Machine Learning for Cybersecurity: Threat Detection and Mitigation

Machine Learning for Cybersecurity: Threat Detection and Mitigation

Guest Editors

Abdussalam Elhanashi
Pierpaolo Dini

Basel • Beijing • Wuhan • Barcelona • Belgrade • Novi Sad • Cluj • Manchester

Guest Editors

Abdussalam Elhanashi
Department of Information
Engineering
University of Pisa
Pisa
Italy

Pierpaolo Dini
Department of Information
Engineering
University of Pisa
Pisa
Italy

Editorial Office
MDPI AG
Grosspeteranlage 5
4052 Basel, Switzerland

This is a reprint of the Special Issue, published open access by the journal *Electronics* (ISSN 2079-9292), freely accessible at: www.mdpi.com/journal/electronics/special_issues/DC2F2R1RZL.

For citation purposes, cite each article independently as indicated on the article page online and using the guide below:

Lastname, A.A.; Lastname, B.B. Article Title. *Journal Name* **Year**, *Volume Number*, Page Range.

ISBN 978-3-7258-2794-7 (Hbk)
ISBN 978-3-7258-2793-0 (PDF)
https://doi.org/10.3390/books978-3-7258-2793-0

© 2024 by the authors. Articles in this book are Open Access and distributed under the Creative Commons Attribution (CC BY) license. The book as a whole is distributed by MDPI under the terms and conditions of the Creative Commons Attribution-NonCommercial-NoDerivs (CC BY-NC-ND) license (https://creativecommons.org/licenses/by-nc-nd/4.0/).

Contents

About the Editors vii

Preface ix

Richard Holdbrook, Olusola Odeyomi, Sun Yi and Kaushik Roy
Network-Based Intrusion Detection for Industrial and Robotics Systems: A Comprehensive Survey
Reprinted from: *Electronics* 2024, *13*, 4440, https://doi.org/10.3390/electronics13224440 1

David Herranz-Oliveros, Marino Tejedor-Romero, Jose Manuel Gimenez-Guzman and Luis Cruz-Piris
Unsupervised Learning for Lateral-Movement-Based Threat Mitigation in Active Directory Attack Graphs
Reprinted from: *Electronics* 2024, *13*, 3944, https://doi.org/10.3390/electronics13193944 24

Fariza Sabrina, Shaleeza Sohail and Umair Ullah Tariq
A Review of Post-Quantum Privacy Preservation for IoMT Using Blockchain
Reprinted from: *Electronics* 2024, *13*, 2962, https://doi.org/10.3390/electronics13152962 50

Jayanthi Ramamoorthy, Khushi Gupta, Ram C. Kafle, Narasimha K. Shashidhar and Cihan Varol
A Novel Static Analysis Approach Using System Calls for Linux IoT Malware Detection
Reprinted from: *Electronics* 2024, *13*, 2906, https://doi.org/10.3390/electronics13152906 69

Mingjie Cheng, Kailong Zhu, Yuanchao Chen, Guozheng Yang, Yuliang Lu and Canju Lu
MSFuzz: Augmenting Protocol Fuzzing with Message Syntax Comprehension via Large Language Models
Reprinted from: *Electronics* 2024, *13*, 2632, https://doi.org/10.3390/electronics13132632 88

Weronika Dracewicz and Mariusz Sepczuk
Detecting Fake Accounts on Social Media Portals—The X Portal Case Study
Reprinted from: *Electronics* 2024, *13*, 2542, https://doi.org/10.3390/electronics13132542 107

Jayanthi Ramamoorthy, Khushi Gupta, Narasimha K. Shashidhar and Cihan Varol
Linux IoT Malware Variant Classification Using Binary Lifting and Opcode Entropy
Reprinted from: *Electronics* 2024, *13*, 2381, https://doi.org/10.3390/electronics13122381 125

Luka Ilić, Aleksandar Šijan, Bratislav Predić, Dejan Viduka and Darjan Karabašević
Research Trends in Artificial Intelligence and Security—Bibliometric Analysis
Reprinted from: *Electronics* 2024, *13*, 2288, https://doi.org/10.3390/electronics13122288 143

Maria Balega, Waleed Farag, Xin-Wen Wu, Soundararajan Ezekiel and Zaryn Good
Enhancing IoT Security: Optimizing Anomaly Detection through Machine Learning
Reprinted from: *Electronics* 2024, *13*, 2148, https://doi.org/10.3390/electronics13112148 160

Michael J. Smith, Michael A. Temple and James W. Dean
Effects of RF Signal Eventization Encoding on Device Classification Performance
Reprinted from: *Electronics* 2024, *13*, 2020, https://doi.org/10.3390/electronics13112020 179

Lampis Alevizos and Martijn Dekker
Towards an AI-Enhanced Cyber Threat Intelligence Processing Pipeline
Reprinted from: *Electronics* 2024, *13*, 2021, https://doi.org/10.3390/electronics13112021 202

Chunlai Du, Yanhui Guo, Yifan Feng and Shijie Zheng
HotCFuzz: Enhancing Vulnerability Detection through Fuzzing and Hotspot Code Coverage Analysis
Reprinted from: *Electronics* **2024**, *13*, 1909, https://doi.org/10.3390/electronics13101909 . . . . . **221**

Zongwen Fan, Shaleeza Sohail, Fariza Sabrina and Xin Gu
Sampling-Based Machine Learning Models for Intrusion Detection in Imbalanced Dataset
Reprinted from: *Electronics* **2024**, *13*, 1878, https://doi.org/10.3390/electronics13101878 . . . . . **233**

Minxiao Wang, Ning Yang, Yanhui Guo and Ning Weng
Learn-IDS: Bridging Gaps between Datasets and Learning-Based Network Intrusion Detection
Reprinted from: *Electronics* **2024**, *13*, 1072, https://doi.org/10.3390/electronics13061072 . . . . . **252**

Rawan Bukhowah, Ahmed Aljughaiman and M. M. Hafizur Rahman
Detection of DoS Attacks for IoT in Information-Centric Networks Using Machine Learning: Opportunities, Challenges, and Future Research Directions
Reprinted from: *Electronics* **2024**, *13*, 1031, https://doi.org/10.3390/electronics13061031 . . . . . **270**

Yixin Yang, Wen Shen, Qian Guo, Qiuhong Shan, Yihan Cai and Yubo Song
EPA-GAN: Electric Power Anonymization via Generative Adversarial Network Model
Reprinted from: *Electronics* **2024**, *13*, 808, https://doi.org/10.3390/electronics13050808 . . . . . . **295**

About the Editors

Abdussalam Elhanashi

Abdussalam Elhanashi is a researcher at the Università di Pisa, Italy, specializing in advanced applications of deep learning and video imaging processing. He holds an M.Sc. in Electronics and Electrical Engineering from the University of Glasgow in Scotland and an MBA from the University of Nicosia in Cyprus. He earned his Ph.D. in Information Engineering from the Università di Pisa, funded by a prestigious merit-based scholarship from the Islamic Bank Development (IsDB) as Libya's top candidate in 2019-2020. Dr. Elhanashi was a Research Fellow at the University of Strathclyde in 2021, where he applied deep learning models to analyze CT scans and X-ray images for medical diagnostics. In 2022, he was a visiting researcher at Hiroshima University in Japan, focusing on advanced video analysis techniques. With over 16 years of industry experience, he has successfully managed engineering projects, conducted system maintenance, and performed root cause analyses to address technical challenges. He authored the first Arabic-language book on artificial intelligence in Libya and has contributed to numerous peer-reviewed articles in international conferences and journals. He is a developer at the Society for Imaging Informatics in Medicine (SIIM) in the USA. His work focuses on real-world AI applications, lightweight model development, video surveillance, IoT-based low-cost embedded systems, designing AI-driven solutions for medical imaging, efficient coding techniques for imaging and video processing systems, machine learning, and deep learning development for cybersecurity.

Pierpaolo Dini

Pierpaolo Dini (Member, IEEE) received his Master's degree in Electrical & Automation Engineering and Ph.D. in Smart Industry from UniPi. He is currently an Assistant Professor of Electronics at UniPi. He is actively involved in the European project Hi-Efficient as an expert in predictive control and maintenance systems for automotive power converters and electric drives. He has also contributed to the European Processor Initiative (EPI) and various regional/national projects, supporting companies in advanced algorithm and system design. He collaborates with international companies such as Marelli Electronics, Magna Mechatronics, Continental, and Toyota Industries. He is the co-inventor of two international patents on automotive cybersecurity (US Patent App. 18/163,488 and US Patent App. 17/929,370) and the co-inventor of a national patent on smart wheelchair electrification (Italy Patent App. 16515979). He is also the co-author of 40+ scientific papers and has an H-index of 16 with over 570 citations (Scopus).

Preface

The rapid evolution of technology has revolutionized our interconnected world, driving innovation across diverse domains such as the Internet of Things (IoT), industrial automation, and critical infrastructure systems. However, this growth has been accompanied by increasingly sophisticated cybersecurity threats, including denial-of-service attacks, malware, and adversarial exploits. These challenges highlight the urgent need for advanced and proactive security measures. Machine learning (ML) has emerged as a transformative tool, reshaping the cybersecurity landscape by enabling intelligent threat detection, anomaly identification, and mitigation strategies.

This reprint, "Machine Learning for Cybersecurity: Threat Detection and Mitigation", explores how ML techniques, such as deep learning, generative adversarial networks, and unsupervised learning, are addressing complex challenges in securing digital ecosystems. For instance, methods like EPA-GAN provide innovative solutions for anonymizing sensitive data while maintaining utility. Similarly, advancements in intrusion detection systems, including sampling-based ML models, bridge the gap between imbalanced datasets and real-world applications, offering robust defense against evolving threats. This reprint also emphasizes pressing concerns such as privacy preservation, secure IoT frameworks, and the mitigation of adversarial attacks. Case studies on topics like Linux IoT malware detection, the integration of AI-enhanced cyber threat intelligence, and fuzzing techniques for vulnerability detection provide a comprehensive perspective on current methodologies and future research directions. Emerging technologies like blockchain and post-quantum cryptography are also explored as pivotal tools for strengthening security in healthcare IoT systems and beyond.

Recognizing the broader implications, this work delves into ethical considerations and the development of security-aware AI systems. The fusion of bibliometric analyses and practical implementations offers readers a balanced view of the opportunities and challenges at the intersection of AI and cybersecurity.

Contributions from leading experts provide actionable insights for researchers, practitioners, and policymakers. By addressing critical themes such as network-based intrusion detection for industrial systems and fake account detection on social media, this reprint aims to foster innovation, inspire resilience, and shape the future of secure digital environments. Together, these efforts ensure that ML remains at the forefront of safeguarding our increasingly interconnected world.

Abdussalam Elhanashi and Pierpaolo Dini
Guest Editors

Article

Network-Based Intrusion Detection for Industrial and Robotics Systems: A Comprehensive Survey

Richard Holdbrook [1], Olusola Odeyomi [1,*,†], Sun Yi [2,*,†] and Kaushik Roy [1]

1 Department of Computer Science, North Carolina Agricultural and Technical State University, Greensboro, NC 27411, USA; rholdbrook@aggies.ncat.edu (R.H.); kroy@ncat.edu (K.R.)
2 Department of Mechanical Engineering, North Carolina Agricultural and Technical State University, Greensboro, NC 27411, USA
* Correspondence: otodeyomi@ncat.edu (O.O.); syi@ncat.edu (S.Y.); Tel.: +1-3362853695 (O.O.) +1-3362853753 (S.Y.)
† These authors contributed equally to this work.

Abstract: In the face of rapidly evolving cyber threats, network-based intrusion detection systems (NIDS) have become critical to the security of industrial and robotic systems. This survey explores the specialized requirements, advancements, and challenges unique to deploying NIDS within these environments, where traditional intrusion detection systems (IDS) often fall short. This paper discusses NIDS methodologies, including machine learning, deep learning, and hybrid systems, which aim to improve detection accuracy, adaptability, and real-time response. Additionally, this paper addresses the complexity of industrial settings, limitations in current datasets, and the cybersecurity needs of cyber–physical Systems (CPS) and Industrial Control Systems (ICS). The survey provides a comprehensive overview of modern approaches and their suitability for industrial applications by reviewing relevant datasets, emerging technologies, and sector-specific challenges. This underscores the importance of innovative solutions, such as federated learning, blockchain, and digital twins, to enhance the security and resilience of NIDS in safeguarding industrial and robotic systems.

Keywords: robotics security; industrial control systems; network-based intrusion detection systems; anomaly detection; machine learning

Citation: Holdbrook, R.; Odeyomi, O.; Yi, S.; Roy, K. Network-Based Intrusion Detection for Industrial and Robotics Systems: A Comprehensive Survey. *Electronics* **2024**, *13*, 4440. https://doi.org/10.3390/electronics 13224440

Academic Editors: Abdussalam Elhanashi and Pierpaolo Dini

Received: 2 October 2024
Revised: 1 November 2024
Accepted: 7 November 2024
Published: 13 November 2024

Copyright: © 2024 by the authors. Licensee MDPI, Basel, Switzerland. This article is an open access article distributed under the terms and conditions of the Creative Commons Attribution (CC BY) license (https:// creativecommons.org/licenses/by/ 4.0/).

1. Introduction

Network-based intrusion detection systems (NIDS) are essential for providing security to industrial and robotic systems in today's cyber threats. An NIDS is a specific type of intrusion detection system (IDS) focused on monitoring and analyzing network traffic for suspicious activity, potentially detecting unauthorized access or malicious behaviors within a network. On the other hand, IDS is a security solution designed to detect malicious activities in a system and can be applied to multiple areas of a system, including web applications, packet filtering, enterprise applications, and database systems. In an NIDS, malicious activity is identified by monitoring the various interconnections between computers, with NIDS typically deployed on routers or switches within the network. The main objective of NIDS is to detect and flag suspicious or harmful activity and report this malicious information to the network administrator. The need for reliable intrusion detection systems (IDS) is increasing as industries adopt digital technologies to stay competitive. Intrusion detection technologies have evolved from simple signature-based methods to complex anomaly detection systems that rely on machine learning and artificial intelligence (AI). Although traditional techniques are effective against known threats, they have struggled to keep pace with the dynamic nature of modern cyberattacks, such as zero-day exploits and advanced persistent threats, which are increasingly common in industrial settings. Machine learning has been a powerful tool in advancing NIDS, enabling them to learn from past data and identify anomalies without explicit programming for every possible threat. This capability is vital in industrial settings, where disruptions can lead to substantial operational delays or safety risks [1]. Studies have shown that incorporating machine learning methods can significantly improve detection accuracy and reduce the occurrence of false positives, which is a common issue in traditional NIDS implementation [2]. Nevertheless, implementing these systems in industrial and robotic settings presents distinct challenges. These systems must possess not only a high level of accuracy but also the ability to operate in real-time, frequently under strict limitations on available resources.

Integrating NIDS in industrial and robotic systems faces significant challenges, including securing vulnerable industrial control system (ICS) protocols, addressing cyber–physical systems (CPS) vulnerabilities, and countering sophisticated advanced persistent threats (APTs). Real-time monitoring limitations, insider threats, and legacy system complexities add to the security demands. Supply chain integrity, regulatory compliance, new and evolving threats, and resource constraints further complicate implementation. Thus, adaptive lightweight NIDS approaches are essential to handle the evolving cyber threat landscape in these critical infrastructures.

The absence of balanced and specialized datasets designed specifically for ICS and robotic systems impedes efficient NIDS adoption. Public datasets, such as UNSW-NB15 and ADFA IDS, typically lack industrial-specific network traffic patterns, protocols, and device interactions [3,4]. The absence of uncommon, crucial attack types causes class imbalances, which leads to more false positives and negatives in detection. Models trained on imbalanced data frequently misinterpret typical operations for abnormalities, resulting in operational interruptions and costly troubleshooting, with large organizations potentially losing an estimate of about $10.3 million yearly and medium-sized companies roughly $11,000 yearly [5].

Therefore, in this survey paper, we aim to answer the following research questions: (1) What strategies can help NIDS detect threats quickly with a low false alarm rate for real-time use in industrial and robotic applications? (2) How can the lack of balanced and specialized datasets for industrial and robotic systems be addressed in NIDS research? (3) How can NIDS integrate well with emerging technologies for industrial and robotic use? To address these questions holistically, we divide the rest of this paper into several sections. Section 2 summarizes the research strategy used to organize this survey paper. Section 3 discusses related work on NIDS. Section 4 gives a brief overview of various existing approaches used to detect network intrusions in industrial and robotic systems, such as signature-based, anomaly-based, machine-learning-based, and deep-learning techniques. Section 5 discusses the integration challenges of NIDS within industrial and robotic systems, emphasizing cybersecurity issues due to increased connectivity and complexity. Section 6 discusses the challenges of existing datasets for industrial and robotic systems and raises suggestions to address these challenges. Section 7 discusses how to integrate NIDS with emerging technologies such as federated learning, quantum computing, blockchain, digital twins, and large language models (LLM). Lastly, Section 8 concludes the survey. The list of abbreviations used in this paper can be found in the Abbreviations section.

2. Research Strategy

We mainly focus on recent developments in IDS and NIDS using research papers published between 2022 and 2024. We specifically focus on industrial and robotic systems since no previous survey papers filled this research gap. We searched for articles on famous online databases, including Google Scholar, IEEE Xplore, and ResearchGate. Articles published in high-ranking journals and conferences are selected for inclusion. On the other hand, we minimized the inclusion of preprints. We excluded papers not written in good English, papers with low-impact factors, and papers from unreputable journals. We use various combinations of keywords to search for articles on the aforementioned online repositories. Some of these keywords include "Network Intrusion Detection Systems", "Network Intrusion in Robotic Systems", "Network Intrusion in Industrial Systems", "Intrusion Detection System", and "Machine Learning-Based Intrusion Detection".

3. Related Work

While several research papers exist on NIDS and intrusion detection systems (IDS), none directly addresses the unique requirements of industrial and robotic environments. Some of the relevant survey papers are discussed below.

Reference [6] provides a comprehensive and systematic review of network-based intrusion detection systems (IDS). Reference [7] is a review paper that focuses strictly on machine learning-based IDS. It discusses critical challenges and provides future research. Reference directions [8] provides an overview of NIDS types, techniques, and advancements. Although these works cover IDS extensively, they lack specific discussion on the demands and challenges associated with industrial and robotic systems. Similarly, Ref. [9] is a systematic review paper that discusses techniques to improve IDS efficiency, such as reducing false positives, increasing detection accuracy, and enhancing system performance. Additionally, Ref. [10] reviews artificial intelligence (AI)-based techniques for network-based intrusion detection systems. Ref. [11] explores learning-based techniques, particularly machine learning, applied to NIDS. Ref. [12] is a systematic literature review focusing on NIDS. It reviews various models used for network intrusion detection systems. Ref. [13] is a comparative study research paper focusing on intrusion detection in IoT networks. It compares the effectiveness of neural network-based intrusion detection with traditional signature-based detection, specifically within IoT network environments. Ref. [14] is another survey paper focusing on NIDS for IoT and non-IoT environments. However, it overlooks IDS and NIDS for robotic and industrial systems. Ref. [15] is yet another survey paper focusing on NIDS with combined deep-learning models. However, it does not consider the challenges unique to industrial and robotic applications. Finally, Ref. [16] is a survey paper focusing on IDS using machine learning and deep learning. However, this paper does not place a strong emphasis on industrial and robotic systems.

Therefore, there is a research gap on NIDS applications in industrial and robotic systems. Moreover, as cyberattacks are getting advanced, there is a need for a survey paper that focuses on how recent emerging technologies can be leveraged in NIDS to improve security in robotic and industrial settings. Thus, this presents a comprehensive overview of the state-of-the-art NIDS for industrial and robotic systems. Table 1 highlights the gaps in existing survey papers by comparing existing survey papers with the unique contributions and focus areas addressed in this study.

Table 1. Comparison of survey papers.

Research Papers	Published Year	Focus on Industrial and Robotic Systems	ICS and Robotics Specific Datasets	Integrates NIDS with Emerging Technologies
[5]	2023	NO	NO	NO
[6]	2024	NO	NO	NO
[9]	2023	NO	NO	NO
[10]	2024	NO	NO	NO
[11]	2023	NO	NO	NO
[7]	2024	NO	NO	NO
[8]	2023	NO	NO	NO
[12]	2024	NO	NO	NO
[13]	2024	NO	NO	NO
[15]	2024	NO	NO	NO
[14]	2023	NO	NO	NO
Our Paper	2024	YES	YES	YES

4. Methodologies of Intrusion Detection

A NIDS operates in several stages to ensure comprehensive threat detection in industrial and robotic systems. As shown in Figure 1, the architecture of an NIDS consists of multiple interconnected components, from network traffic monitoring and data preprocessing to detection processing and response. Each phase plays a crucial role in identifying and mitigating the potential threats. The architecture begins with network traffic monitoring, where data flowing between devices such as robots, computers, or sensors is constantly observed. The goal is to gather data on what is happening within the network. These data are then cleaned and organized in the data preprocessing phase, where irrelevant information is filtered out, making the data ready for further analysis. In feature extraction, key characteristics are extracted from the data to identify potential signs of malicious behavior or anomalies. This is followed by detection processing, in which algorithms analyze the extracted features to detect suspicious activities. Once a potential threat is identified, decision-making takes place to classify detection as either a legitimate threat or a false alarm. If confirmed, the system triggers a response that may involve alerting human operators, blocking certain actions, or taking measures to contain threats. This process ensures the security and smooth operation of industrial and robotic systems in real-time.

The rest of this section will discuss specific methodologies employed by IDS, including signature-based, anomaly-based, machine learning-based, and hybrid systems.

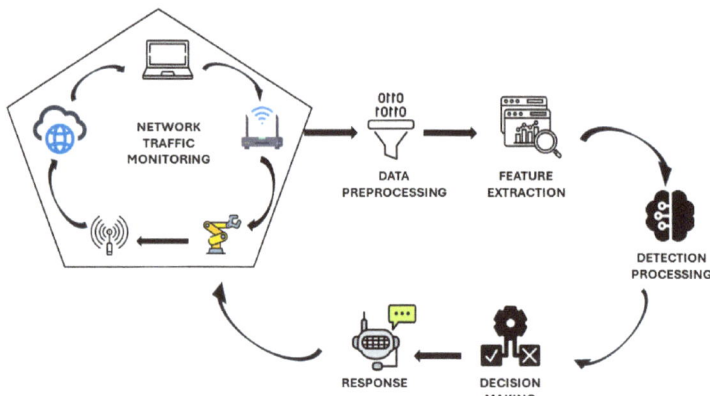

Figure 1. Architecture of NIDS in industrial and robotic systems.

4.1. Signature-Based

An IDS uses databases of known threat signatures to efficiently detect and prevent malicious activities. Signature-based intrusion detection has been widely applied in robotics and ICS. Kaouk et al. [17] discussed the significance of intrusion detection systems in defending these systems from cyberattacks, emphasizing the importance of IDS in protecting systems from cyberattacks. However, a major drawback of these systems is the requirement for frequent updates to the signature database to address new threats. In addition, Ring et al. [18] highlighted the challenges of maintaining up-to-date signature datasets to ensure robust defense against constantly evolving cybersecurity threats. Another aspect shown in [18] is how to make signature-based IDS work better in static environments with stable attack vectors. The industrial and robotics sectors continue to benefit from signature-based intrusion detection, offering a straightforward and effective way to identify malicious activities.

4.2. Anomaly-Based

Anomaly-based IDS excels in identifying zero-day vulnerabilities and complex cyberattacks that do not match known signatures, as they detect anomalies that differ from typical operational activities. Numerous applications of anomaly-based intrusion detection exist in industrial and robotic-related environments. Gillen et al. [19] emphasized the importance of customized security measures in these settings, specifically focusing on the impact of various configurations on safety. Two reviews, one by Kabore et al. [20] and the other by Rosa et al. [21], discussed how anomaly detection in ICS makes use of deep learning techniques such as convolutional neural networks (CNN) and long short-term memory (LSTM). Ganeshan et al. [22] and Alsoufi et al. [23] offered thorough assessments of IDS that depend on abnormalities. Ganeshan et al. categorized IDS into two primary classifications: characteristic-based and methodological. Characteristic-based classification focuses on the location and operation of an IDS within a system, encompassing host-based, network-based, stack-based, signature-based, and anomaly-based approaches. On the other hand, methodological classification emphasizes the techniques employed by the IDS, including tree-based, Adaboost algorithm-based, probabilistic-based, support vector machine (SVM)-based, neural network-based, and unsupervised classification-based methods. Tree-based methods use decision trees to differentiate normal and abnormal network packets. AdaBoost combines multiple weak classifiers to create a more accurate classifier. Probabilistic-based methods employ probabilistic models to handle noisy data and determine optimal solutions. The SVM is a supervised learning algorithm that identifies patterns by finding the optimal hyperplane to separate different classes. Neural networks can analyze network traffic in real-time and learn complex patterns. Unsupervised classification techniques handle large datasets without prior labeling, making them useful when the labeled data are limited or unavailable. The authors emphasized the need for improved detection algorithms to manage unidentified attacks within an IDS effectively. These enhancements can optimize performance and address specific cybersecurity challenges. Alsoufi et al. [23] conducted a detailed analysis of deep learning methods for Internet of Things (IoT) anomaly detection. IDS are increasingly employing deep learning owing to their superior data processing and accuracy. They also emphasized IoT security issues and recommended prioritizing the development of efficient deep-learning models for IoT devices with limited resources.

4.3. Machine Learning-Based

A machine learning-based IDS uses algorithms to learn from data and identify potentially harmful patterns. They can adapt to new cyber threats over time. Numerous studies have examined the application of machine learning for detecting intrusions in robots and industrial systems. Flaus and Georgakis [24] and Haripriya and Jabbar [25] explored the potential benefits of machine learning in this context. In particular, Haripriya and Jabbar discussed how it could help achieve a low false alarm rate and a high detection rate. Unfortunately, Chattopadhyay et al. [26] pointed out problems such as the need for multiple detection methods and how challenging it is to define and use machine-learning techniques. Mishra and Mishra [27] provided more specific information about the challenges in intrusion detection and how to implement them in ICS. They include handling large network traffic volumes, uneven data distribution, continuous adaptation, distinguishing between faults and attacks, handling encrypted data, and ensuring the performance of IDS in large and complex systems. The authors also highlight the need for machine learning and other collaborative IDS solutions to address these challenges.

Kaouk et al. [17] presented a detailed exploration of IDS for ICS, emphasizing the role of machine learning in advancing IDS for ICS and highlighting associated challenges. They discussed the role of machine learning in advancing IDS for ICS, the challenges associated with machine learning-based IDS, such as high false alarm rates, and the difficulty in distinguishing between faults and attacks. They also outlined future research directions, including methods to reduce false positives, improve accuracy, and adapt IDS to new technologies such as the Industrial Internet of Things (IIoT) and advanced encryption methods.

4.4. Deep Learning-Based

Deep learning-based IDS uses advanced neural networks capable of analyzing extensive data to detect complex and subtle threat patterns. This approach is particularly efficient for applications requiring large amounts of data, such as industrial and robotic systems, in which traditional detection approaches may be insufficient. Various studies have investigated the application of deep learning in IDS for industrial and robotic systems. Alani et al. [28] introduced DeepIIoT, an IDS that uses deep learning techniques to identify intrusions. Furthermore, DeepIIoT explains its decision-making process using Shapley additive explanation (SHAP) values. The system uses a multi-layer perceptron (MLP) classifier trained on the WUSTL-IIOT-2021 dataset. This dataset encompasses a range of cyberattacks, including command injection, Denial of Service (DoS), reconnaissance, and backdoor assaults. The model trained on the dataset provided enhanced IIoT security against cyberattacks. Notwithstanding, the increased connectivity and data interchange of IIoT devices make them vulnerable to assaults, especially in critical infrastructure such as power grids and water treatment facilities. The model attained a precision of 99 percent, exhibiting minimal rates of false-positive and false-negative outcomes. Using SHAP values allows the model to provide comprehensive insights into the primary factors that impact classification decisions, thereby improving the transparency and reliability of the system. In future studies, the authors aim to investigate how standardizing the feature set affects the model's accuracy. The goal was to analyze the connection between IDS classification parameters and feature extraction methods to enhance the performance.

Mohamed et al. [29] conducted an extensive analysis of deep learning techniques employed in cyber intrusion detection. They discussed ten deep learning models employed for intrusion detection and applicable in robotics: deep neural networks (DNN), feed-forward deep neural networks (FFDNN), recurrent neural networks (RNN), convolutional neural networks (CNN), restricted Boltzmann machines (RBM), deep belief networks (DBN), deep auto-encoders, deep migration learning, self-taught learning, and replicator neural networks. They classified these deep learning models into two categories: deep discriminative and generative/unsupervised. Their research report categorized 35 well-recognized cyber datasets into seven distinct groups: network traffic-based, electrical network-based, inter-traffic-based, virtual private network-based, Android app-based, IoT traffic-based, and Internet-connected device-based datasets. The authors assessed the performance of these deep learning models using two actual traffic datasets: CSE-CIC-IDS2018 and BoT-IoT. They highlighted that deep learning models generally surpass conventional machine learning techniques, such as Naive Bayes and SVM, in terms of accuracy, detection rate, and false alarm rate. Their results showed that deep learning techniques, specifically CNN and RNN, demonstrated exceptional performance in managing data with many dimensions and capturing time-related patterns in network traffic. These results resulted in improved accuracy and reduced false positive rates. However, the research acknowledged notable obstacles, such as the requirement for extensive labeled datasets and substantial processing resources, which restrict the ability to use the technology in real-time situations and for resource-limited applications such as IoT devices. Their study determined that future investigations should prioritize the development of lightweight and efficient models to tackle these issues and improve the practical implementation of deep learning-based IDS in various domains, including ICS and IoT networks. Abdel-Wahab et al. [30] and Lánský et al. [31] conducted comparative and systematic evaluations of machine learning and deep learning models. Abdel-Wahab et al. [30] used the KDD-99 dataset to test different machine learning and deep learning models to identify intrusions based on unusual behavior. The tested models were SVM, K-Nearest Neighbor (KNN), Gaussian Naive Bayes (GNB), Decision Tree (DT), Random Forest (RF), and Gradient Boosting Classifier (GBC). The authors presented, compared, and evaluated the performance of these models. Multiple models have distinct advantages and disadvantages, and no single model is superior. Nevertheless, the RF model demonstrated encouraging outcomes and warrants further evaluation in actual IDS settings. Additionally, they established

Table 2. Comparative evaluation of intrusion detection methodologies for industrial applications.

Methodology	Detection Accuracy	Adaptability to New Threats	Computational Efficiency	False Positive Rate	Suitability for Industrial Use
Signature-Based	High	Low	High	Low	Good for stable environments
Anomaly-Based	Moderate	High	Moderate	High	Excellent for dynamic environments
Machine Learning-Based	High	High	Varies	Moderate	Very effective for diverse threats
Deep Learning-Based	Very High	Very High	Low	Low	Best for complex data analysis
Hybrid Based	Very High	High	Moderate	Lower than anomaly-only	Optimal for comprehensive coverage

5. Practical Implementation and Challenges of Integrating Network-Based Intrusion Detection in Industrial and Robotic Systems

The integration of advanced networking and computing technologies into industrial and robotic systems has resulted in numerous benefits, including increased efficiency and automation. However, this integration introduces significant cybersecurity challenges. This section summarizes the primary challenges specific to industrial and robotic systems. These challenges are summarized in Figure 2.

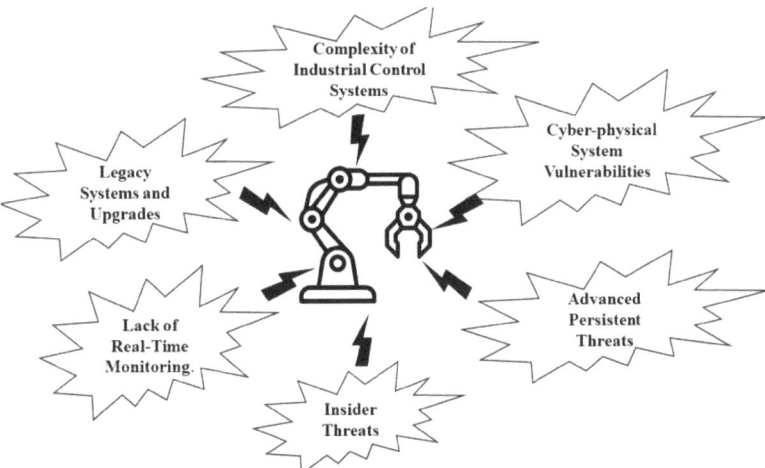

Figure 2. Challenges in industrial and robotics systems for NIDS.

5.1. Complexity of ICS

ICS is becoming increasingly complicated and interconnected with industrial and robotic systems, which has led to increased security concerns. Using the Modbus, PROFINET, S7COMM, and DNP3 protocols in critical infrastructure has posed security risks in many environments. For instance, the U.S. Department of Homeland Security demonstrated the vulnerability of Modbus communication in ICS in simulated attacks on power grid setups, showing how attackers could intercept or alter communication between devices [37,38]. Researchers and innovators have developed a wide range of approaches, including specification-based methodologies [39], process awareness [40], and anomaly detection, to address this issue. Research has shown conflicting opinions on how best to monitor and control network traffic and process semantics. Some studies recommend concentrating on the last connection in the network, whereas others highlight the importance of doing both at once. Applying statistical models and machine learning is promising for anomaly identification in ICS setups. Much work remains before we have an IDS that can handle the unique requirements of complicated industrial and robotic systems, even with current advancements [41].

5.2. Cyber–Physical System (CPS) Vulnerabilities

The combination of cyber and physical components in CPS poses a growing security problem. Cyberattacks on CPS can have real-world effects, such as breaking equipment, stopping production, or creating safety risks. CPS are considered dual because they integrate both physical and cyber components to provide real-time data processing and decision-making capabilities, thereby rendering hacking breaches more dangerous. One prominent example is the 2014 cyberattack on a German steel mill, where attackers compromised network systems, leading to the failure of production controls and significant physical damage. This event was often cited in discussions on CPS vulnerabilities by [42]. Protecting CPS, especially in industrial settings, requires an IDS [43]. Many recent studies have focused on improving the performance of IDS in CPS using machine learning [44]. Cybercriminals use advanced techniques to exploit vulnerabilities in CPS, particularly robotic systems. Concerning identifying suspicious actions and breaches in CPS, deep learning algorithms have demonstrated encouraging results [45]. Control theory has been used to investigate many attacks and methods for detecting them [46]. Identifying hazards in many CPS applications relies heavily on anomaly detection strategies, but there are also obstacles, such as different CPS settings, limited resources, and an absence of established standards [47].

5.3. Advanced Persistent Threats (APTs)

Industrial and robotic systems, particularly those involved in intrusion detection over networks, face severe threats from APTs. They are sophisticated cyberattacks carried out by skilled professionals, often infiltrating high-value systems, stealing sensitive data, or disrupting operations over extended periods. Unlike typical cyber threats, APTs involve a "low-and-slow" approach, where attackers focus on stealth and persistence, avoiding detection as they gradually establish control within a network [48,49]. APTs are typically state-sponsored or highly organized criminal enterprises, and they target critical infrastructure, government agencies, large corporations, and defense systems. A real-world example of an APT is the Stuxnet attack, widely regarded as one of the most prominent APTs in recorded history. Unearthed in 2010, Stuxnet was an exceptionally advanced computer worm created for the specific purpose of attacking Iran's nuclear program. It deliberately targeted the software of industrial control systems, causing significant damage to approximately 20 percent of Iran's nuclear centrifuges. The worm spreads across Microsoft Windows machines and can self-replicate [50]. The SolarWinds Attack of 2020 is another prominent example of an APT. Suspected to be state-sponsored, this attack targeted SolarWinds, a U.S. IT management company, by injecting malicious code into their Orion software updates. When customers installed these updates, attackers gained backdoor access to their systems, compromising numerous high-profile U.S. government agencies such as Homeland Security, Treasury, and State. Microsoft and FireEye were among the affected private sector companies. The attackers used stealthy methods to move laterally within the affected networks, exfiltrating sensitive information and maintaining access over several months. The incident highlighted APT's ability to operate undetected within a supply chain, demonstrating how such threats exploit trusted software and vendors. This event reinforced the need for enhanced cybersecurity measures in government and industry. Traditional security assessment methods are extremely difficult to deal with because they are crafty and effective at evading detection [51]. Artificial immune systems, recurrent neural networks [52], graph theory, and deep learning are some of the machine learning algorithms that have been recently examined as possible ways to find APTs [53]. These strategies aim to improve the detection capacities of both the IT and OT systems. The analysis of network traffic patterns, event correlation, and multi-stage detection frameworks are common components of the proposed solutions [54].

5.4. Lack of Real-Time Monitoring

Monitoring cyber dangers in real-time and promptly responding to them is crucial for ensuring the safety of industrial systems. However, owing to the need for low latency and high reliability in industrial settings, it is not always possible to use full real-time monitoring systems [55]. The lack of real-time monitoring in industrial and robotic systems poses a notable obstacle because of the strict demands for minimal delay and exceptional reliability. These systems require immediate reactions to prevent operational disturbances; however, real-time monitoring can cause delays and put pressure on the limited computational resources. The large amount and speed at which data are produced, along with the complicated nature of diverse networks and the requirement for cyber–physical integration, make it even more challenging to implement efficient monitoring solutions. Therefore, in the absence of strong and immediate detection and response capabilities, cyber threats can remain undetected, which can result in significant risks to operational safety and efficiency. The Colonial Pipeline attack in 2021 highlighted the critical need for real-time monitoring in ICS networks. This ransomware attack led to a major fuel supply disruption in the U.S. and underscored challenges around latency and real-time response in critical infrastructures.

5.5. Insider Threats

Industrial and robotic systems are very vulnerable to insider attacks, whether deliberate or unintentional. Employees who compromise credentials or commit malicious acts can significantly harm an organization's systems. It is still difficult to effectively identify and eliminate insider risks. In 2019, a former Tesla employee disclosed proprietary data and system credentials to outsiders, potentially endangering system integrity and safety. The incident reflects the difficulty of identifying insiders who have legitimate system access, as discussed in [56]. Identifying and managing insider threats in industrial and robotic systems pose significant challenges owing to multiple considerations. First, individuals with privileged access, such as employees, contractors, or partners, frequently possess real access to vital systems, which complicates the differentiation between regular and harmful activities. This access gives them the ability to skip security procedures intended to prevent external threats. Furthermore, insider attacks may manifest as either planned or accidental actions, thus complicating the process of identifying potential threats. Malicious insiders can take advantage of their understanding of system weaknesses, while well-intentioned staff may unintentionally cause harm due to negligence or human error. Moreover, insider threats may disguise themselves during normal activities, rendering the identification of abnormalities challenging without the use of sophisticated monitoring and analytic techniques. The implementation of intrusive monitoring technologies is made more complex by issues related to privacy and the necessity of establishing trust among employees. The complex and diverse structure of industrial networks, including outdated and contemporary technologies, contributes to the challenge of implementing consistent security protocols. The combination of these characteristics poses significant difficulties in accurately identifying and addressing insider dangers in industrial and robotic environments.

5.6. Legacy Systems and Upgrades

The complexity of upgrades and old systems makes it difficult to integrate network-based IDS in robotic and industrial settings. The lack of security considerations during their design makes many industrial control systems vulnerable to cyberattacks. When it comes to linked settings, legacy systems can be a real pain due to their antiquated design patterns and lack of modern security measures. Many SCADA systems still in use today lack encryption, making them liable to attacks such as the sandworm attack in 2014, which targeted Ukraine's infrastructure and exploited SCADA system vulnerabilities. Similar challenges are discussed in [57]. Many industrial systems rely on outdated technology without considering modern cybersecurity concerns. Upgrades are disruptive and expensive, making it difficult to secure these systems. It is extremely difficult to guarantee the safety of older systems while minimizing operational interruptions. The lack of experts trained in outdated control systems coupled with outdated hardware and software makes upgrading these systems difficult [43]. To address these challenges, researchers are developing emerging techniques such as holistic intrusion detection ideas [58], machine learning-based detection methods [59], and last-line defense intrusion detection systems [60]. In addition, scientists are trying to figure out how to incorporate old systems into new IoT designs without destroying any working hardware [61].

5.7. Supply Chain Security

There is a significant security risk associated with integrating a network-based intrusion IDS into robotic and industrial settings. Various cyber dangers may penetrate supply chains owing to their growing complexity and interconnections. In 2020, the SolarWinds cyberattack compromised thousands of organizations globally by inserting malware through a widely-used IT monitoring product. This highlights the potential vulnerabilities across a supply chain reliant on third-party software. Refs. [51,62] explore supply chain threats that can be referenced. Researchers have developed and investigated modern intrusion detection system (IDS) frameworks based on reconfigurable hardware technology to reduce these threats and explore ways to secure geographically distributed industrial networks [63]. Industrial robots are CPSs; therefore, their security is particularly vulnerable to threats that could affect their physical operations. Deploying effective security measures is challenging because of issues related to knowledge, limited testbeds, proprietary systems, and limited resources. Studies have shown the importance of dealing with supply chain integrity in the ICS domain [64] and how important it is to use methods that consider both the physical and virtual parts of critical infrastructure [65]. Additionally, the safety of industrial systems is frequently dependent on security measures taken by vendors and providers outside organizations. Ensuring the security of the entire supply chain is crucial because industrial systems may be compromised by compromised components or software from vendors and suppliers [66,67].

5.8. Regulatory Compliance

When NIDS are used in robotics and factories, they create significant problems when it comes to following rules, finding cyber threats in industrial control systems, and making robots work together. However, company cultural issues, rapidly evolving regulations, and resource limitations complicate compliance management [60]. Regulatory compliance and cybersecurity interact in a complicated way; compliance usually drives security measures but may also leave weaknesses in the general security posture. The General Data Protection Regulation (GDPR) in the EU has introduced complexities for global companies integrating NIDS due to stringent data protection and monitoring requirements [68]. Cybersecurity policies and standards apply to industrial systems; however, they may differ greatly between sectors and locations. It can be especially difficult for global corporations operating in several regulatory ecosystems to meet these standards [69]. Researchers have suggested creative solutions to these difficulties, such as data streaming methods for industrial networks and distributed IDS for robotic systems. Furthermore, thorough risk assessment and mitigation rely on incorporating safety and security concerns into the industrial systems [43].

5.9. Resource Constraints

Resources severely limit the integration of industrial and robotic systems with network-based IDS. Given the constraints of limited processing capacity and energy resources, it is essential to have efficient scheduling of IDS jobs [70] together with lightweight detection techniques [71]. Researchers are currently exploring various methodologies to address these challenges. Some of these methods are developing decision tree-based solutions, deep learning with computational offloading, and the best way to spread IDS functions across network nodes [72]. For mobile CPS, integrating cyber and physical elements in an IDS design has shown the potential to enhance detection accuracy. Researchers have examined policy-based intrusion detection systems (IDS) for wireless sensor networks configured with IPv6 [73]. They also investigated the impact of the training set size on the use of resources for an IDS based on the IoT [74]. Despite significant advancements that have been made, there are still challenges in developing an appropriate IDS for environments with restricted resources [75,76].

5.10. Evolving Threat Lanscape

Industrial and robotic systems are vulnerable to developing threats such as APTs, zero-day vulnerabilities, and malware suited specifically for ICS protocols. Recent attacks, such as the Colonial Pipeline ransomware outbreak [77], demonstrate the widespread disruptions that these threats may create by targeting redundant or traditional security systems that struggle against sophisticated, covert approaches. Similarly, recent attacks on sensor integrity and control signal authenticity in robotic systems, such as the sensor spoofing attack on a pharmaceutical production robot in 2022 [78], which resulted in considerable financial losses owing to product faults, demonstrate the enormous physical hazards involved. The expansion of IoT and IIoT devices increases the attack surface, complicating security by offering more possible entry points. Traditional NIDS are frequently unprepared for such fast changes in attack vectors, making adaptive, learning-based techniques critical for future resilience.

6. BenchMark Datasets

In this survey, we reviewed relevant datasets commonly utilized in the development and evaluation of network-based intrusion detection systems (NIDS) for industrial and robotic systems. These datasets vary from older ones such as NSL-KDD to SWaT and BATADAL, which are concerned with cybersecurity in Industrial Control Systems (ICS) and the Industrial Internet of Things. Researchers can use these open-source datasets for simulations in industrial and robotic systems. Below are the descriptions of the datasets.

6.1. NSL-KDD

The University of Brunswick developed this dataset in 2009 to address redundancies and inaccuracies [79], improving upon the KDD Cup 1999 data. It focused on Denial of Service (DoS), Remote-to-Local (R2L), User-to-Root, and Probes as the attack types. It contains 125,973 records with reduced redundancy, making it more manageable for researchers. Each record includes 41 features, such as the protocol type, service, and flag. Despite its limited representation of real-world attacks and lack of industrial-specific features, this dataset remains synthetic and can be applicable in certain industrial and robotic cybersecurity scenarios. This restricts its relevance in modern industrial environments, which demand context-specific data.

6.2. ISCX

The Canadian Institute of Cybersecurity created this dataset in 2012 with HTTP, SMTP, and SSH as the attack types [80]. They designed this well-balanced real-world network traffic dataset for anomaly detection. It includes 18 h of labeled traffic across a wide range of protocols with around 1 million records. Although the dataset imitates real-world traffic, it does not specifically incorporate ICS or IoT protocols. This makes it less relevant for NIDS research in robotic or ICS settings.

6.3. SWaT

Singapore University of Technology created this dataset in 2016 with a focus on the security of water treatment systems and ICS [81]. The attack types are ICS-specific, such as command injection and data manipulation. The data creation process involved six days of attack data against an operational water treatment plant. SWaT contains 946,722 data points and 51 unique attributes, such as sensor readings and actuator states. Despite being industry-specific, SWaT only represents a water treatment ICS and lacks diversity across different industrial environments, which limits its applicability to other robotic and industrial systems.

6.4. CICIDS 2017

The Canadian Institute for Cybersecurity developed this dataset in 2017 with the focus of emulating modern real-world attack scenarios in a controlled network [82]. It uses DoS, brute force, web attacks, and infiltration as the attack types. It has over 3.1 million instances, including 83 network flow features and labeled attack types. This dataset primarily targets enterprise networks lacking the attributes of ICS and IoT applications commonly found in some industrial systems.

6.5. UGR'16

The University of Granada developed this dataset in 2017 with a focus on long-term analysis of real network traffic for anomaly detection. DoS, brute force, and botnet were its attack types [83]. The team collected approximately 19 gigabytes of data, capturing daily network traffic for more than a year and including eight basic flow features. It can be applied in certain scenarios but lacks the industrial specificity and IoT-focused protocols needed for a wide range of robotics systems, limiting its applicability in manufacturing and robotic scenarios.

6.6. BETADAL

Different institutions developed this dataset, with Singapore University of Technology and Design (SUTD) and Cornell University being the prominent ones in 2018 [84]. The dataset includes water infrastructure attacks, such as sensor and command spoofing, and focuses on attack scenarios for water distribution ICS. It simulates multiple attacks on water distribution systems, capturing data for sensors and actuators for several weeks. Its scope is limited to water distribution and machinery and does not generalize to other ICS or robotic applications, leaving a gap for other sectors, such as manufacturing and power distribution.

6.7. WUSTL-IIOT

Washington University in St. Louis developed this dataset in 2018 with a focus on intrusion detection for smart manufacturing and IIoT environments [85]. The dataset includes IIoT network attacks such as DDoS and command injection, with approximately 2.5 million labeled packets focusing on IIoT protocols and smart manufacturing scenarios. It is limited in scope and lacks some robotic attack cases essential for testing NIDS in broader industrial systems.

6.8. TON IoT

The University of South Wales developed this dataset in 2020, focusing on the security of IIoT environments that encompass IoT, ICS, and network flows [86]. The dataset includes IoT-specific attacks such as DDoS and data manipulation. The data creation is from network traffic, IoT sensors, and ICS systems and includes 22 million entries across 29 features. Even though this dataset covers a lot of IIoT protocols, it might not cover all of the attack vectors that are unique to robotics. Table 3 summarizes the datasets, their properties, and the research publications where they were initially introduced or cited.

Table 3. Key datasets for network-based intrusion detection in industrial and robotics systems.

Dataset	Developer	Year	Attack Types	Focus	Reference
NSL-KDD	University of New Brunswick	2009	DoS, R2L,U2L,Probe	Improved Version of KDD CUP 1999	[79]
ISCX	Canadian Institute for Cybersecurity	2012	HTTP,SMTP,SSH attacks	Well-balanced real-world network traffic	[80]
SWaT	Singapore University of Technology	2016	ICS-specific attacks	Cybersecurity in Industrial Control	[81]
CICIDS 2017	Canadian Institute of Cybersecurity	2017	DoS,brute force	Real-world attack scenarios	[82]
UGR'16	University of Granada	2017	DoS,brute force	Long-term network traffic analysis	[83]
BATADAL	Singapore University of Technology and Design (SUTD) and Cornell University	2018	Water Infrastructure	ICS attack on water distribution systems	[84]
WUSTL-IIOT	Washington University in St.Louis	2018	IIoT network attacks	Intrusion detection in smart manufacturing	[85]
TON IoT	University of New South Wales	2020	IoT-Specific attacks	Industrial IoT (IIoT) environment	[86]

The limitations of existing datasets reveal the need for more comprehensive data to support effective NIDS research in industrial and robotic environments. Most datasets only cover specific sectors and lack the cross-sector applicability necessary to generalize NIDS for various industrial systems. Also, only a few datasets comprehensively cover the wide array of attacks faced by industrial and robotic systems, such as APTs, sensor spoofing, and ICS-specific malware. On the other hand, there are numerous cases of data imbalance, where many datasets contain limited instances of attack data. This can lead to biased models that are ineffective for real-time industrial NIDS applications. Solving these challenges will require holistic approaches, such as expanding datasets with simulated data from diverse environments. Simulation data from smart grids, automated factories, manufacturing assembly plants, and water treatment plants can improve generalizability and model performance in real-world applications. Combining real-world data with high-fidelity simulated attacks, such as from digital twins of industrial systems, can enhance dataset diversity without compromising data authenticity. Robotics-specific datasets that cover sensor spoofing, command injection, and other unique attack vectors can better support NIDS research for robotic applications. Integrating a broader range of cyber threats will enable researchers to train models that respond effectively to evolving threats in industrial and robotic environments. Researchers can also reduce the risks of model bias, which results in inaccurate and ineffective NIDS, by generating attack data to balance normal and malicious instances. Implementing these solutions will support the development of robust NIDS capable of protecting modern industrial and robotic systems from increasingly sophisticated cyber threats.

7. Future Directions

Researchers and innovators can infuse or combine NIDS with emerging technologies to ensure robust and effective security for industrial and robotic systems. We will elaborate on some of the best of these technologies.

7.1. Federated Learning

Federated learning (FL) enables decentralized model training, allowing devices to learn collaboratively without sharing raw data. In this approach, data remains localized, and only model updates are shared, ensuring data privacy and reducing communication overhead [87]. Studies demonstrate that FL-based IDS can achieve comparable or superior accuracy to traditional deep learning models while maintaining data confidentiality [88,89], and its applications in IDS span various domains, including IoT and industrial networks [90]. Due to its decentralized nature, federated learning holds promise for intrusion detection in industrial and robotic systems. By enabling local NIDS training across industrial endpoints, FL will ensure that sensitive factory data remains on-site, safeguarding proprietary processes and sensitive information while building robust, location-specific models. FL can also adaptively enhance NIDS by leveraging local threat intelligence without centralizing data, making it ideal for factories with high privacy demands. For example, automotive manufacturers are testing FL to improve security in connected vehicles by continuously adapting to regional threat profiles without sharing proprietary data externally [91,92].

7.2. Large Language Models

Large language models (LLMs) are advanced AI models designed to understand, generate, and manipulate human-like language based on large datasets. These models are trained on billions of words or tokens from sources like books, articles, websites, and technical documents, allowing them to understand context, semantics, and grammar at a high level. Recent research has explored the application of large language models (LLMs) in network intrusion detection systems and broader cybersecurity contexts. LLMs demonstrate potential in various tasks, including vulnerability detection, malware analysis, and DDoS attack identification [93,94]. Studies show that LLMs, particularly BERT-based models, can effectively extract features from network data and outperform traditional machine learning methods in intrusion detection [95]. Frameworks like HuntGPT integrate LLMs with explainable AI to enhance user understanding and interaction in threat detection [96].

LLMs have shown promising results in DDoS detection, achieving high accuracy rates through fine-tuning and few-shot learning [97]. In NIDS, LLMs can parse and analyze vast amounts of industrial communication data, identifying anomalies based on deviations in patterns. Their ability to understand contextual semantics enables them to detect unusual activities and generate explanations for identified threats. For instance, an LLM-powered NIDS could monitor machine-to-machine communications in factories and detect unusual command sequences that might indicate an intrusion. This functionality is increasingly relevant in smart factories, where real-time, language-based anomaly detection aids in quick response to potential threats, reducing downtime.

7.3. Quantum Computing

Quantum computing is a revolutionary computing paradigm based on the principles of quantum mechanics, which studies the behavior of matter and energy at extremely small scales, like atoms and subatomic particles. Unlike classical computers, which process information in binary bits (0s and 1s), quantum computers use quantum bits, or qubits. These qubits can exist in multiple states simultaneously due to a property called superposition, enabling quantum computers to handle complex calculations with immense efficiency and speed. Quantum machine learning (QML) approaches, including quantum support vector machines and quantum neural networks, have shown improved accuracy and reduced training time compared with classical methods in recent research [98]. Novel approaches like quantum-enhanced machine learning Security Systems [99] and Quantum Generative Adversarial Networks [100] have shown promise in enhancing network security. Quantum computing offers unprecedented processing power, which can enhance NIDS by facilitating the rapid analysis of vast data streams typical in industrial environments. Its potential for breaking classical encryption also pushes the need for quantum-resistant cybersecurity measures within NIDS frameworks. Currently, large manufacturers and logistics firms are researching quantum-based encryption to secure industrial networks, preparing for a future where quantum threats may become viable. Quantum-enhanced NIDS could analyze network traffic at speeds unattainable by classical systems, significantly advancing detection capabilities in time-sensitive environments like robotic assembly lines.

7.4. Blockchain

Blockchain is a decentralized, distributed ledger technology that records transactions in a way that is transparent, secure, and resistant to tampering. Originally developed to support cryptocurrencies like Bitcoin, blockchain is now applied across various industries due to its ability to create trust in environments without a central authority. Blockchain provides an immutable audit trail and fosters collaboration among stakeholders for faster threat detection [101]. The technology's application extends to various domains, including IoT networks, where it strengthens security against emerging attacks [102]. Researchers have proposed frameworks for blockchain-based IDS, analyzing their effectiveness in decentralized environments [103]. Blockchain can be integrated with NIDS to enhance data transmission and logging trustworthiness. By using blockchain, NIDS can store records of detected threats and responses in an immutable ledger, ensuring traceability and preventing tampering. This is particularly valuable in industrial supply chains, where multiple vendors access factory networks. For instance, pharmaceutical manufacturers employ blockchain to verify the authenticity of equipment and software components, thereby mitigating risks from third-party suppliers. Blockchain-based NIDS can also validate data integrity across distributed industrial networks, improving trust and reducing vulnerabilities from insider threats.

7.5. Digital Twin

A digial twin is a virtual replica of a physical asset, system, or process that continuously updates in real-time based on data from its physical counterpart. This concept enables real-time monitoring, analysis, and simulation, offering a dynamic model that mirrors the physical entity's behavior and conditions and can also facilitate data sharing and control of security-critical processes. Digital twins are commonly used in industrial applications, smart manufacturing, healthcare, and even urban planning, as they provide a comprehensive view of how a system operates under varying conditions. They can be equipped with intrusion detection algorithms to detect and classify attacks in real time [104]. A novel hybrid model combining deep learning and random forest for intrusion detection in digital twins has shown promising results [105]. Digital twins can empower NIDS by providing a mirrored environment to simulate attacks and assess the impact of intrusion without affecting actual operations. NIDS models trained on digial twins can proactively identify and counteract attacks based on simulated threat scenarios, offering a powerful preemptive security tool. For example, aerospace and automotive industries utilize digial twins to test and validate system robustness under simulated cyberattacks, minimizing risk before deployment. Industrial systems can continuously monitor and detect deviations by integrating real-time digial twin data with NIDS, strengthening cybersecurity within dynamic manufacturing processes.

These technologies, when integrated into NIDS, can significantly advance the capabilities of intrusion detection in industrial and robotic systems. Leveraging FL, LLMs, quantum computing, blockchain, and digial twins opens pathways for robust, efficient, and intelligent NIDS, addressing the unique cybersecurity challenges in the industrial and robotic sectors.

8. Conclusions

As cyber threats continue to escalate, this survey highlights the importance of advanced and specialized Network Intrusion Detection Systems (NIDS) tailored specifically to the needs of industrial and robotic environments. Current solutions, including machine learning and deep learning-based NIDS, show promise in achieving low false positive rates, high detection accuracy, and operational efficiency. However, these advancements face challenges related to legacy systems, complex network protocols, real-time monitoring needs, and resource limitations. To address these, emerging technologies such as federated learning, blockchain, quantum computing, and digital twins are poised to provide enhanced data security, privacy, and adaptability within NIDS. By integrating these technologies, future NIDS can evolve into more resilient, scalable, and intelligent systems, effectively protecting industrial and robotic systems against advanced cyber threats. This study emphasizes the ongoing need for robust, sector-specific NIDS solutions, offering a pathway for continued research and development to fortify industrial cybersecurity.

Author Contributions: Conceptualization, S.Y., O.O., and K.R.; writing—original draft preparation, R.H.; writing—review and editing, S.Y., O.O. and K.R.; supervision, S.Y., O.O., and K.R. All authors have read and agreed to the published version of the manuscript.

Funding: This research was funded by the National Science Foundation (NSF) Engineering Research Center (EEC-2133630), Hybrid Autonomous Manufacturing Moving from Evolution to Revolution (HAMMER).

Data Availability Statement: No new data were created or analyzed in this study. Data sharing is not applicable to this article.

Conflicts of Interest: The authors declare no conflicts of interest.

Abbreviations

The following abbreviations are used in this manuscript:

NIDS	Network-based Intrusion Detection System
ICS	Industrial Control Systems
AI	Artificial Intelligence
IDS	Intrusion Detection Systems
DoS	Denial of Service
CNN	Convolutional Neural Networks
LSTM	Long Short-Term Memory

SVM	Support Vector Machine
IoT	Internet of Things
IIoT	Industrial Internet of Things
SHAP	Sharply Adaptive Explanations
MLP	Multi-Layer Perception
DNN	Deep Neural Networks
FFDNN	Feed-Forward Deep Neural Networks
RNN	Recurrent Neural Network
RBM	Restricted Boltzmann Machines
DBN	Deep Belief Networks
KNN	K-Nearest Neighbor
GNB	Gaussian Naive Bayes
DT	Decision Tree
RF	Random Forest
GBC	Gradient Boosting Classifier
DRL	Deep Reinforcement Learning
GANs	Generative Adversarial Networks
MES	Manufacturing Executive System
ANNs	Artificial Neural Networks
FFNN	Feedforward Neural Networks
RBNN	Radial Basis Neural Networks
PFCM	Probabilistic Fuzzy C-Means
IDE	Integrated Development Environment
IT	Information Technology
OT	Operational Technologies
PCA	Principal Component Analysis
PPO2	Proximal Policy Optimization
APCs	Automate Process Control Systems
GNNs	Graph Neural Networks
DLDA	Direct Linear Discriminant Analysis
SSS	Small Sample Size
NN	Nearest-Neighbor
LDA	Linear Discriminant Analysis
CPS	Cyber–Physical System
APTs	Advanced Persistent Threats

References

1. Kheddar, H.; Himeur, Y.; Awad, A.I. Deep transfer learning for intrusion detection in industrial control networks: A comprehensive review. *J. Netw. Comput. Appl.* **2023**, *220*, 103760. [CrossRef]
2. Jin, K.; Zhang, L.; Zhang, Y.; Sun, D.; Zheng, X. A network traffic intrusion detection method for industrial control systems based on deep learning. *Electronics* **2023**, *12*, 4329. [CrossRef]
3. Rajasa, M.C.; Rahma, F.; Rachmadi, R.F.; Pratomo, B.A.; Purnomo, M.H. A Review of Imbalanced Datasets and Resampling Techniques in Network Intrusion Detection System. In Proceedings of the 2023 8th International Conference on Information Technology and Digital Applications (ICITDA), Yogyakarta, Indonesia, 4–5 October 2023; pp. 1–6.
4. Balla, A.; Habaebi, M.H.; Elsheikh, E.A.; Islam, M.R.; Suliman, F.E. The Effect of Dataset Imbalance on the Performance of SCADA Intrusion Detection Systems. *Sensors* **2023**, *23*, 758. [CrossRef] [PubMed]
5. Keserwani, P.K.; Govil, M.C.; Pilli, E.S. An Effective NIDS Framework Based on a Comprehensive Survey of Feature Optimization and Classification Techniques. *Neural Comput. Appl.* **2023**, *35*, 4993–5013. [CrossRef]
6. Abdulganiyu, O.H.; Ait Tchakoucht, T.; Saheed, Y.K. A systematic literature review for network intrusion detection system (IDS). *Int. J. Inf. Secur.* **2023**, *22*, 1125–1162. [CrossRef]
7. Thakkar, A.; Lohiya, R. A review on challenges and future research directions for machine learning-based intrusion detection system. *Arch. Comput. Methods Eng.* **2023**, *30*, 4245–4269. [CrossRef]
8. Suleiman, F.; Iliyasu, U.; Abubakar, M. Review on the Network Intrusion Detection Systems (NIDS). *BIMA J. Sci. Technol.* **2024**, *8*, 141–155.
9. Abdulganiyu, O.H.; Tchakoucht, T.A.; Saheed, Y.K. Towards an efficient model for network intrusion detection system (IDS): Systematic literature review. *Wirel. Netw.* **2024**, *30*, 453–482. [CrossRef]

10. Gala, Y.; Vanjari, N.; Doshi, D.; Radhanpurwala, I. AI based techniques for network-based intrusion detection system: A review. In Proceedings of the 2023 10th International Conference on Computing for Sustainable Global Development (INDIACom), New Delhi, India, 15–17 March 2023; pp. 1544–1551.
11. Panchal, R.K.; Snehkunj, R.; Panchal, V.V. A Survey on Network-based Intrusion Detection System using Learning Techniques. In Proceedings of the 2024 5th International Conference on Image Processing and Capsule Networks (ICIPCN), Dhulikhel, Nepal, 3–4 July 2024; pp. 740–747.
12. Yogesh; Goyal, L.M. A Systematic Literature Review of Network Intrusion Detection System Models. In Proceedings of the International Conference on Paradigms of Communication, Computing and Data Analytics, Delhi, India, 22–23 April 2023; Springer Nature: Singapore, 2023; pp. 453–468.
13. Schrötter, M.; Niemann, A.; Schnor, B. A Comparison of Neural-Network-Based Intrusion Detection against Signature-Based Detection in IoT Networks. *Information* **2024**, *15*, 164. [CrossRef]
14. Abdulkareem, S.A.; Foh, C.H.; Shojafar, M.; Carrez, F.; Moessner, K. Network Intrusion Detection: An IoT and Non IoT-Related Survey. *IEEE Access* **2024**, *12*, 147167–147191. [CrossRef]
15. Idrissi, H.K.; Kartit, A. Network Intrusion Detection using Combined Deep Learning Models: Literature Survey and Future Research Directions. *IAENG Int. J. Comput. Sci.* **2024**, *51*, 998–1010.
16. Momand, A.; Jan, S.U.; Ramzan, N. A systematic and comprehensive survey of recent advances in intrusion detection systems using machine learning: Deep learning, datasets, and attack taxonomy. *J. Sens.* **2023**, *2023*, 6048087. [CrossRef]
17. Kaouk, M.; Flaus, J.; Potet, M.; Groz, R. A review of intrusion detection systems for industrial control systems. In Proceedings of the 2019 6th International Conference on Control, Decision and Information Technologies (CoDIT), Paris, France, 23–26 April 2019; pp. 1699–1704.
18. Ring, M.; Wunderlich, S.; Scheuring, D.; Landes, D.; Hotho, A. A survey of network-based intrusion detection data sets. *Comput. Secur.* **2019**, *86*, 147–167. [CrossRef]
19. Gillen, R.E.; Carter, J.M.; Craig, C.; Johnson, J.A.; Scott, S.L. Assessing anomaly-based intrusion detection configurations for industrial control systems. In Proceedings of the 2020 IEEE 21st International Symposium on "A World of Wireless, Mobile and Multimedia Networks" (WoWMoM), Cork, Ireland, 31 August–3 September 2020; pp. 360–366.
20. Kabore, R.; Kouassi, A.; N'goran, R.; Asseu, O.; Kermarrec, Y.; Lenca, P. Review of anomaly detection systems in industrial control systems using deep feature learning approach. *Engineering* **2021**, *13*, 30–44. [CrossRef]
21. Rosa, L.; Cruz, T.R.; Freitas, M.B.; Quitério, P.; Henriques, J.; Caldeira, F.; Monteiro, E.; Simões, P. Intrusion and anomaly detection for the next-generation of industrial automation and control systems. *Future Gener. Comput. Syst.* **2021**, *119*, 50–67. [CrossRef]
22. Ganeshan, R.; Daniya, T. A systematic review on anomaly-based intrusion detection system. *IOP Conf. Ser. Mater. Sci. Eng.* **2020**, *981*, 022010. [CrossRef]
23. Alsoufi, M.A.; Razak, S.B.; Siraj, M.M.; Nafea, I.T.; Ghaleb, F.A.; Saeed, F.; Nasser, M. Anomaly-based intrusion detection systems in IoT using deep learning: A systematic literature review. *Appl. Sci.* **2021**, *11*, 8383. [CrossRef]
24. Flaus, J.; Georgakis, J. Review of machine learning-based intrusion detection approaches for industrial control systems. In Proceedings of the Computer & Electronics Security Applications Rendez-vous (C&ESAR) Conference, Rennes, France, 19–21 November 2018.
25. Haripriya, L.; Jabbar, M.A. Role of machine learning in intrusion detection system. In Proceedings of the 2018 Second International Conference on Electronics, Communication and Aerospace Technology (ICECA), Coimbatore, India, 29–31 March 2018; pp. 925–929.
26. Chattopadhyay, M.; Sen, R.; Gupta, S. A comprehensive review and meta-analysis on applications of machine learning techniques in intrusion detection. *Australas. J. Inf. Syst.* **2018**, *22*. [CrossRef]
27. Mishra, N.; Mishra, S. A review of machine learning-based intrusion detection system. *EAI Endorsed Trans. Internet Things* **2024**, *10*. [CrossRef]
28. Alani, M.M.; Damiani, E.; Ghosh, U. DeepIIoT: An explainable deep learning-based intrusion detection system for industrial IoT. In Proceedings of the 2022 IEEE 42nd International Conference on Distributed Computing Systems Workshops (ICDCSW), Bologna, Italy, 10 July 2022; pp. 169–174.
29. Yadav, S.; Kalpana, R. A survey on network intrusion detection using deep generative networks for cyber-physical systems. In *Artificial Intelligence Paradigms for Smart Cyber-Physical Systems*; IGI Global: Hershey, PA, USA, 2021; pp. 137–159.
30. Abdel-Wahab, M.S.; Neil, A.M.; Atia, A. A comparative study of machine learning and deep learning in network anomaly-based intrusion detection systems. In Proceedings of the 2020 15th International Conference on Computer Engineering and Systems (ICCES), Cairo, Egypt, 15–16 December 2020; pp. 1–6.
31. Lansky, J.; Ali, S.; Mohammadi, M.; Majeed, M.K.; Karim, S.H.T.; Rashidi, S.; Hosseinzadeh, M.; Rahmani, A.M. Deep learning-based intrusion detection systems: A systematic review. *IEEE Access* **2021**, *9*, 101574–101599. [CrossRef]
32. Aldweesh, A.; Derhab, A.; Emam, A.Z. Deep learning approaches for anomaly-based intrusion detection systems: A survey, taxonomy, and open issues. *Knowl.-Based Syst.* **2020**, *189*, 105124. [CrossRef]
33. Khacha, A.; Saadouni, R.; Harbi, Y.; Aliouat, Z. Hybrid deep learning-based intrusion detection system for industrial internet of things. In Proceedings of the 2022 5th International Symposium on Informatics and its Applications (ISIA), M'sila, Algeria, 29–30 November 2022; pp. 1–6.

34. Maseno, E.M.; Wang, Z.; Xing, H. A systematic review on hybrid intrusion detection system. *Secur. Commun. Netw.* **2022**, *2022*, 9663052. [CrossRef]
35. Zhang, L.; Zhang, J.; Chen, Y.; Lao, S. Hybrid intrusion detection based on data mining. In Proceedings of the 2018 11th International Conference on Intelligent Computation Technology and Automation (ICICTA), Changsha, China, 22–23 September 2018; pp. 299–301.
36. Hasan, M. A Hybrid Real-Time Intrusion Detection System for an Internet of Things Environment with Signature and Anomaly-Based Intrusion Detection. Ph.D. Thesis, National College of Ireland, Dublin, Ireland, 2019.
37. Sverko, M.; Grbac, T.G. Complex systems - network component security of SCADA systems. In Proceedings of the 2021 44th International Convention on Information, Communication and Electronic Technology (MIPRO), Opatija, Croatia, 24–28 May 2021; pp. 1630–1635.
38. Duo, W.; Zhou, M.; Abusorrah, A. A survey of cyber attacks on cyber physical systems: Recent advances and challenges. *IEEE/CAA J. Autom. Sin.* **2022**, *9*, 784–800. [CrossRef]
39. Hotellier, E.; Sicard, F.; Francq, J.; Mocanu, S. Standard specification-based intrusion detection for hierarchical industrial control systems. *Inf. Sci.* **2024**, *659*, 120102. [CrossRef]
40. Ayodeji, A.; Liu, Y.; Chao, N.; Yang, L. A new perspective towards the development of robust data-driven intrusion detection for industrial control systems. *Nucl. Eng. Technol.* **2020**, *52*, 2687–2698. [CrossRef]
41. Hu, Y.; Yang, A.; Li, H.; Sun, Y.; Sun, L. A survey of intrusion detection on industrial control systems. *Int. J. Distrib. Sens. Netw.* **2018**, *14*, 1550147718794615. [CrossRef]
42. Pu, H.; He, L.; Cheng, P.; Sun, M.; Chen, J. Security of Industrial Robots: Vulnerabilities, Attacks, and Mitigations. *IEEE Netw.* **2023**, *37*, 111–117. [CrossRef]
43. Bonagura, V.; Foglietta, C.; Panzieri, S.; Pascucci, F. Advanced Intrusion Detection System for Industrial Cyber-Physical Systems. *IFAC-PapersOnLine* **2022**, *55*, 265–270. [CrossRef]
44. Ali, J. Intrusion Detection Systems Trends to Counteract Growing Cyber-Attacks on Cyber-Physical Systems. In Proceedings of the 2021 22nd International Arab Conference on Information Technology (ACIT), Muscat, Oman, 21–23 December 2021; pp. 1–6.
45. Umer, M.; Sadiq, S.; Karamti, H.; Alhebshi, R.M.; Alnowaiser, K.; Eshmawi, A.A.; Song, H.; Ashraf, I. Deep Learning-Based Intrusion Detection Methods in Cyber-Physical Systems: Challenges and Future Trends. *Electronics* **2022**, *11*, 3326. [CrossRef]
46. Ding, D.; Han, Q.-L.; Xiang, Y.; Ge, X.; Zhang, X.-M. A survey on security control and attack detection for industrial cyber-physical systems. *Neurocomputing* **2018**, *275*, 1674–1683. [CrossRef]
47. Jeffrey, N.; Tan, Q.; Villar, J.R. A Review of Anomaly Detection Strategies to Detect Threats to Cyber-Physical Systems. *Electronics* **2023**, *12*, 3283. [CrossRef]
48. Khalid, A.; Zainal, A.; Maarof, M.A.; Ghaleb, F.A. Advanced Persistent Threat Detection: A Survey. In Proceedings of the 2021 3rd International Cyber Resilience Conference (CRC), Virtual, 29–31 January 2021; pp. 1–6.
49. Eke, H.N.; Petrovski, A. Advanced Persistent Threats Detection based on Deep Learning Approach. In Proceedings of the 2023 IEEE 6th International Conference on Industrial Cyber-Physical Systems (ICPS), Wuhan, China, 8–11 May 2023; pp. 1–10.
50. Collins, S.; McCombie, S. Stuxnet: The emergence of a new cyber weapon and its implications. *J. Polic. Intell. Count. Terror.* **2012**, *7*, 80–91. [CrossRef]
51. Farooq, M.J.; Zhu, Q. IoT Supply Chain Security: Overview, Challenges, and the Road Ahead. *arXiv* **2019**, arXiv:1908.07828.
52. Eke, H.N.; Petrovski, A.; Ahriz, H. The use of machine learning algorithms for detecting advanced persistent threats. In Proceedings of the 2019 International Conference on Computational Science and Computational Intelligence (CSCI), Las Vegas, NV, USA, 5–7 December 2019.
53. Gan, C.; Lin, J.; Huang, D.-W.; Zhu, Q.; Tian, L. Advanced Persistent Threats and Their Defense Methods in Industrial Internet of Things: A Survey. *Mathematics* **2023**, *11*, 3115. [CrossRef]
54. Ghafir, I.; Prenosil, V. Proposed Approach for Targeted Attacks Detection. In *Advanced Computer and Communication Engineering Technology*; Springer: Berlin/Heidelberg, Germany, 2016; pp. 73–80.
55. Gan, C.; Lin, J.; Huang, D.-W.; Zhu, Q.; Tian, L. Robotics cyber security: Vulnerabilities, attacks, countermeasures, and recommendations. *Int. J. Inf. Secur.* **2022**, *21*, 1–14.
56. Dudek, W.; Szynkiewicz, W. Cyber-security for mobile service robots–challenges for cyber-physical system safety. *J. Telecommun. Inf. Technol.* **2019**, *2*, 29–36. [CrossRef]
57. Marali, M.; Sudarsan, S.D. Graceful Reincarnation of Legacy Industrial Control Systems. In Proceedings of the 2018 IEEE 3rd International Conference on Computing, Communication and Security (ICCCS), Kathmandu, Nepal, 25–27 October 2018; pp. 224–229.
58. Duque Antón, S.D.; Schotten, H.D. Putting Together the Pieces: A Concept for Holistic Industrial Intrusion Detection. *arXiv* **2019**, arXiv:1905.11701.
59. Rakas, S.V.B.; Stojanović, M.D.; Marković-Petrović, J.D. A Review of Research Work on Network-Based SCADA Intrusion Detection Systems. *IEEE Access* **2020**, *8*, 93083–93108. [CrossRef]
60. Abdullah, N.S.; Sadiq, S.; Indulska, M. Information systems research: Aligning to industry challenges in management of regulatory compliance. In Proceedings of the Pacific Asia Conference on Information Systems (PACIS), Taipei, Taiwan, 9–12 July 2010; p. 36.

61. Cunha, J.; Batista, N.; Cardeira, C.; Melicio, R. Upgrading a Legacy Manufacturing Cell to IoT. *J. Sens. Actuator Netw.* **2021**, *10*, 65. [CrossRef]
62. Papadogiannaki, E.; Chrysos, G.; Georgopoulos, K.; Ioannidis, S. A Reconfigurable IDS Framework for Encrypted and Non-Encrypted Network Data in Supply Chains. In Proceedings of the 2023 International Conference on Engineering and Emerging Technologies (ICEET), Istanbul, Turkiye, 27–28 October 2023; pp. 1–6.
63. de Moura, R.L.; Gonzalez, A.; Franqueira, V.N.L.; Maia Neto, A.L. A Cyber-Security Strategy for Internationally-dispersed Industrial Networks. In Proceedings of the 2020 International Conference on Computational Science and Computational Intelligence (CSCI), Las Vegas, NV, USA, 16–18 December 2020; pp. 62–68.
64. Nygård, A.R.; Katsikas, S. *SoK: Combating Threats in the Digital Supply Chain*; Association for Computing Machinery: New York, NY, USA, 2022; p. 128.
65. Ciocarlie, G.F.; Zhou, J. Securing Critical Infrastructure Across Cyber and Physical Dimensions. *IEEE Secur. Priv.* **2023**, *21*, 9. [CrossRef]
66. Nuaimi, M.; Chaari, L.; Hamed, B.B. Intelligent approaches toward intrusion detection systems for Industrial Internet of Things: A systematic comprehensive review. *J. Netw. Comput. Appl.* **2023**, *215*, 103637. [CrossRef]
67. Mayoral-Vilches, V. Robot Cybersecurity, a Review. *Int. J. Cyber Forensics Adv. Threat Investig.* **2022**, *3*, 11.
68. Marotta, A.; Madnick, S. *Analyzing the Interplay Between Regulatory Compliance and Cybersecurity (Revised)*; MIT Sloan School of Management, Cybersecurity Interdisciplinary Systems Laboratory (CISL): Cambridge, MA, USA, 2020.
69. Santoso, F.; Finn, A. An In-Depth Examination of Artificial Intelligence-Enhanced Cybersecurity in Robotics, Autonomous Systems, and Critical Infrastructures. *IEEE Trans. Serv. Comput.* **2024**, *17*, 1293–1310. [CrossRef]
70. Abbas, W.; Laszka, A.; Vorobeychik, Y.; Koutsoukos, X. Scheduling Intrusion Detection Systems in Resource-Bounded Cyber-Physical Systems. In Proceedings of the First ACM Workshop on Cyber-Physical Systems-Security and/or Privacy, Denver, CO, USA, 16 October 2015; pp. 55–66.
71. Vuong, T.P.; Loukas, G.; Gan, D.; Bezemskij, A. Decision tree-based detection of denial of service and command injection attacks on robotic vehicles. In Proceedings of the 2015 IEEE International Workshop on Information Forensics and Security (WIFS), Rome, Italy, 16–19 November 2015; pp. 1–6.
72. Hassanzadeh, A.; Xu, Z.; Stoleru, R.; Gu, G.; Polychronakis, M. PRIDE: Practical Intrusion Detection in Resource Constrained Wireless Mesh Networks. In *Information and Communications Security*; Springer: Berlin/Heidelberg, Germany, 2013; pp. 213–228.
73. Asulba, B.A.; Schumacher, N.; Souto, P.F.; Almeida, L.; Santos, P.M.; Martins, N.; Sousa, J. Impact of Training Set Size on Resource Usage of Machine Learning Models for IoT Network Intrusion Detection. In Proceedings of the 2023 19th International Conference on Distributed Computing in Smart Systems and the Internet of Things (DCOSS-IoT), Pafos, Cyprus, 19–21 June 2023; pp. 330–337.
74. Amaral, J.P.; Oliveira, L.M.; Rodrigues, J.J.P.C.; Han, G.; Shu, L. Policy and network-based intrusion detection system for IPv6-enabled wireless sensor networks. In Proceedings of the 2014 IEEE International Conference on Communications (ICC), Sydney, Australia, 10–14 June 2014; pp. 1796–1801.
75. Hamouda, D.; Ferrag, M.A.; Benhamida, N.; Seridi, H. Intrusion Detection Systems for Industrial Internet of Things: A Survey. In Proceedings of the 2021 International Conference on Theoretical and Applicative Aspects of Computer Science (ICTAACS), Skikda, Algeria, 15–16 December 2021; pp. 1–8.
76. Arnaboldi, L.; Morisset, C. A review of intrusion detection systems and their evaluation in the IoT. *arXiv* **2021**, arXiv:2105.08096.
77. Kang, Q.; Gu, Y. A Survey on Ransomware Threats: Contrasting Static and Dynamic Analysis Methods. *J. Cybersecur. Res.* **2023**, 2023110798. [CrossRef]
78. Abed, M.S.; Al-Doori, Q.F.; Abdullah, A.T.; Abdallah, A.A. Security Vulnerabilities and Threats in Robotic Systems: A Comprehensive Review. *Int. J. Saf. Secur. Eng.* **2023**, *13*, 3. [CrossRef]
79. Tavallaee, M.; Bagheri, E.; Lu, W.; Ghorbani, A.A. A detailed analysis of the KDD CUP 99 data set. In Proceedings of the 2009 IEEE Symposium on Computational Intelligence for Security and Defense Applications (CISDA), Ottawa, ON, Canada, 8–10 July 2009; pp. 1–6.
80. Shiravi, A.; Shiravi, H.; Tavallaee, M.; Ghorbani, A.A. Toward developing a systematic approach to generate benchmark datasets for intrusion detection. *Comput. Secur.* **2012**, *31*, 357–374. [CrossRef]
81. Mathur, A.P.; Tippenhauer, N.O. SWaT: A water treatment testbed for research and training on ICS security. In Proceedings of the 2016 International Workshop on Cyber-Physical Systems for Smart Water Networks (CySWater), Vienna, Austria, 11 April 2016; pp. 31–36.
82. Sharafaldin, I.; Lashkari, A.H.; Ghorbani, A.A. Toward generating a new intrusion detection dataset and intrusion traffic characterization. In Proceedings of the 2018 International Conference on Information Systems Security and Privacy (ICISSp), Funchal, Portugal, 22–24 January 2018; pp. 108–116.
83. Maciá-Fernández, G.; Camacho, J.; Magán-Carrión, R.; García-Teodoro, P.; Therón, R. UGR '16: A new dataset for the evaluation of cyclostationarity-based network IDSs. *Comput. Secur.* **2018**, *73*, 411–424. [CrossRef]
84. Taormina, R.; Galelli, S. Deep-learning approach to the detection and localization of cyber-physical attacks on water distribution systems. *J. Water Resour. Plan. Manag.* **2018**, *144*, 04018065. [CrossRef]
85. Zolanvari, M.; Teixeira, M.A.; Gupta, L.; Khan, K.M.; Jain, R. Machine learning-based network vulnerability analysis of industrial Internet of Things. *IEEE Internet Things J.* **2019**, *6*, 6822–6834. [CrossRef]

86. Alsaedi, A.; Moustafa, N.; Tari, Z.; Mahmood, A.; Anwar, A. TON_IoT telemetry dataset: A new generation dataset of IoT and IIoT for data-driven intrusion detection systems. *IEEE Access* **2020**, *8*, 165130–165150. [CrossRef]
87. Belenguer, A.; Navaridas, J.; Pascual, J.A. A review of Federated Learning in Intrusion Detection Systems for IoT. *arXiv* **2022**, arXiv:2204.12443.
88. Alazab, A.; Khraisat, A.; Singh, S.; Jan, T.; Alazab, M. Enhancing Privacy-Preserving Intrusion Detection through Federated Learning. *Electronics* **2023**, *12*, 3382. [CrossRef]
89. Rashid, M.M.; Khan, S.U.; Eusufzai, F.; Redwan, M.A.; Sabuj, S.R.; Elsharief, M. A Federated Learning-Based Approach for Improving Intrusion Detection in Industrial Internet of Things Networks. *Network* **2023**, *3*, 158–179. [CrossRef]
90. Fedorchenko, E.; Novikova, E.; Shulepov, A. Comparative Review of the Intrusion Detection Systems Based on Federated Learning: Advantages and Open Challenges. *Algorithms* **2022**, *15*, 247. [CrossRef]
91. Shibly, K.H.; Hossain, M.D.; Inoue, H.; Taenaka, Y.; Kadobayashi, Y. Personalized Federated Learning for Automotive Intrusion Detection Systems. In Proceedings of the 2022 IEEE Future Networks World Forum (FNWF), Montreal, QC, Canada, 12–14 October 2022; pp. 544–549.
92. Han, W.; Peng, J.; Yu, J.; Kang, J.; Lu, J.; Niyato, D. Heterogeneous Data-Aware Federated Learning for Intrusion Detection Systems via Meta-sampling in Artificial Intelligence of Things. *IEEE Internet Things J.* **2023**, *11*, 13340–13354. [CrossRef]
93. Xu, H.; Wang, S.; Li, N.; Wang, K.; Zhao, Y.; Chen, K.; Yu, T.; Liu, Y.; Wang, H. Large Language Models for Cyber Security: A Systematic Literature Review. *arXiv* **2024**, arXiv:2405.04760.
94. Lira, O.G.; Marroquin, A.; To, M.A. Harnessing the Advanced Capabilities of LLM for Adaptive Intrusion Detection Systems. In Proceedings of the International Conference on Advanced Information Networking and Applications, Fukuoka, Japan, 25–28 March 2024.
95. Lai, H. Intrusion Detection Technology Based on Large Language Models. In Proceedings of the 2023 International Conference on Evolutionary Algorithms and Soft Computing Techniques (EASCT), Wuhan, China, 15–17 May 2023; pp. 1–5.
96. Ali, T.; Kostakos, P. HuntGPT: Integrating Machine Learning-Based Anomaly Detection and Explainable AI with Large Language Models (LLMs). *arXiv* **2023**, arXiv:2309.16021.
97. Guastalla, M.; Li, Y.; Hekmati, A.; Krishnamachari, B. Application of Large Language Models to DDoS Attack Detection. In Proceedings of the SmartSP Conference, Paris, France, 6–8 June 2023.
98. Nicesio, O.K.; Leal, A.G.; Gava, V.L. Quantum Machine Learning for Network Intrusion Detection Systems, a Systematic Literature Review. In Proceedings of the 2023 IEEE 2nd International Conference on AI in Cybersecurity (ICAIC), Los Angeles, CA, USA, 14–16 March 2023; pp. 1–6.
99. Shaik, N.; Harichandana, D.B.; Chitralingappa, D.P. Quantum Computing and Machine Learning: Transforming Network Security. *Int. J. Adv. Res. Sci. Commun. Technol.* **2024**, 500–509. [CrossRef]
100. Rahman, M.A.; Shahriar, H.; Clincy, V.A.; Hossain, M.F.; Rahman, M.A. A Quantum Generative Adversarial Network-based Intrusion Detection System. In Proceedings of the 2023 IEEE 47th Annual Computers, Software, and Applications Conference (COMPSAC), Turin, Italy, 20–24 June 2023; pp. 1810–1815.
101. Issac K Varghese, E.A. Blockchain Technology in the Intrusion Detection Domain. *Int. J. Recent Innov. Trends Comput. Commun.* **2023**, *11*, 1731–1740. [CrossRef]
102. Shalabi, K.; Al-Haija, Q.A.; Al-Fayoumi, M. A Blockchain-based Intrusion Detection/Prevention Systems in IoT Network: A Systematic Review. In *Procedia Computer Science*; Elsevier: Amsterdam, The Netherlands, 2024.
103. Kaur, A.; Pawar, S.; Jore, N.; Chavan, V.; Mule, N. Intrusion Detection System using Blockchain. *Int. J. Res. Appl. Sci. Eng. Technol.* **2024**. [CrossRef]
104. Varghese, S.A.; Dehlaghi Ghadim, A.; Balador, A.; Alimadadi, Z.; Papadimitratos, P. Digital Twin-based Intrusion Detection for Industrial Control Systems. In Proceedings of the 2022 IEEE International Conference on Pervasive Computing and Communications Workshops and Other Affiliated Events (PerCom Workshops), Pisa, Italy, 21–25 March 2022; pp. 611–617.
105. Lipsa, S.; Dash, R.K. A novel intrusion detection system based on deep learning and random forest for digital twin on IoT platform. *Int. J. Sch. Res. Eng. Technol.* **2023**, *2*, 51–64.

Disclaimer/Publisher's Note: The statements, opinions and data contained in all publications are solely those of the individual author(s) and contributor(s) and not of MDPI and/or the editor(s). MDPI and/or the editor(s) disclaim responsibility for any injury to people or property resulting from any ideas, methods, instructions or products referred to in the content.

Article

Unsupervised Learning for Lateral-Movement-Based Threat Mitigation in Active Directory Attack Graphs

David Herranz-Oliveros [1], Marino Tejedor-Romero [1], Jose Manuel Gimenez-Guzman [2,*] and Luis Cruz-Piris [1]

1 Departamento de Automática, Universidad de Alcalá, 33,600, 28805 Madrid, Spain; david.herranz@uah.es (D.H.-O.); marino.tejedor@uah.es (M.T.-R.); luis.cruz@uah.es (L.C.-P.)
2 Departamento de Comunicaciones, Universitat Politècnica de València, 46022 Valencia, Spain
* Correspondence: jmgimenez@upv.es

Abstract: Cybersecurity threats, particularly those involving lateral movement within networks, pose significant risks to critical infrastructures such as Microsoft Active Directory. This study addresses the need for effective defense mechanisms that minimize network disruption while preventing attackers from reaching key assets. Modeling Active Directory networks as a graph in which the nodes represent the network components and the edges represent the logical interactions between them, we use centrality metrics to derive the impact of hardening nodes in terms of constraining the progression of attacks. We propose using Unsupervised Learning techniques, specifically density-based clustering algorithms, to identify those nodes given the information provided by their metrics. Our approach includes simulating attack paths using a snowball model, enabling us to analytically evaluate the impact of hardening on delaying Domain Administration compromise. We tested our methodology on both real and synthetic Active Directory graphs, demonstrating that it can significantly slow down the propagation of threats from reaching the Domain Administration across the studied scenarios. Additionally, we explore the potential of these techniques to enable flexible selection of the number of nodes to secure. Our findings suggest that the proposed methods significantly enhance the resilience of Active Directory environments against targeted cyber-attacks.

Keywords: cybersecurity; lateral movement; threat mitigation; unsupervised learning; attack graphs; active directory; hardening placement

Citation: Herranz-Oliveros, D.; Tejedor-Romero, M.; Gimenez-Guzman, J.M.; Cruz-Piris, L. Unsupervised Learning for Lateral-Movement-Based Threat Mitigation in Active Directory Attack Graphs. *Electronics* **2024**, *13*, 3944. https://doi.org/10.3390/electronics13193944

Academic Editors: Abdussalam Elhanashi and Pierpaolo Dini

Received: 4 September 2024
Revised: 30 September 2024
Accepted: 4 October 2024
Published: 6 October 2024

Copyright: © 2024 by the authors. Licensee MDPI, Basel, Switzerland. This article is an open access article distributed under the terms and conditions of the Creative Commons Attribution (CC BY) license (https://creativecommons.org/licenses/by/4.0/).

1. Introduction

The digital security landscape is increasingly threatened by sophisticated attackers who navigate and manipulate enterprise networks to compromise critical assets. In this context, preventing or significantly impeding lateral movement within networks is a crucial defense strategy for cybersecurity professionals. The goal is to limit the reach of such incursions and slow down adversaries as they attempt to breach the targeted infrastructure. This task is particularly challenging in complex network environments, where maintaining a balance between effective security protocols and optimal network performance is essential.

Networks incorporating Microsoft Active Directory (AD) are especially vulnerable to these challenges. AD, predominantly used in Windows-based environments to manage user permissions and resources, is a prime target for cybercriminals due to its central role in network administration [1]. A breach in AD infrastructure can lead to severe disruptions, as evidenced by several high-profile cyber incidents in recent years.

Examples include operations by the self-named Ransomware as a Service (RaaS) group, Black Basta, between 2022 and 2024, which compromised over 500 organizations. They exploited AD environments to perform domain enumeration and expand their attack surface within compromised networks. Another recent case is the ransomware attack on

Ascension Health by the Akira group in May 2024, disrupting critical healthcare operations across the United States. These incidents underscore the significance of such threats that can compromise AD, as well as the serious consequences that follow. Both of these threats have recently been highlighted by the U.S. Cybersecurity and Infrastructure Security Agency (CISA) in their *#StopRansomware* communications series [2,3].

Given this context, there is an urgent need for innovative defense mechanisms that are both effective and minimally disruptive to network performance. Our research proposes the use of Unsupervised Learning (UL) techniques, specifically density-based clustering algorithms based on centrality metrics, to enhance the security of AD infrastructures. This approach leverages concepts from graph theory to model the logical architectures that govern permissions and trusts in these environments, as well as from epidemiological models to simulate threat propagation within them [4].

As shown in Figure 1, our proposed methodology utilizes graph-based models of AD infrastructure, where network actors (computers, users, groups, domains, etc.) are represented as nodes connected by edges that denote different relationships between them.

We model the progression of attackers from their initial access point to critical organizational assets using an Attack-Path-Based (APB) approach [5]. This involves simulating a snowball attack [6,7], where attackers advance by chaining credentials and permissions from the network's entry point towards their targets, progressively compromising more AD components. In network graph terms, attackers compromise more nodes as they traverse the connecting edges. The resulting paths from the network's periphery to its central critical elements form the attack paths, which are the focus of this research.

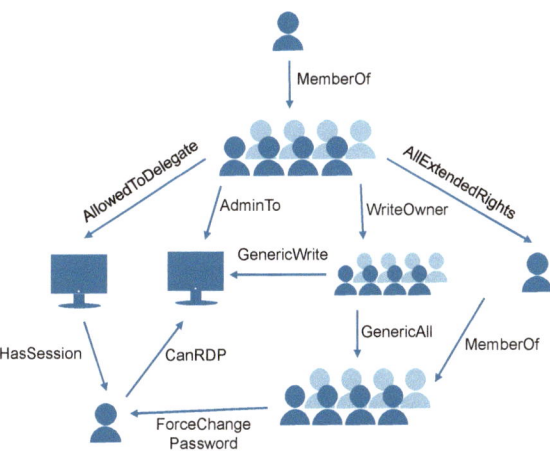

Figure 1. Example of an AD graph model illustrating some typical relationships between network actors.

We apply centrality metrics to assess the strategic importance of each node in potential attack paths to domain management. By employing unsupervised density-based clustering on these metrics, we identify and prioritize critical points where hardening (i.e., applying safeguards) can significantly mitigate the risk of lateral movement, thus protecting the organization's critical assets while minimizing operational disruption.

This strategic immunization approach mimics epidemiological interventions, targeting nodes whose security enhancements yield significant network-wide benefits. Through extensive simulations of potential attack scenarios, we assess the effectiveness of our immunization strategies in curbing the success of cyber threats. These scenarios include evaluations on both a real network graph and three more generated using various widely referenced tools (BadBlood, BloodHoundDBCreator, AD Simulator).

In parallel, we evaluate two well-known clustering algorithms (DBSCAN [8] and HDBSCAN-GLOSH [9]) to determine their respective benefits based on the nature of the analyzed networks.

We adopt a Time to Compromise (TTC)-based evaluation framework [10], which highlights the impact of our measures on the time it takes for attackers to compromise the network, specifically targeting the Domain Administration (DA). In summary, the novel contributions of our research that advance this goal are as follows:

- Network graph modeling is applied to AD infrastructures, focusing on generating subgraphs that capture the dynamics of attack paths that lateral-movement-based threats might follow from the network periphery to the DA. This approach optimizes the subsequent application of UL techniques. Unlike other studies, our research analyzes four varied AD graphs, including one from a real network infrastructure. Additionally, related studies limit their scope to a few specific types of edges in the AD attack graph (typically *MemberOf*, *AdminTo*, and *HasSession*), whereas our model encompasses all possible enumerated edges.
- UL techniques combined with centrality metrics commonly applied in epidemiology are used to identify network elements whose hardening can delay the time it takes for threats to reach the DA in the AD, thereby reducing the likelihood of full network compromise. In contrast to most related studies, we do not consider these metrics in isolation but instead employ clustering to detect global anomalies throughout the attack graph.
- HDBSCAN-GLOSH is analyzed and evaluated as a method for identifying candidate network points for hardening, which does not depend on clustering groupings. This includes a comparative accuracy analysis against DBSCAN for the same purpose. This method enables the flexible determination and prioritization of the desired number of candidates based on a continuous scoring system.
- Inspired by epidemiology, we assess the impact of hardening identified network points through a tailored compartmental Susceptible–Infected (SI) propagation model, which realistically simulates all possible lateral movement dynamics from the network's periphery (where threats typically originate) to the DA. Our model stands out by considering multidirectional propagation, allowing us to analyze the global impact of intrusions on the attack graph without requiring the attacker to iteratively choose a specific path.
- Yielding on the stochastic nature of the aforementioned propagation model, we construct a Continuous-Time Markov Chain (CTMC) to reflect, probabilistically, the impact of countermeasures on slowing down attack progress (i.e., a higher success rate per edge decreases the time it takes for an attacker to traverse it).
- We address the mitigation of targeted threats in AD environments as a node-hardening–placement problem rather than one of node-blocking or edge-blocking. We achieve this by modeling hardening a node as a delay factor α on the average time required to compromise it rather than completely preventing the attacker from being able to achieve it (i.e., $\alpha = \infty$). This enables us to propose a set of nodes for hardening (node budget) that are not restricted to distinct paths towards the DA but are instead distributed across the entire graph.
- Regarding the temporal nature of our proposed model, we define a novel TTC-based metric that quantifies the delay imposed on attackers reaching the DA after applying countermeasures. This is measured throughout extensive simulations as the Median Time to Compromise DA (MTCDA).

Our approach aims to enhance the resilience of AD environments by providing a defense mechanism primarily intended for rapid and effective protection following the detection or suspicion of an ongoing intrusion. Beyond this, our ultimate goal is to offer a scalable framework for securing all types of complex and hierarchical network infrastructures against lateral movement in targeted cyber-attacks. Our experimental results demonstrate that the use of UL-based hardening–placement techniques can delay threats by 2 to 7 times in reaching the central DA of the AD, depending on the analyzed graph and the delay factor α associated with the application of countermeasures. Additionally, the findings suggest that depending on the topology of the graphs, in most cases, comparable or even superior benefits can be obtained by employing techniques that provide greater flexibility in selecting the number of nodes to harden.

This paper is structured as follows: Section 2 discusses the most relevant works related to our research. Section 3 presents the system model, including how we model lateral movement processes on AD graphs, the graphs we analyze, and the considerations regarding the origin and targets of the threats we study. Section 4 introduces the UL-based algorithms that underpin our proposed node identification strategies for threat mitigation, the centrality metrics on which they are applied, and the subgraphs on which these metrics are calculated. Section 5 presents our study's results. Finally, Section 6 summarizes our contributions and outlines further research avenues.

2. Related Work

Recent advancements in cybersecurity have leveraged graph-based models, Machine Learning (ML), and Reinforcement Learning (RL) to enhance threat detection and mitigation in multiple environments. Modeling networks as graphs to study lateral movement has been extensively utilized in previous research [4,11,12]. These models help visualize and analyze the complex relationships within a network, allowing for effective detection and mitigation of lateral movement threats. For instance, Ref. [13] employed graph theory to propose a centrality-based methodology for analyzing and reducing potential attack vectors in Microsoft cloud environments. Unlike our approach, their proposal focuses exclusively on Microsoft Azure capabilities and uses centrality metrics isolated from each other. In [14], a methodology to build a threat model based on multilayer graphs is proposed, alongside a set of techniques to reconfigure the network in order to mitigate the risk over assets.

At the same time, ML and RL have emerged as powerful tools for enhancing cybersecurity and network resilience. Ref. [15] developed ML-driven anomaly detection systems to protect against zero-day attacks, and Ref. [16] explored deep learning techniques for threat detection and anomaly analysis. Refs. [17,18] proposed the use of Graph Neural Networks (GNNs) for detecting Structural Hole Spanners (SHS) in generic dynamic networks. These studies highlight the crucial role of advanced learning algorithms in improving the detection and response capabilities against sophisticated cyber threats.

Deepening in graph-based models, the concept of APB analysis is fundamental to understanding and mitigating lateral movement in networks. Ref. [5] introduced APB analysis, where network relationships are viewed as "hops" forming attack paths within a graph. Ref. [7] developed Heat-ray, a system combining ML and combinatorial optimization to reduce the potential of identity snowball attacks within large networks. Ref. [19] further contributed to this field by modeling hops between machines in parallel graphs, utilizing centrality metrics to rank nodes for immunization (once again, unlike us, using these metrics in isolation from one another).

Building on the concept of immunization, epidemiological approaches have been adapted from public health to various fields, including cybersecurity, to model and mitigate the spread of threats. Refs. [20,21] discussed the application of these epidemiological models in public health, which have also been applied to social networks [22,23] and financial analysis [24]. The authors of [25] conducted an in-depth study on the dynamics of epidemiological infection processes in generic networks, leading to the development of a framework for simulating these processes [26], which enables the evaluation of the impact of mitigation strategies in real infrastructures. In the area of epidemiology and cybersecurity, the works of [4,10] stand out by modeling the spread of cyber threats as analogous to epidemiological infections. Notably, Ref. [4] proposed the use of clustering for identifying key nodes in lateral-movement-based threats spreading within AD networks. However, unlike our approach, their methodology did not experimentally employ a tailored epidemiological model and relied solely on DBSCAN applied to authentication graphs between computers. In [10], the authors further extended these methodologies to whole AD network graphs, integrating centrality metrics with k-shell decomposition to enhance node selection for immunization and network reconfiguration. Unlike our current approach, their proposal focuses on mitigating non-targeted threats (i.e., without a defined target) and does not explore the use of more flexible node-ranking strategies for hardening–placement, such as those based on HDBSCAN-GLOSH.

Focusing on the defense against threats in AD attack graphs, several recent works are relevant to our study, employing both edge-blocking strategies [6,27–31] and node-blocking strategies [32,33]. Refs. [27–30] primarily focus on using RL-based strategies to generate attack policies, which are then used to apply Evolutionary Diversity Optimization (EDO) techniques to identify optimal edge budgets for defense. In contrast, Refs. [6,31] adopt Fixed-Parameter Tractable (FPT)-based approaches to achieve optimal defense. However, scalability issues lead them to tightly link their proposals to the ideal characteristics of AD networks (assuming high tree-likeness and absence of cycles) or resort to GNN-based heuristics. All of these proposals are inapplicable to the problem we address, as they focus on identifying edges rather than nodes. Interestingly, although these studies use probabilistic models as we do, they distinguish between the failure rate, detection rate, and success rate for each edge, except for the authors of [31]. Similar to this work, and since we do not tackle threat detection, we focus exclusively on the success rate.

In [32,33], while the focus is indeed on node identification, their approach focuses on detection via decoy placement rather than threat mitigation. Specifically, Ref. [33] seeks to ensure intrusion detection by positioning decoys that, within a given budget, cover the maximum number of shortest paths (SPs) between the DA and a defined set of entry points to the network. Ref. [32] extends this approach by adopting a temporal-based approach, similar to our methodology (TTC-based), but with the aim of maximizing the time between threat detection and the attacker reaching the target.

These edge-blocking and node-blocking strategies also differ from our work in their use of a Stackelberg Game model, where the attacker and defender make iterative decisions based on each other's actions. This contrasts with our focus on hardening–placement strategies, as these approaches rely on blocking strategies, either by removing links (i.e., failure rate = 1) or assuming detection as the primary means of mitigation. Additionally, they limit their scope to a few specific types of edges in the AD attack graph (typically *MemberOf*, *AdminTo*, and *HasSession*), whereas our model encompasses all possible enumerated edges. Finally, none of these studies incorporate real AD attack graphs into their experiments, further distinguishing our approach.

In conclusion, the integration of graph-based models, ML, RL, and epidemiological techniques provides a comprehensive framework for enhancing cybersecurity in AD environments. The synergy between these approaches facilitates effective threat detection, mitigation, and overall network security, underscoring the importance of continuous advancements and interdisciplinary methodologies in this critical field.

3. System Model

Our proposal relies on a network model using directed graphs designed to efficiently and accessibly aggregate all information regarding an organization's various assets and the interdependencies among them. We specifically target Microsoft AD, the predominant tool for network management at the enterprise or organizational level [34]. During our research, we evaluated our proposal using four distinct AD graphs—one derived from a real, anonymized network infrastructure (referred to as RS) and three generated using different synthetic tools: BadBlood https://github.com/davidprowe/BadBlood (accessed on 1 January 2024), AD Simulator https://github.com/nicolas-carolo/adsimulator (accessed on 1 January 2024), and BloodHound DB Creator https://github.com/BloodHoundAD/BloodHound-Tools/tree/master/DBCreator (accessed on 1 January 2024) (referred to as BB, AS, and BH, respectively). RS and BB graphs were collected using SharpHound https://github.com/BloodHoundAD/SharpHound (accessed on 1 January 2024). AD

Simulator and BloodHound DB Creator are widely used in AD security research [27–30,32,33] due to the lack of studies using real graphs because of their sensitive data. To enrich our analysis, we also incorporate BadBlood, a well-regarded tool within the AD community. Unlike typical synthetic graph generators, it automates the deployment of a whole simulated AD domain, which we subsequently enumerate as a real graph. As we will discuss later, the structure of the BB graph closely resembles that of the RS graph, especially concerning the low exposure of the DA. See Table 1 for detailed information on the nodes and edges of all these graphs.

Table 1. An overview of node and edge categories in the studied graphs based on AD elements. *AD Simulator generated a graph with four domains: one primary and three empty ones. All analyses in this paper focus solely on the primary domain.

Graph		RS	BB	AS	BH
Node Type	User	102,589	99,957	10,004	99,957
	Computer	3236	3236	341	3237
	Group/Container	984	1026	154	1005
	OU	399	223	41	21
	GPO	159	2	18	22
	Domain	1	1	4 *	1
		107,368	104,445	10,562	104,243
Edge Type	AddKeyCredentialLink	0	206,382	0	0
	AddMember	959	0	3	1
	AdminTo	85	5	1467	11,181
	AllExtendedRights	734	0	10,070	0
	AllowedToDelegate	0	0	64	646
	CanPSRemote	0	3	64	0
	CanRDP	18	3	64	645
	Contains	106,180	104,390	10,386	1021
	DCSync	0	11	0	0
	ExecuteDCOM	0	2	64	646
	ForceChangePassword	204,728	0	3	1
	GenericAll	535,998	598,000	31,416	104,191
	GenericWrite	1471	104,219	10,528	2
	GetChanges	4	1	3	2
	GetChangesAll	2	1	2	3
	GetChangesInFilteredSet	0	1	0	0
	GpLink	198	2	33	42
	HasSession	449	0	15,086	99,827
	MemberOf	636,600	118,128	125,419	453,667
	Owns	105,936	158	10,527	0
	ReadLAPSPassword	0	0	2	0
	SQLAdmin	1	0	0	0
	TrustedBy	0	0	4	0
	WriteDACL	108,271	0	10,636	4
	WriteOwner	53	261	10,632	4
		1,701,687	1,131,567	236,473	671,883

Our primary objective is to assess the risk exposure of critical network elements. One prevalent risk factor is lateral movement, which enables attackers to compromise one or more low-value network elements and subsequently navigate through the network by exploiting multiple security vulnerabilities until critical assets are reached. To represent this behavior in our network graphs, we model a snowball attack [6,7] on AD. This entails the attacker's initial access to various elements of the AD network (e.g., users, computers) and the progressive expansion of the attack, element by element, through the chained use of stolen credentials and the exploitation of existing privileges among the network's entities. For cybersecurity purposes, this type of attack involving the spread of malware is categorized as a lateral-movement-based attack.

3.1. Lateral Movement through Active Directory as an Epidemiological Process

Most proposals related to AD defense cited in Section 2 utilize a Stackelberg Game model. In this framework, the attacker, starting from a defined set of entry points, makes decisions based on the defensive strategy while considering various conditions (such as prior knowledge of the network, the risk of detection, the ability to reattempt exploitation of edges, etc.).

In contrast, our approach evaluates hardening–placement in a much broader and demanding context. We simulate an attacker who moves indiscriminately throughout the network, similar to an epidemiological process, where any prior knowledge of the network structure is irrelevant and any technique for exploiting an edge in the attack graph is applicable. This model enables us to accurately capture all possible lateral-movement-based attack dynamics on AD.

We utilize an SI epidemiological model [25], which is extensively employed across various research fields, including cybersecurity. We model lateral movement as a probabilistic infection mechanism within a compartmental framework. Each simulation begins with a single initially infected node and concludes once the DA is compromised. This approach allows us to evaluate the effectiveness of threat mitigation by strategically hardening specific nodes, thereby slowing the progression toward the DA takeover.

We utilize an SI model rather than a Susceptible–Infected–Recovered (SIR) model, as the SI model allows for both preventive and reactive countermeasures. This is because the countermeasures considered in our research can be applied preventively or reactively without altering the dynamics of the SI model. These countermeasures influence the infection transmission rate between nodes in the graph rather than the state transitions (i.e., recovery) of already infected nodes. Here, S represents the set of susceptible nodes and I the infected nodes. At any time t, $S(t)$ and $I(t)$ denote their respective counts. The infection rate is β, with N being the total number of nodes. The dynamics of infection are captured by the following equations:

$$\frac{dS}{dt} = -\beta I(t) \frac{S(t)}{N}; \frac{dI}{dt} = \beta I(t) \frac{S(t)}{N} \tag{1}$$

This model assumes nodes transition from susceptible to infected. The infection rate β is expressed as τN, where τ is the transmission rate per edge. Thus, the equation for $\frac{dI}{dt}$ becomes:

$$\frac{dI}{dt} = \tau I(t)(N - I(t)) \tag{2}$$

The success rate per edge inherent in node-blocking and edge-blocking models is analogous to this transmission rate τ, carrying over its meaning to the hardening–placement problem we address. This factor allows us to model the difficulty an attacker faces in compromising a specific relationship in the graph. The higher the transmission rate of an edge, the shorter the expected time for the attacker to compromise it. The impact of hardening on this transmission rate will be discussed further in Section 5.1.

To incorporate randomness, we model the process as a CTMC using Kolmogorov equations to describe the probability of different numbers of infected nodes over time. Building on this model, we created a tailored Discrete Event Simulator (DES) to study the infection process based on the models implemented in [26]. The simulation begins with an initially infected node located as far as possible from the DA and proceeds to compromise it. Initially, all nodes are marked as susceptible, and infection events for the initial compromised node are scheduled at $t = 0$. The simulator processes the event queue iteratively, with infection events governed by an exponentially distributed ($X \sim \text{Exp}(\lambda)$) time interval Δt, determined by the rate $\lambda = \tau^{-1}$. At each iteration of the simulator at a given time t, the node(s) scheduled for infection will transition to an infected state (i.e., $S \to I$). New infection events will then be queued for all newly reachable nodes from those just infected, provided they remain susceptible. Notably, if a node has an infection event scheduled in the queue, but a new event for the same node is queued for an earlier time, the original event will be replaced by the incoming one.

This behavior of our DES is key to understanding the multidirectional nature of the attacker we model. It allows us to assess the effectiveness of our approach under attack conditions that go far beyond an attacker simply choosing the most optimal available path into the network while remaining undetected. Consider, for instance, an Advanced Persistent Threat (APT) whose presence in the network is suspected or even detected, but the full extent of the intrusion remains unknown.

Our main goal would be to avoid network collapse by minimizing the time between threat detection and complete remediation or, conversely, by delaying DA compromise as long as possible. In this scenario, we would not know from what distance to which the DA countermeasures should be applied without severely affecting the network's functionality. Additionally, the attacker might pivot through suboptimal paths to ensure persistence. Therefore, a hardening–placement strategy that proposes nodes distributed throughout the network (regardless of their distance from the DA, and even including those along the same attack paths) by prioritizing hardening over blocking (i.e., reinforcing paths rather than splitting them) emerges as a critical defense mechanism since maximizes network coverage by securing the largest possible segment [35,36].

Our DES is designed to evaluate different hardening–placement strategies by simulating a multidirectional attacker as described, subjecting them to all potential lateral movement dynamics, including the aforementioned scenarios.

3.2. Domain Administration as the Target of the Attackers

While our previous research focused on the analysis of lateral movement spreading horizontally in AD networks [10], our current contribution aims to conduct an analogous analysis regarding the vertical spreading of attackers (i.e., privilege escalation). But what does this horizontal–vertical duality mean in terms of lateral movement propagation?

To understand this differentiation, it is essential to recall the network model we are working with. In this model, we find numerous elements of the organization's network, as well as all the trust relationships, privilege grants, and memberships among them. It is no secret that the ultimate goal of attackers is to compromise the element with the highest functional capacity over the network (i.e., the highest level of privilege). Since we are dealing with environments orchestrated by AD, this critical asset is the main server or Domain Controller (DC), which is responsible for the overall DA. Based on the data model inherent to the graphs we analyze, we observe that, in all cases, administrative users or groups, as well as the computer acting as DC (if enumerated in the graph), ultimately have a direct outgoing link to the Domain-typed node. From now on, we will refer to this single node as the DA. Notice that, for simplicity, our study does not consider multi-domain AD networks.

Furthermore, keeping in mind Microsoft's privilege grant pyramid for AD-based environments (AD Tier Model), we observe a well-defined vertical hierarchy with several levels or tiers. According to this hierarchy, the DC and other assets capable of compromising the entire network are located in tier 0. These critical assets are logically the least numerous. Conversely, non-administrator user accounts and common workstations are the most numerous elements at the base of this pyramid (i.e., tier 2). Their lower value within the organization typically confers a lower degree of cybersecurity hardening. When considering that these elements are often operated by non-IT users or individuals without security knowledge, it becomes evident that one or more elements from this tier are likely to be initially compromised during an attack. This is usually where lateral movement attempts to progress from. In real-world scenarios, this often happens through social engineering techniques, phishing or spear phishing, mail spoofing, etc.

With this hierarchical pyramid in mind, it becomes much easier to understand what we mean by horizontal or vertical progression of lateral movement. When discussing points in the network that, if compromised, enable attackers to access a broad segment of the network (regardless of the value or privilege level of those exposed elements), we refer to the horizontal expansion of the lateral movement. Conversely, when discussing points in the network that allow attackers to access higher privilege elements, whether few or many, we refer to vertical progression. This is analogous to malware spreading across an AD environment horizontally along the general privilege pyramid or escalating from the bottom to the top.

In [4], the author explains how the spread of cyberattacks is fundamentally due to three types of nodes in the affected organization, based on the role they play in lateral movement progression: spreaders, gatekeepers, and escalators. Considering the network as a directed graph to model a snowball attack, spreaders are nodes that allow access to many others, i.e., horizontal propagation. On the other hand, we have nodes involved in the vertical propagation of threats, known as escalators and gatekeepers.

Escalators are nodes that enable access to higher privilege nodes, i.e., performing cross-tier logins. A clear example would be the exposure of an administrator's credentials (tier 0 or 1) on a machine regularly used by non-administrator accounts (tier 2). This could occur due to the residual presence of privileged credentials from a previous session, such as one initiated for maintenance tasks. In the attack graph, this scenario would manifest as a *HasSession* edge between the two nodes, allowing an attacker, after compromising the machine, to retrieve those credentials and, therefore, escalate.

Gatekeepers, on the other hand, are topologically close nodes or immediate ancestors of the escalators within the attack graph. This means that an attacker must pass through them to reach the escalators. Their role is specifically defined to designate them as detection agents where monitoring measures can be applied. However, since our work focuses on a post-detection scenario (or one where the risk of detection does not influence the attacker's behavior), we can operationally consider them as escalators since they also contribute indirectly to the attacker's vertical progression through the network.

Our objective is to identify this latter group (escalators and gatekeepers) related to the attackers' progression from the most mundane elements in tier 2 to the top of the pyramid in tier 0, the DA.

3.3. Identifying Entry Points to the Network

In AD environments, attackers typically aim to compromise the DA. To model a directional snowball attack, we must consider not only the targets of intrusions but also their origins. As previously mentioned, attackers usually gain initial access to the network by compromising one or more non-administrative elements with low privilege levels and, consequently, low value to the organization.

Following this premise, to evaluate the effects of threat mitigation in AD during our research, we define the set of entry points P as the network elements (nodes in the network graph) from which the attacker begins lateral movement during simulations. Inspired by existing research [6,31], we select these nodes based on their distance from the DA in the network graph. These nodes must satisfy the following conditions:

- Each entry point $p \in P$ must be part of the node set V in the analyzed graph G.
- Each entry point $p \in P$ must have at least one path leading to the DA (i.e., $p \to \text{DA}$).
- The SP from any p to the DA ($\text{SP}(p, \text{DA})$) must have the maximum path length (i.e., the number of hops) among all nodes $v \in V$ in the graph G that have a path to the DA. Note that nodes v without any path to the DA (i.e., $v \not\to \text{DA}$) are excluded from this calculation.

Considering these conditions, we formally define our selection of the entry points for attackers into the network as follows:

$$P = \{p \in V \mid (p \to \text{DA}) \wedge \text{SP}(p, \text{DA}) = \max(\{\text{SP}(v, \text{DA}) \mid v \in V \wedge (v \to \text{DA})\})\} \quad (3)$$

Additionally, we define the set R of non-reaching nodes as those nodes from which there is no path to the DA in G.

$$R = \{r \in V \mid (r \not\to \text{DA})\} \quad (4)$$

4. Unsupervised Learning for Active Directory Threat Mitigation

4.1. Domain Reachability Graphs

In the literature on attack graph studies, both generally and specifically for AD, attack graphs typically represent the paths an attacker might traverse to achieve their objectives, given one or more initial starting points [6]. In our study, we begin with network graphs representing all interactions related to privilege grants and memberships among network elements indiscriminately. We use the previously defined entry points P as access points for attackers, and the node representing the DA as their ultimate target. The subgraph extracted from the original network graph—considering the starting point, tentative attack destination, and the routes in between—can already be considered an attack graph. However, the method for extracting this attack graph, as well as the volume of data or routes it contains, is not arbitrary.

Initially, we considered composing an attack graph that strictly contains all $SP(p \in P \to DA)$. This graph would exhibit a Directed Acyclic Graph (DAG) nature, ensuring that all optimal paths an attacker would take to achieve their goal are reflected. If we also ensure a tree-like structure by guaranteeing only one path $i \to j$ between each pair of vertices (i, j), the resulting attack graph could accurately reflect the attack dynamics leading attackers optimally to the DA. But what should be the nature of the attack graph?

To answer this question, we must consider the analytical purpose of our desired attack graph. Our goal is to analyze the centrality of the attack graph nodes via UL (specifically using density-based clustering algorithms) to determine nodes where applying security safeguards would drastically reduce attackers' success chances. Therefore, referencing an attack graph that only reflects the optimal route from each node p to the DA ensures threat mitigation for those specific routes, without considering all other possible routes. These include other routes of equal length if we have taken only one $SP(p \to DA)$ between each pair (p, DA), as well as longer routes (e.g., only one hop longer), even if we calculated all $SP(p \to DA)$ of equal length between each pair. Additionally, the tree-like nature of the optimal SP-based attack graph conceals another possibility for attackers during analysis: the ease with which attackers pivot from one route to another.

Thus, we conclude that we are interested in the attack graph capturing the maximum number of possible routes and their interactions without cycles. Logically, this graph would contain not only the SPs but also all paths mediating between P and the DA (i.e., $\{(p \to DA) \mid p \in P, (p \to DA) \subseteq G\}$). The granularity of the solution we seek will depend on this. If we aim to mitigate the most critical network routes effectively and in isolation, the attack graph will contain only one or a subset of SPs between each pair (p, DA). If we seek a solution to apply safeguards in a general way throughout the network, the subset of SPs will grow (first taking, one by one, all the SPs of equal length, then starting to take, one by one, the shorter ones of greater length than the first ones, and so on) until finally resulting in the graph of all paths between each pair (p, DA).

Unfortunately, calculating the attack graph based on all routes or a subset of SPs between P and the DA is computationally impractical. Therefore, our proposed approach introduces an attack graph we call the Domain Reachability Graph (DRG). Our interest is to cover as much of the network as possible when determining where to apply safeguards to evaluate our proposal under the most demanding scenario. Thus, the DRG theoretically seeks to approximate that graph encompassing all paths of any length between each pair (p, DA) by finding the graph that contains all nodes involved in at least one route to the DA.

Recalling the previously defined set R as non-reaching nodes, the graph we seek contains all nodes $v \in V$ in the network graph G that are not part of the set R. Additionally, the set E of edges in the original graph is reduced to E', containing only the edges mediating between nodes also included in the resulting DRG. Thus:

$$\text{DRG}(V', E'): \begin{cases} V' = \{v \in V \mid v \notin R\} \\ E' = \{(u,v) \in E \mid u \in V' \wedge v \in V'\} \end{cases} \quad (5)$$

Finally, it is worth noting that this DRG generation mechanism does not exempt us from encountering cycles among different routes to the DA. While this might compromise the precision of the subsequent centrality-based analysis, we must consider the inherent tree-like nature of logical AD graphs like those we work with [6,31]. Since we generate the DRG through a mere node elimination process, the original graph's tree-like nature will remain. This property ensures the minimal presence of cycles along the routes between different node pairs in the graph and, thus, the impact of ignoring them when calculating node centrality. Notably, given the characteristics of the simulation process carried out to evaluate the proposal's effectiveness, detailed in Section 5.1, the presence of cycles in the graph will not affect the obtained results.

4.2. Centrality Analysis

Below, we present the centrality metrics calculated prior to applying UL-based techniques to them. We draw on multiple previous works, particularly those focused on cybersecurity and attack graphs [4,10,13,19], to select a set of six widely used centrality metrics that, in our view, provide valuable insights from sufficiently orthogonal perspectives when evaluating the role nodes play in the overall connectivity of the aforementioned DRGs. We decided to use a set of well-known metrics that, although widely used in the security domain, are as general-purpose as possible. A deeper exploration remains to assess the performance of more specialized metrics tailored to specific contexts.

It is worth noting that [4,10] have already demonstrated the effectiveness of using these metrics, among others, to identify key nodes relevant to lateral movement on AD through the application of density-based clustering. Ref. [13] also addresses AD security using some of the metrics we propose. Additionally, Ref. [19] highlights the benefits of adapting the calculation of these metrics to specific attack subgraphs when dealing with threats targeting a defined objective, rather than considering the entire network. In our case, we achieve this by calculating these metrics on our DRGs. Let V be the set of nodes, and E the set of edges of a given DRG:

- The betweenness centrality B_i of a node i is a measure of the node's influence over the flow of information in the network:

$$B_i = \sum_{\substack{a \neq i \neq b \\ a \neq b}} \frac{\delta_{ab}(i)}{\delta_{ab}}, \quad (6)$$

Here, δ_{ab} denotes the total number of SPs from node a to node b, and $\delta_{ab}(i)$ represents the count of those paths that pass through node i, normalized by dividing by $(V-1)(V-2)$.

- Closeness centrality C_i for a node i indicates how close the node is to all other nodes in the network:

$$C_i = \sum_{j \neq i} \frac{1}{d_{ij}}, \quad (7)$$

where d_{ij} represents the SP distance between i and j. If no path exists, $d_{ij} \to \infty$. The values are normalized by dividing by $(V-1)$.

- The eigenvector centrality e_i of a node i is determined by the principal eigenvector of the adjacency matrix:

$$\mathbf{Me} = \mu \mathbf{e}, \quad (8)$$

where **M** is the adjacency matrix, and μ is the largest eigenvalue. Normalization is achieved by dividing by the Euclidean norm $\|e\|_2$:

$$\|e\|_2 = \sqrt{\sum_{i=1}^{V} e_i^2}. \tag{9}$$

- Katz centrality z_i for node i incorporates both the number and the quality of connections:

$$z_i = \beta \mathbf{M} y_i + \gamma, \tag{10}$$

Here, **M** is the adjacency matrix, and γ is a vector representing initial centrality values, usually set to **1**. The attenuation factor β adjusts the influence of nearby versus distant nodes and must be less than the inverse of the largest eigenvalue of **M**. Normalization uses the Euclidean norm $\|y\|_2$.

- The degree centrality g_i of a node i is its degree, defined as:

$$g_i = |\{(i,j) \in E : i \neq j\}| + |\{(j,i) \in E : i \neq j\}|, \tag{11}$$

where (i,j) is a directed edge from i to j. Values are normalized by dividing by $2(V-1)$.

- The number of descendants D_i of node i represents the count of reachable nodes:

$$D_i = |\{j \in V \setminus \{i\} : i \rightarrow j\}|, \tag{12}$$

Path $i \rightarrow j$ is calculated using the Breadth First Search (BFS) algorithm [37]. The values are normalized by dividing by $(V-1)$.

4.3. Density-Based Clustering Techniques

The use of centrality metrics to evaluate the significance of elements within a network (i.e., nodes in graph-based models) is a well-established method in both general epidemiology [38,39] and cybersecurity research applied to ICT networks [4,10,13,19]. Many of these studies [13,19,38,39] explore various strategies for achieving this through ranking based on individual centrality metrics, which often limits the scope of the analysis. Ref. [19] emphasizes the importance of combining multiple metrics into a unified ranking system to better prioritize safeguards in attack graphs. However, it stops short of exploring more advanced methods for effectively aggregating these metrics beyond simple weighting and combining individual rankings.

We based our proposal on the works in [4,10]. Identifying key nodes whose hardening can significantly delay an attacker's progress toward their objectives (i.e., the DA in the case of an AD attack graph) is feasible through the use of UL-based techniques, specifically clustering. Ref. [4] introduces a novel approach by applying density-based clustering on centrality metrics in AD authentication graphs between computers, allowing the aggregation of information from multiple metrics simultaneously. The proposal is grounded in the observation that centrality values, especially when considering multiple metrics per node, tend to show high concentrations in a small subset of nodes. In Section 5.1, we explore how this phenomenon also occurs in the metrics we calculate. This leads to the idea of identifying these potentially critical nodes as outliers (i.e., nodes that remain unclustered after applying density-based clustering) in an n-dimensional metric space. This aligns with the concept highlighted in [19] (p. 3), "highly ranked hosts are more likely to be used in an attack", but offers a more comprehensive perspective centered on identifying global maxima within the examined metric space. Our previous work on detecting superspreaders in non-targeted AD reachability graphs [10], which differs from our current focus on targeted attacks as explained in Section 3.2 further supports the validity of applying UL techniques to the problem at hand. Additionally, in this work, we explore performance not only in a global anomaly detection framework but also in a combined global and local context.

We have considered two different UL algorithms based on clustering techniques: Density-Based Spatial Clustering of Applications with Noise (DBSCAN) [8] and a variant of DBSCAN called Hierarchical DBSCAN (HDBSCAN) [40]. Next, we briefly describe both clustering techniques.

The foundations of DBSCAN are based on the principle that for each node in a cluster, the neighborhood with radius ϵ must contain at least a minimum number of nodes *minsamples*. This is equivalent to stating that the cardinality of the neighborhood must exceed a certain threshold. Specifically, the concept of a cluster corresponds to the set of all nodes that are density-reachable from a core node within the cluster. DBSCAN examines the ϵ-neighborhood of each node in the graph, and if it contains more than *minsamples* nodes, a new cluster is formed. It then checks the ϵ-neighborhood of all nodes that have not yet been processed. In our study, the neighborhood is defined within the 6-dimensional space determined by the metrics specified in Section 4.2. It is important to note that both ϵ and *minsamples* are tunable hyperparameters. Given that our goal is to identify nodes with atypical centrality as candidates for safeguard application (i.e., outliers in the clustering), the *minsamples* parameter will be the one we modify. The higher the desired number of outliers, the greater the value of the *minsamples* should be.

On the other hand, HDBSCAN follows Hartigan's model of density–contour clusters and is capable of producing a complete clustering hierarchy that encompasses all possible DBSCAN-like clusterings for an infinite range of density thresholds. From this result, a simplified cluster tree can be easily extracted using Hartigan's concept of rigid clusters [41].

The hyperparameters of HDBSCAN are also tunable. In this case, the two main parameters are *minsamples* and *minclustersize*. Once again, adjusting *minsamples* will govern the number of outliers resulting from the clustering process, while *minclustersize* determines the granularity of the clustering (i.e., the minimum size a cluster must reach to be recognized as such, rather than being merged with another existing bigger cluster).

Moreover, and crucial to our work, the density-based hierarchy produced by HDBSCAN can be utilized for outlier detection using Global–Local Outlier Scores from Hierarchies (GLOSH) [40]. In our research, we focused on using this algorithm specifically for outlier detection rather than relying on clustering results in HDBSCAN. Nevertheless, from now on, we refer to this approach as HDBSCAN-GLOSH, as its execution depends on HDBSCAN. Its distinctive features are as follows:

- The ability to consider both local and global outliers during detection. This allows for the identification of anomalous values in the 6-dimensional metrics space both unidimensionally and multidimensionally.
- The result of outlier detection is independent of the number and size of the resulting clusters. HDBSCAN-GLOSH thus provides a general scoring for all nodes in the graph based on their degree of anomaly (i.e., a score indicating how anomalous each node is).

This second aspect is what specifically motivates us to include HDBSCAN-GLOSH in our study as a UL-based alternative to DBSCAN. The resulting scoring offers a much more flexible framework for outlier detection since, when selecting candidate nodes for hardening, we can choose the desired number of nodes without being constrained by the number of nodes left unassigned to any cluster after detection. Once a specific node budget has been determined, this approach also allows for the optimal application or planning of countermeasures. Within the identified set of nodes, we retain an outlier score that enables us to rank them based on the priority of action, ensuring efficient resource allocation.

5. Experiments

Below, we detail the outcomes derived from the application of UL analysis on the DRGs extracted from the different graphs whose results we compared: RS, BB, BH, and AS (see Table 2). Finally, we present and compare the simulation results obtained after applying security safeguards based on the findings from this analysis.

Table 2. A summary of the number of nodes and edges per original graph and its associated DRG.

Graph	Original		DRG			
	#Nodes	#Edges	#Nodes	%	#Edges	%
RS	107,368	1,701,687	150	0.14%	1001	0.06%
BB	104,445	1,131,567	374	0.36%	3251	0.29%
AS	10,562	236,473	2474	23.42%	49,674	21.01%
BH	104,243	671,883	10,011	9.60%	53,737	8.00%

5.1. Experimental Settings

The first step involves calculating the centrality metrics mentioned in Section 4.2 for all nodes in each of the DRGs under study. To analyze the obtained values, we refer to Figure 2. This figure presents the descending progression of values for each metric across all nodes in each graph. It is important to note that, for better data visualization, the values shown for each metric are min-max normalized, and the progression of nodes on the X-axis is logarithmic. We observe that in all graphs, to a greater or lesser extent, a relatively small subset of nodes exhibits centrality values that significantly deviate from the overall trend in at least one or more of the calculated metrics.

This distribution of values is precisely what ensures that identifying nodes with atypical centrality (i.e., outliers in the clustering) will allow us to pinpoint network elements that play a critical role in network connectivity, as discussed in Section 4.3. Here, we have to demonstrate that the identified elements play a role, such that, when security safeguards are applied to them, we achieve a significant impact on the potential intrusions based on lateral movement, both in terms of delaying the expected time to reach the DA and, therefore, the risk to which it is subjected.

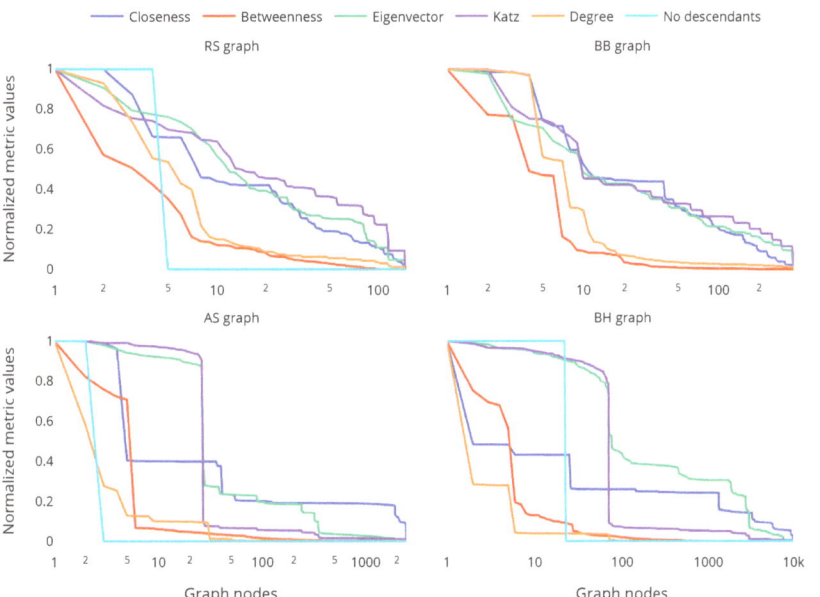

Figure 2. The distribution of centrality metrics along the DRG nodes.

Within the six-dimensional space of centrality metrics generated for each graph, we employed UL methods to identify anomalies. This approach involved the use of DBSCAN and HDBSCAN-GLOSH algorithms. To simplify the hyperparameter tuning and considering that we do not seek an excessively large number of candidate nodes for safeguard application (relative to the total number of nodes), we varied the possible values for the *minsamples* parameter in DBSCAN. For each graph, we selected the value that best suited our analysis based on the number of candidate nodes identified.

For HDBSCAN, the goal was to perform detection under conditions as similar as possible to those used in DBSCAN to allow for a fair comparison of the results obtained from both algorithms. Therefore, we determined the *minsamples* value analogously to the method used for DBSCAN in each case. As for the *minclustersize* parameter, we set it by default to be equal to the *minsamples* value.

As shown in Table 3, given the varying sizes of the DRGs associated with each network graph, and to ensure that we can analyze the results across a broader range of scenarios, we selected different *minsamples* values for each graph. This approach allowed us to generate cases with varying rates of candidate nodes for hardening while consistently maintaining a relatively small number of candidates in relation to the DRG node count.

Table 3. Overview of candidate node number per graph for applying security countermeasures.

Graph	#DRG Nodes	Minsamples	#Candidate Nodes	Candidates Rate
RS	150	5	13	8.67%
BB	374	5	19	5.08%
AS	2474	20	60	2.43%
BH	10,011	45	128	1.28%

With this in mind, the following scenarios were considered during the experimental phase for hardening–placement techniques aimed at mitigating lateral-movement-based threats:

- *Baseline*: This scenario involves no hardening techniques. The infection process will proceed under the normal conditions inherent to the network graph being simulated. This scenario serves as a baseline to observe the impact of the various strategies applied in the subsequent scenarios on the infection process.
- *ULdb*: In this scenario, hardening is applied to the nodes identified as candidates through the execution of the DBSCAN algorithm. Here, the candidate nodes are those not belonging to any of the identified clusters (i.e., those considered outliers by the clustering algorithm).
- *ULgl*: In this scenario, hardening is applied to the nodes identified as candidates through the execution of the HDBSCAN-GLOSH algorithm. For each graph, if DBSCAN identifies n candidate nodes, HDBSCAN-GLOSH will identify the n nodes with the highest anomaly scores, ensuring that the candidate sets have equal cardinality in both cases.
- *RNDngb*: In this scenario, hardening is applied to the nodes identified as candidates using the immunization algorithm proposed in [42], which favors the random selection of nodes with high degree centrality. As in the previous scenario, the number of candidate nodes selected using this strategy will match the number identified by DBSCAN for each graph. This scenario serves as a reference to evaluate whether the effect achieved by UL-based strategies represents a significant improvement. To ensure a fair comparison of results, random node selection will be performed on the DRGs, ensuring at least one path exists from any proposed node to the DA.

Given the characteristics of our simulator, as detailed in Section 3.1, it is important to note that throughout the entire simulation phase, there will be no practical difference between using the original network graphs or their corresponding DRGs. Since the simulations will start from the various entry points P identified for each graph and converge upon reaching the DA, any nodes in the original graphs that are not included in the associated DRG will not play a role in the simulated infection process. To optimize resource usage, such as memory during simulations, we will conduct the simulations on the DRGs extracted from each original graph. However, whether we use the original graphs or the DRGs will not affect the outcomes of the simulations.

To measure the effectiveness of the safeguards applied to the identified nodes, we introduce a TTC-based metric called MTCDA. It is important to remember that the simulations involve a sequence of infection events that unfold over a certain period until the DA is compromised, meaning the attacker has traversed the DRG from start to finish. The simulation ends at this point. The critical data point is when the DA is compromised or, in other words, when the simulation concludes. Naturally, our objective is to delay this moment as much as possible by implementing countermeasures. Thus, the MTCDA represents the median time it takes for the attacker to reach the DA across all iterations of the simulation. The number of iterations performed by the simulator will be determined by three factors:

- **Entry Points (*P*)**: We conducted multiple simulations, initiating the infection process from each of the entry points $p \in P$ that were initially identified for each graph. The count of these entry points can be found in Table 4.

Table 4. Number of nodes acting as entry points P for each graph.

Graph	RS	BB	AS	BH
$|P|$	9	1	3	6

- **Hardening–placement strategy**: We carried out various simulations where hardening measures were applied to mitigate lateral movement on different sets of nodes identified through the strategies described earlier (*ULdb*, *ULgl*, and *RNDngb*), as well as a scenario where no mitigation strategy was applied (*Baseline*).
- **Equal setup simulations**: Given specific initial conditions (i.e., a particular infection source node and a selected mitigation strategy, if any), and to account for the inherent randomness of the stochastic process governing the simulations, we performed 500 iterations of the simulator under these conditions. This approach ensures that the confidence intervals (CIs) for the MTCDA are precise enough.

Taking all of this into account, the total number of simulator runs I performed for each graph is given by the following expression:

$$I = |P| \times 4 \times 500 \tag{13}$$

where $|P|$ represents the cardinality of the set of entry points, the factor 4 corresponds to the number of hardening–placement strategies applied, and the factor 500 accounts for the number of iterations with the same initial setup in each case. To execute all these simulations, we used multi-threaded Python 3.10 code (optimized with Numpy and Numba), running on an Intel Xeon server with 48 cores and 128 GB of RAM.

Finally, we need to clarify how we model the application of security countermeasures (i.e., hardening) to the various network nodes identified through the aforementioned strategies. We model this by introducing a mitigation factor α that takes effect whenever the infection attempts to reach a node designated for hardening. This factor α implies that each time the infection process tries to reach one of these nodes, the expected average time for that transition to occur (derived from the mean λ^{-1} of the probability distribution $X \sim \text{Exp}(\lambda)$ governing the time between infections) will be increased by a factor of α (i.e, redefining the distribution mean as $\lambda = \alpha \tau^{-1}$). Without loss of generality, we set a uniform value of $\tau = 50\%$ for all edges in the graph and a value of $\alpha = 10$ for the incoming edges to hardened nodes. It is important to emphasize that our model aims to establish a framework for estimating the impact on the compromise time of the hardened nodes. This will enable us to assess the area of the network that we can secure through the application of countermeasures. However, a quantitative analysis of the specific impact of implementing these countermeasures remains to be conducted.

5.2. Experimental Evaluation

In this section, we present the experimental results obtained from applying the proposed methodology developed throughout this work, following the preliminary considerations detailed in Section 5.1.

We focus on the results derived from the TTC-based evaluation we propose, which aims to determine the impact of hardening by implementing security countermeasures at specific points in the network identified via UL techniques (i.e., DBSCAN and HDBSCAN-GLOSH) to mitigate lateral-movement-based threats. This mitigation is reflected in the delay imposed on potential attackers in reaching the organization's critical assets (i.e., the DA or Domain node), assuming that they initiate the compromise by lateral movement from the farthest point in the network from which this objective can be achieved.

The objective is to make it as difficult as possible for attackers to traverse the network when attempting to compromise the entire affected AD environment. This approach not only reduces the likelihood of a successful attack but also extends the available reaction time for incident response teams to take reactive measures, if necessary, before the network is fully compromised.

Figure 3 illustrates the distribution of time values (TTC-DA) that attackers, modeled as an infection process, took in each of the proposed scenarios to reach the DA, across all iterations conducted by the simulator. The figure also highlights the median value of these times (MTCDA) in each scenario. Note that the time in these figures is dimensionless, as we are focused on a consistent comparison among techniques.

At first glance, it is evident that the *ULdb* strategy yields outstanding results, not only significantly outperforming the baseline scenario but also surpassing the *RNDngb* strategy, which serves as a reference. A deeper analysis of these results can be found in Table 5, where the median values (MTCDA) for each scenario are presented along with the corresponding CI-95%. Additionally, the gain indicator $\mathcal{G}_{ULdb}$ is defined as the ratio between the MTCDA using the *ULdb* strategy and the MTCDA for the baseline scenario where no hardening has been applied to the network. For simplicity, this gain is not calculated for the other scenarios since *ULdb* clearly stands out as the most advantageous.

Table 5. MTCDA and CI-95% (lower and upper bounds) for the different graphs and hardening–placement strategies.

	Graph											
	RS			BB			AS			BH		
	Med	CI_{lo}	CI_{up}	Med	CI_{lo}	CI_{up}	Med	CI_{lo}	CI_{up}	Med	CI_{lo}	CI_{up}
Baseline	10.47	0.144	0.178	20.10	0.671	0.909	2.43	0.060	0.057	8.39	0.122	0.127
ULdb	41.87	1.214	1.195	144.60	6.751	5.116	4.65	0.095	0.111	21.91	0.326	0.234
ULgl	21.04	0.475	0.449	69.93	3.149	3.432	4.04	0.092	0.076	11.63	0.138	0.140
RNDngb	24.98	0.884	0.882	81.66	4.442	3.344	4.32	0.092	0.085	15.24	0.296	0.260
$\mathcal{G}_{ULdb}$	3.9970	-	-	7.1925	-	-	1.9168	-	-	2.6101	-	-

The results show that the expected time for attackers to reach the DA, after applying countermeasures to the nodes identified through DBSCAN, increases by approximately 2 to 7 times, depending on the graph. Furthermore, a clear trend is observed: graphs where attackers would naturally take longer to reach the DA due to the network's inherent topological characteristics are the cases where the gain is even greater. This makes logical sense, as the more of the network the attacker must traverse through lateral movement, the more likely they are to encounter the points where we have implemented the appropriate countermeasures (i.e., the application of safeguards according to our proposal has a broader effective action area).

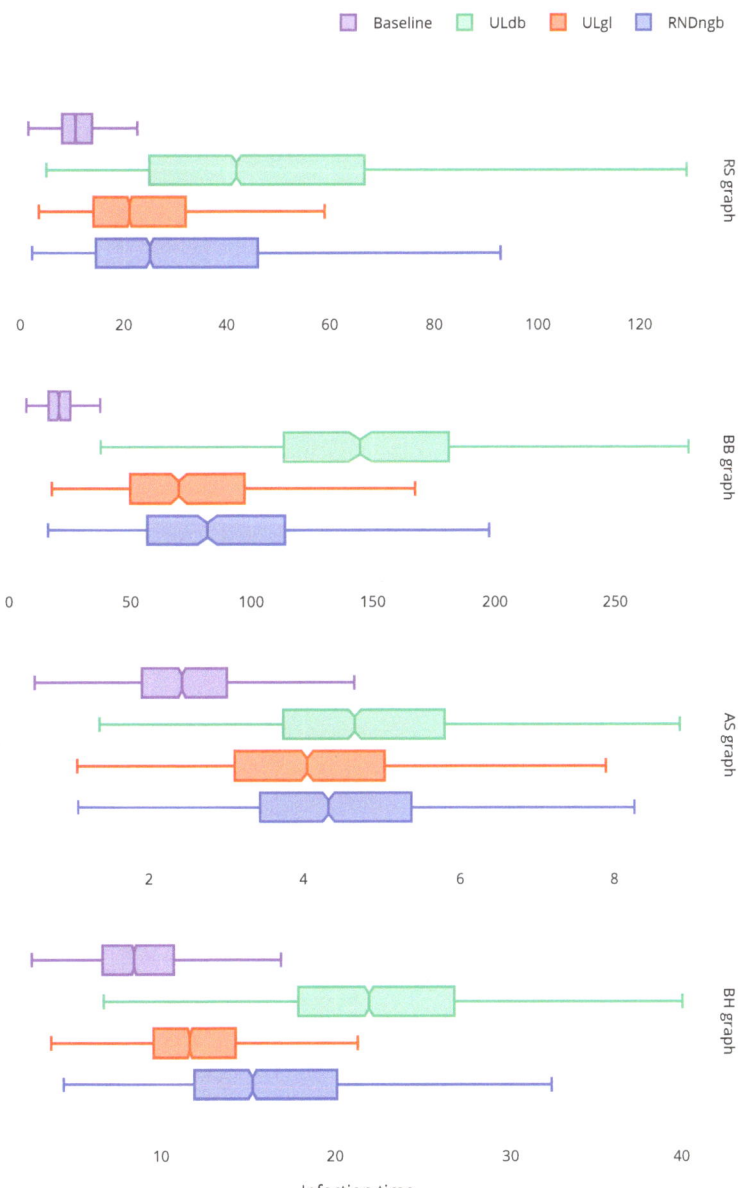

Figure 3. TTC-DA for different hardening–placement techniques.

On the other hand, we observed less favorable results with our secondary proposal, which used the *ULgl* strategy for hardening–placement. While the MTCDA did increase compared to the scenario without countermeasures, the improvement was not significant. In fact, the *RNDngb* strategy showed greater benefits in all study cases. The goal of applying the *ULgl* strategy, based on HDBSCAN-GLOSH, was to explore an alternative to *ULdb* that could provide a general anomaly score for all nodes in the graph. This would allow for flexible selection of the number of candidate nodes for hardening. Although the initial results did not meet expectations, we hypothesize that the issue may stem from incorrect hyperparameter tuning. This would imply that although HDBSCAN-GLOSH accounts for both global and local outliers, this distinction from DBSCAN is not the cause of the problem. This makes sense, as DBSCAN effectively identifies nodes crucial to network connectivity across multiple factors in the six-dimensional centrality metric space. However, a node's local importance in just one of these factors might also indicate its relevance.

We then assessed whether varying HDBSCAN-GLOSH hyperparameters could improve the performance of the *ULgl* strategy on each graph. The basic hyperparameters for HDBSCAN-GLOSH were *minsamples* and *minclustersize*. By default, we set both values to be equal, but this time we decoupled them from the *minsamples* parameter in DBSCAN.

We kept the other hardening–placement strategies (*baseline*, *ULdb*, and *RNDngb*) unchanged. We then redefined a new GLOSH-based strategy called *ULgl'*. For this strategy, the execution of HDBSCAN-GLOSH remained the same; however, we adjusted the reference values of its hyperparameters and recalculated the outlier scores. Since our goal was to match the results of applying *ULgl'* with those of *ULdb*, we selected the parameter values to identify a set of nodes for each graph with the highest possible overlap with the set previously identified by DBSCAN. Table 6 shows the parameter values used for this configuration, as well as the number and proportion of nodes identified that coincide between both strategies (i.e., *ULdb* and *ULgl'*).

Table 6. Overview of #nodes per graph identified by both DBSCAN and HDBSCAN-GLOSH, and their rate relative to DBSCAN's original candidates.

Graph	#Candidate Nodes	*Ulgl'* Hyperparameters	#Coincident Nodes	Overlap Rate
RS	13	10	12	92.31%
BB	19	25	18	94.74%
AS	60	500	30	50%
BH	128	1750	29	22.66%

In Figure 4 and Table 7, we present the extended results for each graph, including the new *ULgl'* strategy. Since the new *ULgl'* parameters were obtained by seeking to maximize node overlap with *ULdb*, it is expected that cases with higher overlap rates (i.e., RS and BB) show closer results between both strategies. Conversely, cases with lower overlap (i.e., AS and BH) are more likely to produce differing results.

Table 7. MTCDA and CI-95% (lower and upper bounds) for the different graphs and hardening–placement strategies, including *ULgl* hyperparameter variation (*ULgl'*).

	Graph											
	RS			BB			AS			BH		
	Med	CI_{lo}	CI_{up}	Med	CI_{lo}	CI_{up}	Med	CI_{lo}	CI_{up}	Med	CI_{lo}	CI_{up}
Baseline	10.47	0.144	0.178	20.10	0.671	0.909	2.43	0.060	0.057	8.39	0.122	0.127
ULdb	41.87	1.214	1.195	144.60	6.751	5.116	4.65	0.095	0.111	21.91	0.326	0.234
ULgl	21.04	0.475	0.449	69.93	3.149	3.432	4.04	0.092	0.076	11.63	0.138	0.140
ULgl'	41.39	1.260	1.122	136.07	4.913	6.859	9.93	0.234	0.300	14.90	0.224	0.204
RNDngb	24.98	0.884	0.882	81.66	4.442	3.344	4.32	0.092	0.085	15.24	0.296	0.260
$\mathcal{G}_{ULdb}$	3.9970	-	-	7.1925	-	-	1.9168	-	-	2.6101	-	-
$\mathcal{G}_{ULgl'}$	3.9513	-	-	6.7679	-	-	4.0911	-	-	1.7753	-	-

Overall, we observed a positive effect. With a simple adjustment to HDBSCAN-GLOSH hyperparameters, the TTC-DA distribution for *ULgl'* and the median value MTCDA increased significantly in three out of the four scenarios studied. This effect can be summarized by noting that the gain from this new strategy ($\mathcal{G}_{ULgl'}$) is nearly identical to that originally achieved by DBSCAN ($\mathcal{G}_{ULdb}$) for both the RS and BB graphs, with MTCDA increasing nearly 4 and 7 times, respectively, compared to the scenario without countermeasures. This was expected since the set of nodes to harden under both strategies overlapped nearly 100% for these graphs (see Table 6).

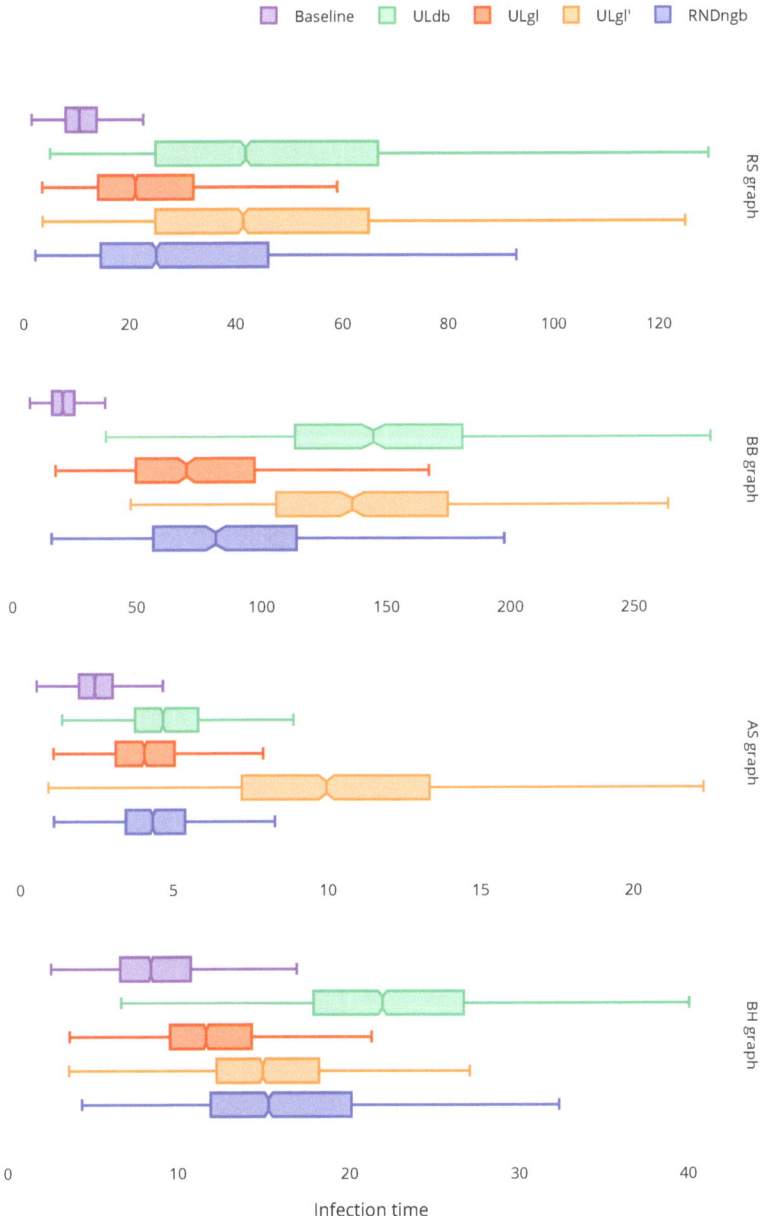

Figure 4. TTC-DA for different hardening–placement techniques, including $ULgl$ hyperparameter variation ($ULgl'$).

Additionally, we explore these results further in Figure 5, which shows, for all graphs, the histogram of outlier scores given by HDBSCAN-GLOSH segmented into 20 equal-sized bins, alongside a complementary CDF plot (i.e., CCDF or $1 - $ CDF) illustrating the ratio of nodes whose outlier score is equal to or higher than a given value. We observe that for the RS graph, the number of nodes is relatively small at high scores, gradually increasing as lower scores are considered. This suggests that there are no significant concentrations of nodes in any high sub-range of outlier scores. In a less regular and gradual manner, we observe a similar trend in the BB graph, too. Thus, by applying the *ULgl'* strategy and selecting the top percentage of nodes (i.e., *Candidates Rate*) with the highest outlier scores, we effectively capture a homogeneous set of the most relevant nodes for hardening, which is almost identical to that obtained by *ULdb*.

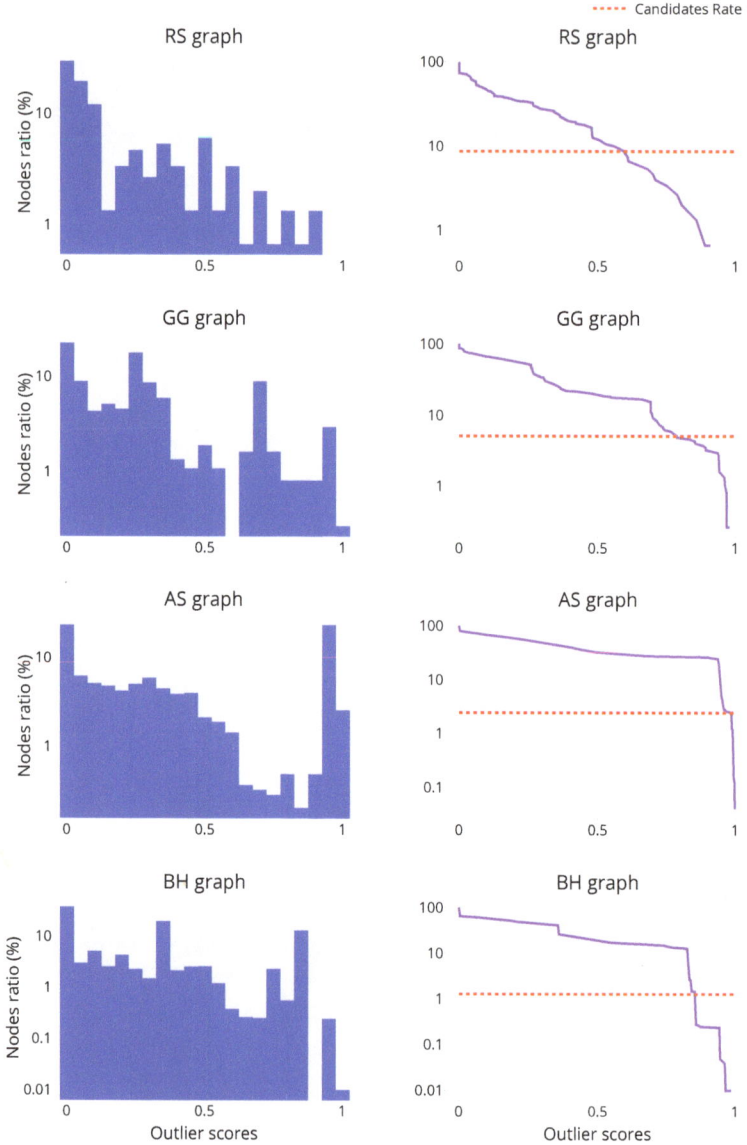

Figure 5. Histogram (**left**) and CCDF (**right**) of outlier scores given by HDBSCAN-GLOSH for each graph.

In the case of the AS graph, the overlap between the two strategies was only 50%, resulting in more unequal outcomes. Notably, the *ULgl'* strategy not only matched but significantly exceeded the results obtained by *ULdb*, increasing the MTCDA from approximately 2% to over 4%. Upon examining Figure 5, we observe a distinct trend compared to previous findings. A considerable proportion of the outlier scores given by HDBSCAN-GLOSH are concentrated in the higher ranges. This is particularly evident when considering the top decile (D_9) of the score values, which stands at 0.946 for the AS graph, whereas for the RS, BB, and BH graphs, its value is 0.56, 0.698, and 0.827, respectively. This suggests that by employing the *ULgl'* strategy, we successfully captured a subset of nodes exhibiting markedly anomalous centrality, thus highlighting their critical importance for being hardened. Furthermore, focusing on the differences between the sets of nodes identified by *ULdb* and *ULgl'*, we note that when using DBSCAN, a particularly high *minsamples* value was not configured (i.e., 20, as indicated in Table 3). This may have prevented it from considering "doubtful" outliers. Returning to Figure 5, we observe that the percentage of nodes designated as candidates (i.e., *Candidates Rate*) excludes many other nodes with similarly high scores that are close to those selected. This might explain why *ULdb* identified a distinct subset of nodes whose hardening also offers significant benefits, despite not being as pronounced as those achieved by *ULgl'*.

In the case of the BH graph, we observe that the hyperparameter adjustments applied to *ULgl'* have only a marginal impact, resulting in a slight improvement that merely brings its performance on par with the *RNDngb* strategy, which remains significantly inferior to that of *ULdb*. Notably, the node set identified by DBSCAN in this scenario required a substantially higher *minsamples* parameter to reach the desired cardinality (1.28% of DRG nodes), as compared to the other graphs (i.e., 45, as shown in Table 3). This suggests that there may be a relatively low proportion (relative to the large size of the DRG, as seen in Table 2) of strongly anomalous nodes within the clustered metric space, complicating DBSCAN's ability to detect significant outliers when *minsamples* is low. Figure 5 further supports this, as the difficulty in detecting pronounced anomalies is also evident for HDBSCAN-GLOSH. There are few nodes with high scores (i.e., 0.9 or above), and only below the threshold of 0.85 does the node count increase considerably. Furthermore, when evaluating the *Candidates Rate* threshold cutoff point, we observe that it excludes in the score range (0.8, 0.85) a significant number of nodes with scores very close to those selected. Given the large size of the BH DRG graph, which increases the need for accurate node selection, these factors force *ULgl'* to identify a set of less strongly outlier nodes, making the selection less reliable. Consequently, *ULdb* proves to be a more consistent and effective strategy for selecting nodes to harden in this case.

In summary, we observe that tuning the HDBSCAN-GLOSH hyperparameters does not necessarily provide a viable alternative to DBSCAN. Our analysis shows that in cases like the RS and BB graphs, where the subgraph of nodes capable of reaching the domain (i.e., DRG) is smaller and centrality distribution results in a progressively lower presence of outlier nodes as they become more anomalous, both hardening–placement strategies yield similar results. In cases where there is a significant presence of strongly outlier nodes, such as the AS graph, HDBSCAN-GLOSH has demonstrated its ability to capture them more extensively and reliably than DBSCAN, significantly surpassing its outcomes. Conversely, in cases like the BH graph, which exhibits a very low presence of strongly outlier nodes, HDBSCAN-GLOSH's results fall short of those achieved by DBSCAN.

Given these findings, it becomes clear that the morphology and centrality distribution of the graph play a crucial role in determining which algorithm is more effective. Further exploration across a broader range of AD attack graphs with varying structures is essential to fully understand the conditions under which DBSCAN is suitable, as well as under which HDBSCAN-GLOSH can reach or even outperform it. However, it is evident that while tuning DBSCAN's *minsamples* parameter adjusts the number of identified nodes in a somewhat uncontrolled manner, the proper tuning of HDBSCAN-GLOSH hyperparameters (without needing to control the number of identified nodes) remains essential for accurate anomaly detection.

6. Conclusions

Over time, industries and other organizations have increasingly been affected not only by the growing amount of cyber threats globally but also by the continuous advancement in the complexity and professionalization of these threats. One of the most prevalent techniques employed by these cyber threats, and whose mitigation is crucial, is lateral movement. In this study, we explore threat mitigation possibilities through the application of ML (specifically UL), focusing on network infrastructures orchestrated by Microsoft AD, one of the most widely used technologies in the world, for this purpose.

We defined DRGs that isolate the dynamics of lateral movement propagation, targeting the most critical asset of AD networks, the DA. Various widely used centrality metrics were calculated on these subgraphs, allowing us to apply density-based clustering (DBSCAN) on the resulting sets. This approach identified anomalous nodes where applying security countermeasures could significantly slow lateral movement before the domain is compromised. Additionally, we applied the HDBSCAN-GLOSH algorithm to attempt to match, or even surpass, the results obtained with DBSCAN by providing a continuous anomaly score for all nodes, enabling the flexible determination of the number of network points to harden.

Our analysis was conducted on four AD graphs: one extracted from real, anonymized infrastructure and three others generated synthetically utilizing widely used tools (BadBlood, ADSimulator, and BloodHoundDBCreator). The application of DBSCAN yielded significantly positive results. Depending on the graph and a given delay factor α (used to model node hardening), we were able to delay the compromise of DA by up to seven times compared to scenarios without countermeasures. We also found that HDBSCAN-GLOSH, depending closely on the characteristics of the analyzed graph and its proper hyperparameter tuning, can achieve results similar to or even surpassing those of DBSCAN while also providing the advantage of specifying the desired number of nodes to harden.

Despite these positive outcomes, our research remains open to further advancements. The main ones are outlined below:

- We aim to explore the optimal hyperparameter tuning characteristics for DBSCAN and HDBSCAN-GLOSH to achieve the best possible results, as well as to include a broader and more diverse range of AD graphs within the scope of our study to validate them. This includes the addition of more real attack graphs whenever possible, as well as the use of state-of-the-art synthetic generation tools like ADSynth [43].
- We plan to analyze the performance of these proposals as we adjust the budget for the number of nodes to be hardened, as well as the selection of centrality metrics used, considering both their quantity and complexity.
- We aim to conduct a quantitative analysis of the exploitability of each type of edge within the AD graph. This will enable us to estimate the exploitation difficulty associated with each edge, as well as the effort and impact involved in implementing specific countermeasures for each case.
- We will explore other UL-based clustering algorithms, including classical methods like OPTICS (Ordering Points to Identify the Clustering Structure) [44], as well as more recent ones as proposed in [45]. Additionally, we will examine other anomaly detection techniques, such as Isolation Forest [46], which is particularly suited for outlier detection.
- In addition to exploring a broader range of anomaly detection techniques (both clustering-based and otherwise), we aim to enhance our research by establishing new baselines for evaluating future proposals. To achieve this, we will explore the applicability of untapped solutions in AD security from other fields of study, such as those based on SHS [17,18].

- We are also considering the possibility of developing a general risk reduction framework that integrates our study on identifying key nodes for attackers to reach AD administration (escalators and gatekeepers) with the work in [10] on identifying generic superspreaders in AD infrastructures.

Additionally, this research could extend even further by generalizing analyses to apply to more types of logical graphs from not only AD-based infrastructures [14]. Moreover, we are considering including dynamic graphs [33,47] in our research scope, which can represent the reaction dynamics of incident response teams when a threat is being mitigated.

Author Contributions: Conceptualization, D.H.-O. and M.T.-R.; data curation, D.H.-O.; formal analysis, D.H.-O., M.T.-R. and J.M.G.-G.; funding acquisition, J.M.G.-G. and L.C.-P.; investigation, D.H.-O., M.T.-R., J.M.G.-G. and L.C.-P.; methodology, D.H.-O., M.T.-R., J.M.G.-G. and L.C.-P.; project administration, J.M.G.-G.; resources, D.H.-O., M.T.-R. and L.C.-P.; software, D.H.-O. and L.C.-P.; supervision, M.T.-R. and J.M.G.-G.; validation, D.H.-O., M.T.-R., J.M.G.-G. and L.C.-P.; visualization, D.H.-O., M.T.-R. and J.M.G.-G.; writing—original draft, D.H.-O., M.T.-R. and J.M.G.-G.; writing—review and editing, D.H.-O., M.T.-R. and J.M.G.-G. All authors have read and agreed to the published version of the manuscript.

Funding: This publication is part of project TED2021-131387B-I00 funded by MCIN/AEI/10.13039/501100011033 and by the European Union "NextGenerationEU"/PRTR and of project PID2021-123168NB-I00 funded by MCIN/AEI/10.13039/501100011033/FEDER, UE. Finally, this work is a part of the research project SBPLY/23/180225/000160, which is funded by the EU through FEDER, Spain, and by the JCCM through INNOCAM. David Herranz is also funded by both an FPU grant and a Mobility Grant for Research Staff in Training from the University of Alcalá.

Data Availability Statement: The datasets presented in this article are not readily available because data from real network infrastructures are confidential. Requests to access the other datasets should be directed to the authors.

Conflicts of Interest: The authors declare no conflicts of interest.

Abbreviations

The following abbreviations are used in this manuscript:

AD	Active Directory
APB	Attack-Path-Based
APT	Advanced Persistent Threat
BFS	Breadth First Search
(C)CDF	(Complementary) Cumulative Distribution Function
CI	Confidence Interval
CTMC	Continuous-Time Markov Chain
DAG	Directed Acyclic Graph
DA	Domain Administration
DBSCAN	Density-Based Spatial Clustering of Applications with Noise
DC	Domain Controller
DES	Discrete Event Simulator
DRG	Domain Reachability Graph
EDO	Evolutionary Diversity Optimization
FTP	Fixed-Parameter Tractable
GLOSH	Global–Local Outlier Score from Hierarchies
GNN	Graph Neural Network
HDBSCAN	Hierarchical Density-Based Spatial Clustering of Applications with Noise
ICT	Information and Communication Technology
ML	Machine Learning
MTCDA	Median Time to Compromise Domain Administration
OPTICS	Ordering Points to Identify the Clustering Structure
RaaS	Ransomware as a service
RL	Reinforcement Learning
SHS	Structural Hole Spanners

SI	Susceptible–Infected
SIR	Susceptible–Infected–Recovered
SP	Shortest Path
TTC	Time to Compromise
UL	Unsupervised Learning

References

1. Grillenmeier, G. Now's the time to rethink Active Directory security. *Netw. Secur.* **2021**, *2021*, 13–16. [CrossRef]
2. Cybersecurity and Infrastructure Security Agency. #StopRansomware: Black Basta (AA24-131A). 2024. Available online: https://www.cisa.gov/news-events/cybersecurity-advisories/aa24-131a (accessed on 29 August 2024).
3. Cybersecurity and Infrastructure Security Agency. #StopRansomware: Akira Ransomware (AA24-109A). 2024. Available online: https://www.cisa.gov/news-events/cybersecurity-advisories/aa24-109a (accessed on 29 August 2024).
4. Powell, B.A. The epidemiology of lateral movement: Exposures and countermeasures with network contagion models. *J. Cyber Secur. Technol.* **2020**, *4*, 67–105. [CrossRef]
5. Lambert, J. Defenders Think in Lists. Attackers Think in Graphs. As Long as This Is True, Attackers Win. 2015. Available online: https://perma.cc/6NZ2-A2HY (accessed on 19 June 2023).
6. Guo, M.; Li, J.; Neumann, A.; Neumann, F.; Nguyen, H. Practical fixed-parameter algorithms for defending active directory style attack graphs. In Proceedings of the AAAI Conference on Artificial Intelligence, Online, 22 February–1 March 2022; Volume 36, pp. 9360–9367.
7. Dunagan, J.; Zheng, A.X.; Simon, D.R. Heat-ray: Combating identity snowball attacks using machinelearning, combinatorial optimization and attack graphs. In Proceedings of the ACM SIGOPS 22nd Symposium on Operating Systems Principles, Big Sky, MT, USA, 11–14 October 2009; pp. 305–320.
8. Ester, M.; Kriegel, H.P.; Sander, J.; Xu, X. A density-based algorithm for discovering clusters in large spatial databases with noise. In Proceedings of the KDD, Portland, ON, USA, 2–4 August 1996; Volume 96, pp. 226–231.
9. Campello, R.J.; Moulavi, D.; Sander, J. Density-based clustering based on hierarchical density estimates. In Proceedings of the Pacific-Asia Conference on Knowledge Discovery and Data Mining, Gold Coast, Australia, 14–17 April 2013; Springer: Berlin/Heidelberg, Germany, 2013; pp. 160–172.
10. Herranz-Oliveros, D.; Marsa-Maestre, I.; Gimenez-Guzman, J.M.; Tejedor-Romero, M.; de la Hoz, E. Surgical immunization strategies against lateral movement in Active Directory environments. *J. Netw. Comput. Appl.* **2024**, *222*, 103810. [CrossRef]
11. Powell, B.A. Role-based lateral movement detection with unsupervised learning. *Intell. Syst. Appl.* **2022**, *16*, 200106. [CrossRef]
12. Bowman, B.; Laprade, C.; Ji, Y.; Huang, H.H. Detecting Lateral Movement in Enterprise Computer Networks with Unsupervised Graph AI. In Proceedings of the 23rd International Symposium on Research in Attacks, Intrusions and Defenses (RAID 2020), San Sebastian, Spain, 14–15 October 2020; USENIX Association: Berkeley, CA, USA, 2020; pp. 257–268.
13. Elmiger, M.; Lemoudden, M.; Pitropakis, N.; Buchanan, W.J. Start thinking in graphs: Using graphs to address critical attack paths in a Microsoft cloud tenant. *Int. J. Inf. Secur.* **2024**, *23*, 467–485. [CrossRef]
14. Marsa-Maestre, I.; Gimenez-Guzman, J.M.; Orden, D.; de la Hoz, E.; Klein, M. REACT: Reactive resilience for critical infrastructures using graph-coloring techniques. *J. Netw. Comput. Appl.* **2019**, *145*, 102402. [CrossRef]
15. Chen, E.; Lockey, S.; Khosravi, H.; Baghaei, N. Enhancing Cybersecurity through Machine Learning-Driven Anomaly Detection Systems. *J. Artif. Intell. Res. Appl.* **2024**, *4*, 123–135.
16. Sarker, I.H. Learning Technologies: Toward Machine Learning and Deep Learning for Cybersecurity. In *AI-Driven Cybersecurity and Threat Intelligence: Cyber Automation, Intelligent Decision-Making and Explainability*; Springer: Berlin/Heidelberg, Germany, 2024; pp. 43–59.
17. Goel, D.; Shen, H.; Tian, H.; Guo, M. Discovering Top-k Structural Hole Spanners in Dynamic Networks. *arXiv* **2023**, arXiv:2302.13292.
18. Goel, D.; Shen, H.; Tian, H.; Guo, M. Effective graph-neural-network based models for discovering Structural Hole Spanners in large-scale and diverse networks. *Expert Syst. Appl.* **2024**, *249*, 123636. [CrossRef]
19. Hong, J.B.; Kim, D.S. Scalable security analysis in hierarchical attack representation model using centrality measures. In Proceedings of the 2013 43rd Annual IEEE/IFIP Conference on Dependable Systems and Networks Workshop (DSN-W), Budapest, Hungary, 24–27 June 2013; IEEE: Piscataway, NJ, USA, 2013; pp. 1–8.
20. He, Z.Y.; Abbes, A.; Jahanshahi, H.; Alotaibi, N.D.; Wang, Y. Fractional-order discrete-time SIR epidemic model with vaccination: Chaos and complexity. *Mathematics* **2022**, *10*, 165. [CrossRef]
21. Thomas, D.M.; Sturdivant, R.; Dhurandhar, N.V.; Debroy, S.; Clark, N. A Primer on COVID-19 Mathematical Models. *Obesity* **2020**, *28*, 1375. [CrossRef]
22. Raponi, S.; Khalifa, Z.; Oligeri, G.; Di Pietro, R. Fake news propagation: A review of epidemic models, datasets, and insights. *ACM Trans. Web (TWEB)* **2022**, *16*, 1–34. [CrossRef]
23. Hosseini, S.; Zandvakili, A. Information dissemination modeling based on rumor propagation in online social networks with fuzzy logic. *Soc. Netw. Anal. Min.* **2022**, *12*, 34. [CrossRef]
24. Bucci, A.; La Torre, D.; Liuzzi, D.; Marsiglio, S. Financial contagion and economic development: An epidemiological approach. *J. Econ. Behav. Organ.* **2019**, *162*, 211–228. [CrossRef]

25. Kiss, I.Z.; Miller, J.C.; Simon, P.L. *Mathematics of Epidemics on Networks*; Springer: Cham, Switzerland, 2017; Volume 598, p. 31.
26. Miller, J.C.; Ting, T. EoN (Epidemics on Networks): A fast, flexible Python package for simulation, analytic approximation, and analysis of epidemics on networks. *J. Open Source Softw.* **2019**, *4*, 1731. [CrossRef]
27. Goel, D.; Neumann, A.; Neumann, F.; Nguyen, H.; Guo, M. Evolving Reinforcement Learning Environment to Minimize Learner's Achievable Reward: An Application on Hardening Active Directory Systems. *arXiv* **2023**, arXiv:2304.03998.
28. Goel, D.; Moore, K.; Guo, M.; Wang, D.; Kim, M.; Camtepe, S. Optimizing Cyber Defense in Dynamic Active Directories through Reinforcement Learning. In Proceedings of the European Symposium on Research in Computer Security, Bydgoszcz, Poland, 16–20 September 2024; Springer: Berlin/Heidelberg, Germany, 2024; pp. 332–352.
29. Goel, D.; Ward, M.; Neumann, A.; Neumann, F.; Nguyen, H.; Guo, M. Hardening Active Directory Graphs via Evolutionary Diversity Optimization based Policies. *ACM Trans. Evol. Learn.* **2024**. [CrossRef]
30. Goel, D.; Ward-Graham, M.H.; Neumann, A.; Neumann, F.; Nguyen, H.; Guo, M. Defending active directory by combining neural network based dynamic program and evolutionary diversity optimisation. In Proceedings of the Genetic and Evolutionary Computation Conference, Boston, MA, USA, 9–13 July 2022; pp. 1191–1199.
31. Guo, M.; Ward, M.; Neumann, A.; Neumann, F.; Nguyen, H. Scalable edge blocking algorithms for defending active directory style attack graphs. In Proceedings of the AAAI Conference on Artificial Intelligence, Washington, DC, USA, 7–14 February 2023; Volume 37, pp. 5649–5656.
32. Ngo, H.; Guo, M.; Nguyen, H. Optimizing cyber response time on temporal active directory networks using decoys. In Proceedings of the Genetic and Evolutionary Computation Conference, Melbourne, VIC, Australia, 14–18 July 2024; pp. 1309–1317.
33. Ngo, H.Q.; Guo, M.; Nguyen, H. Near Optimal Strategies for Honeypots Placement in Dynamic and Large Active Directory Networks. In Proceedings of the 2023 International Conference on Autonomous Agents and Multiagent Systems, London, UK, 29 May–2 June 2023; pp. 2517–2519.
34. Dias, J. *A Guide to Microsoft Active Directory (AD) Design*; Technical Report; Lawrence Livermore National Lab. (LLNL): Livermore, CA, USA, 2002.
35. Kang, H.; Liu, B.; Mišić, J.; Mišić, V.B.; Chang, X. Assessing security and dependability of a network system susceptible to lateral movement attacks. In Proceedings of the 2020 International Conference on Computing, Networking and Communications (ICNC), Big Island, HI, USA, 17–20 February 2020; IEEE: Piscataway, NJ, USA, 2020; pp. 513–517.
36. He, D.; Gu, H.; Zhu, S.; Chan, S.; Guizani, M. A Comprehensive Detection Method for the Lateral Movement Stage of APT Attacks. *IEEE Internet Things J.* **2023**, *11*, 8440–8447. [CrossRef]
37. Lawande, S.R.; Jasmine, J.; Anbarasi, J.; Izhar, L.I. A systematic review and analysis of intelligence-based pathfinding algorithms in the field of video games. *Appl. Sci.* **2022**, *12*, 5499. [CrossRef]
38. Sartori, F.; Turchetto, M.; Bellingeri, M.; Scotognella, F.; Alfieri, R.; Nguyen, N.K.K.; Le, T.T.; Nguyen, Q.; Cassi, D. A comparison of node vaccination strategies to halt SIR epidemic spreading in real-world complex networks. *Sci. Rep.* **2022**, *12*, 21355. [CrossRef]
39. Rodrigues, F.A. Network centrality: An introduction. In *A Mathematical Modeling Approach from Nonlinear Dynamics to Complex Systems*; Springer: Berlin/Heidelberg, Germany, 2019; pp. 177–196.
40. Campello, R.J.; Moulavi, D.; Zimek, A.; Sander, J. Hierarchical density estimates for data clustering, visualization, and outlier detection. *ACM Trans. Knowl. Discov. Data (TKDD)* **2015**, *10*, 1–51. [CrossRef]
41. Hartigan, J.A. *Clustering Algorithms*, 99th ed.; John Wiley & Sons, Inc.: Hoboken, NJ, USA, 1975.
42. Cohen, R.; Havlin, S.; Ben-Avraham, D. Efficient immunization strategies for computer networks and populations. *Phys. Rev. Lett.* **2003**, *91*, 247901. [CrossRef] [PubMed]
43. Nguyen, N.L.; Falkner, N.; Nguyen, H. ADSynth: Synthesizing Realistic Active Directory Attack Graphs. In Proceedings of the 2024 54th Annual IEEE/IFIP International Conference on Dependable Systems and Networks (DSN), Brisbane, Australia, 24–27 June 2024; IEEE: Piscataway, NJ, USA, 2024; pp. 66–74.
44. Ankerst, M.; Breunig, M.; Kriegel, H.P.; Ng, R.; Sander, J. Ordering points to identify the clustering structure. In Proceedings of the ACM SIGMOD, Vancouver, BC, Canada, 10–12 June 2008; Volume 99.
45. Shojafar, M.; Taheri, R.; Pooranian, Z.; Javidan, R.; Miri, A.; Jararweh, Y. Automatic clustering of attacks in intrusion detection systems. In Proceedings of the 2019 IEEE/ACS 16th International Conference on Computer Systems and Applications (AICCSA), Abu Dhabi, United Arab Emirates, 3–7 November 2019; IEEE: Piscataway, NJ, USA, 2019; pp. 1–8.
46. Liu, F.T.; Ting, K.M.; Zhou, Z.H. Isolation Forest. In Proceedings of the 2008 Eighth IEEE International Conference on Data Mining, Pisa, Italy, 15–19 December 2008; pp. 413–422. [CrossRef]
47. Khoury, J.; Klisura, D.; Zanddizari, H.; Parra, G.D.L.T.; Najafirad, P.; Bou-Harb, E. Jbeil: Temporal Graph-Based Inductive Learning to Infer Lateral Movement in Evolving Enterprise Networks. In Proceedings of the 2024 IEEE Symposium on Security and Privacy (SP), San Francisco, CA, USA, 19–23 May 2024; IEEE Computer Society: Piscataway, NJ, USA, 2023; p. 9.

Disclaimer/Publisher's Note: The statements, opinions and data contained in all publications are solely those of the individual author(s) and contributor(s) and not of MDPI and/or the editor(s). MDPI and/or the editor(s) disclaim responsibility for any injury to people or property resulting from any ideas, methods, instructions or products referred to in the content.

Review

A Review of Post-Quantum Privacy Preservation for IoMT Using Blockchain

Fariza Sabrina [1,*,†], Shaleeza Sohail [2,†] and Umair Ullah Tariq [1]

1 School of Engineering and Technology, Central Queensland University, Rockhampton, QLD 4701, Australia; u.tariq@cqu.edu.au
2 College of Engineering, Science and Environment, The University of Newcastle, Callaghan, NSW 2308, Australia; shaleeza.sohail@newcastle.edu.au
* Correspondence: f.sabrina@cqu.edu.au
† These authors contributed equally to this work.

Abstract: The Internet of Medical Things (IoMT) has significantly enhanced the healthcare system by enabling advanced patient monitoring, data analytics, and remote interactions. Given that IoMT devices generate vast amounts of sensitive data, robust privacy mechanisms are essential. This privacy requirement is critical for IoMT as, generally, these devices are very resource-constrained with limited storage, computation, and communication capabilities. Blockchain technology, with its decentralisation, transparency, and immutability, offers a promising solution for improving IoMT data security and privacy. However, the recent emergence of quantum computing necessitates developing measures to maintain the security and integrity of these data against emerging quantum threats. This work addresses the current gap of a comprehensive review and analysis of the research efforts to secure IoMT data using blockchain in the quantum era. We discuss the importance of blockchain for IoMT privacy and analyse the impact of quantum computing on blockchain to justify the need for these works. We also provide a comprehensive review of the existing literature on quantum-resistant techniques for effective blockchain solutions in IoMT applications. From our detailed review, we present challenges and future opportunities for blockchain technology in this domain.

Keywords: quantum computing; privacy preservation; blockchain; IoMT; post-quantum

Citation: Sabrina, F.; Sohail, S.; Tariq, U.U. A Review of Post-Quantum Privacy Preservation for IoMT Using Blockchain. *Electronics* **2024**, *13*, 2962. https://doi.org/10.3390/electronics13152962

Academic Editors: Abdussalam Elhanashi and Pierpaolo Dini

Received: 31 May 2024
Revised: 19 July 2024
Accepted: 22 July 2024
Published: 26 July 2024

Copyright: © 2024 by the authors. Licensee MDPI, Basel, Switzerland. This article is an open access article distributed under the terms and conditions of the Creative Commons Attribution (CC BY) license (https://creativecommons.org/licenses/by/4.0/).

1. Introduction

The Internet of Things (IoT) has significantly improved various aspects of our daily lives. However, these devices are vulnerable to security threats due to its resource-constrained nature. Several research efforts have provided security solutions for IoT devices using contemporary and emerging technologies [1]. IoT has a huge impact on the health domain and the rapid evolution of IoMT has enabled unprecedented levels of patient monitoring, data analytics, and remote medical interactions. As these technologies integrate more deeply into healthcare infrastructures, they generate vast amounts of sensitive medical data, necessitating robust privacy preservation mechanisms. Some of the common techniques to secure IoMT-generated sensitive medical data are cryptographic algorithms, access control mechanisms, machine learning approaches, blockchain, and steganography [2]. The advent of quantum computing provides access to new and emerging quantum-enabled security solutions for the healthcare domain [3]. However, it also poses new challenges to the existing security measures, threatening the integrity and confidentiality of IoMT data. These emerging threats require the development of post-quantum solutions to safeguard medical data against future quantum-enabled attacks [4].

One of the recent technologies used for enhancing data security for IoT-based applications is Blockchain due to its decentralisation, transparency, and immutability. For healthcare IoMT applications, this technology is a promising option for enhancing data security and privacy. One of the main strengths is that by leveraging blockchain, it is possible to create a secure and transparent environment for managing IoMT health data, where modifications are traceable and immutable, which makes it very attractive for applications with sensitive data. Additionally, blockchain can facilitate secure, decentralised data exchanges across the network without relying on trusted third parties, thus reducing vulnerability points for healthcare applications. Hence, we only consider blockchain-based security solutions for healthcare IoMT applications in this work.

Recently, quantum computing has emerged as a critical technology with an immense amount of computing power that helps solve problems that cannot be solved by classical computers [5]. However, in addition to the huge benefits that can be provided by this technology some threats to existing techniques are posed as well. A common concern in this area is the vulnerability of existing cryptographic algorithms to the quantum era. Most of these cryptographic algorithms rely on the fact that they cannot be solved by classical computers due to limited computing resources; however, quantum speed-up will break this barrier by making a huge amount of computing power available for decrypting these algorithms.

The rest of the paper is organised as follows. Section 2 discusses the importance of blockchain technology for IoMT privacy preservation. Section 3 discusses the key algorithms in post-quantum cryptography and compares those algorithms based on their strength and utilisation. Section 4 analyses how blockchain technology is affected by quantum computing due to its reliance on cryptographic algorithms. Section 5 provides a performance comparison of post-quantum cryptographic algorithms for IoT devices. in Section 6, we provide a comprehensive review of the existing literature addressing quantum-resistant techniques for effective blockchain solutions for IoMT applications. Section 7 summarises the key approaches against quantum attacks discussed in this paper and future research opportunities in this domain, and Section 8 concludes our paper.

2. Background

In this section, we will briefly look at the emerging use of IoMT in healthcare applications and emphasise the importance of privacy preservation requirements for these applications due to the inherent constraints associated with these devices. We will review some of the common techniques used for providing data privacy for these applications. The last subsection focuses on discussing the architecture and mechanism of blockchain technology that make it suitable for this purpose. With recent advancements in quantum computing, all these blockchain-based approaches can become vulnerable and will not be able to resist quantum attacks. Hence, in this paper, we will analyse all the existing research efforts in the field of post-quantum cryptography that can provide privacy for blockchain solutions for healthcare IoMT applications.

2.1. Overview of Internet of Medical Things

The Internet of Medical Things can be defined as a network comprising medical devices and individuals, utilising wireless communication to facilitate the exchange of healthcare information [6]. IoMT enables the seamless collection, analysis, and transmission of health data, empowering healthcare providers to make better decisions, monitor patients more effectively, and provide personalised care remotely [7]. Ultimately it enhances healthcare delivery by making it more efficient, cost-effective, and patient-centric.

Key aspects of IoMT include its ability to connect various medical devices and sensors that monitor vital patient data in real time. These devices can communicate and share information over the Internet, allowing for immediate medical interventions and continuous patient monitoring without the need for physical presence. This technology is particularly beneficial in managing chronic diseases, improving emergency responses, and optimising hospital operations by reducing unnecessary visits and streamlining processes [7].

Irrespective of the transformative benefit that IoMT brings to healthcare applications, it presents several significant challenges in terms of privacy, security, heterogeneity, interoperability, the availability of data, etc. IoMT devices come with diverse specifications in terms of hardware and software (such as connections, power requirements, processing capabilities, supported protocols, and security measures), and this heterogeneity among the devices which can be part of one network elevates the system's susceptibility cyber attacks. Hence, the use and integration of IoMT devices into healthcare systems present both a technical challenge and a critical opportunity for improving patients' experience [8].

2.2. Importance of Privacy Preservation for IoMT

Privacy preservation refers to the techniques and methods used to protect sensitive and private information from unauthorised access, disclosure, or misuse while maintaining the functionality and quality of service of the application [9]. Privacy preservation in medical data is crucial due to the highly sensitive nature of personal health information. Unauthorised access can lead to significant privacy violations and misuse of data, including identity theft and discrimination. Medical data often include not only health information but also personally identifiable information, which, if exposed, can harm the patient's privacy and security. Additionally, the potential for cyber attacks on healthcare systems necessitates robust security measures to protect these data from being compromised.

IoMT devices collect highly sensitive health and medical data, including medical histories, current health status, and biometric information. Ensuring the privacy of these data from unauthorised access is essential to maintain patient privacy and trust in healthcare applications [10].

2.3. Privacy Preservation Techniques for IoMT

A significant amount of work has been conducted in the existing literature highlighting the technologies that could protect the privacy of medical data. Some of the technologies that could be used to ensure privacy in IoMT are as follows:

- Cryptographic Algorithms: Cryptographic algorithms include methods that are used to secure and protect data and communication by applying complex mathematical and logical techniques. Data encryption is one of the principal applications of cryptography that ensures the data are encrypted at their sources and decrypted only at their destination, and hence it prevents unauthorised access during the transmission of data. Homomorphic encryption and secret sharing ensure that even when data are processed or analysed by third-party systems, the privacy and integrity of the data remain intact, preventing any unauthorised access or interpretation of sensitive information [11].
Putra et al. [12] highlight that centralised medical data repositories, while useful for streamlined data access and efficient healthcare workflows, are vulnerable to security threats such as unauthorised access and cyber attacks. These threats jeopardize the integrity and confidentiality of sensitive patient information. To mitigate these risks, the authors emphasise the necessity of robust security measures, including encryption, access control, and breach detection mechanisms.
- Anonymisation and Pseudonymisation: Anonymisation removes personally identifiable information, making data untraceable to individuals. Pseudonymisation replaces private data identifiers with artificial identifiers or pseudonyms, protecting privacy while allowing data analysis without directly exposing personal information [13].
- Privacy-Preserving Machine Learning: Machine learning has been one of the most used approaches when it comes to data privacy preservation in any domain [14]. Different machine learning approaches like SVM, CNN, deep learning, and multiple ensemble approaches have been used for the detection and mitigation of security and privacy attacks when IoT devices are used [15]. Distributed machine learning approaches like federated learning allow model training on multiple devices with local data, which eliminates the requirement of exchanging data and, hence, minimises privacy concerns [16]. To further mitigate privacy attacks, Differential Privacy can be utilised, which adds noise to the data or to the outputs of data analyses to obscure the presence or absence of individuals in the data set [17].

- Edge Computing: Processing data at the edge (closer to where they are generated) reduces the need to transmit sensitive information over the network, thus enhancing privacy. As mentioned in [12], cloud-edge computing, federated learning, and AI could secure and preserve the privacy of medical data in the IoMT ecosystem.
- Blockchain: Blockchain can significantly enhance privacy preservation in medical data through its decentralised architecture, which eliminates the need for a central authority, reducing the risk of data tampering and unauthorised access. Each transaction on the blockchain is encrypted and linked to the previous one, creating an immutable ledger. This ensures that medical records are secure and can only be accessed by parties who have been granted permission, enhancing patient privacy. Furthermore, the use of smart contracts on blockchain platforms can automate the consent management process, allowing patients to control who can access their data and for what purpose, ensuring compliance with privacy regulations like HIPAA. Additionally, blockchain's transparency feature can be fine-tuned to balance confidentiality with the traceability of access and changes to data, making it a robust solution for securing sensitive health information [18]. For example, [19] proposed quantum blockchain technology called the Quantum Blockchain Integrated Medical Data Processing System (QB-IMD), providing a promising solution to these emerging threats. The QB-IMD system utilises a quantum blockchain structure along with a novel Electronic Medical Record Algorithm (QEMR) to ensure data legitimacy and tamper-proofing through quantum signatures and quantum identity authentication. Elkhodr et al. [20] proposed a blockchain-based framework aimed at secure and privacy-preserving biomedical data sharing. The authors demonstrated that the proposed blockchain-based solution is efficient in enhancing the privacy of data in the sharing of biomedical data.
- Hybrid Approaches: A combination of blockchain technology and federated learning (FL) can also be employed to efficiently tackle privacy issues in IoMT [21]. Ali et al. [22] present a framework that integrates deep learning, homomorphic encryption, and blockchain to enhance the privacy and security of medical data. The consortium blockchain component provides a decentralised and immutable ledger to manage data access, ensuring that only authorised parties can access sensitive information. Smart contracts within the blockchain enforce access control policies, maintaining data integrity and preventing unauthorised access. Together, these technologies create a robust solution for the secure, efficient, and privacy-preserving management of medical data in the Industrial Internet of Medical Things (IIoMT).

Among the above-mentioned approaches, end-to-end encryption is often considered the most foundational and effective for the immediate protection of data in transit and at rest. For comprehensive privacy preservation, combining multiple technologies such as encryption with blockchain for secure data management and federated learning for privacy-preserving analytics can provide a robust solution. However, hybrid approaches come with their challenges and limitations and the integration of these approaches is non-trivial in most application architectures. In this paper, we only consider the use of blockchain for privacy preservation in IoMT-based applications. In the next section, we first look at the workings of blockchain and then further elaborate on how quantum computing can affect this technology and related healthcare applications.

2.4. Overview of Blockchain

Blockchain is a decentralised and distributed digital ledger technology that records transactions across multiple nodes in a peer-to-peer network [23]. In blockchain, once a record is made, it cannot be altered. The structure of blockchain consists of data blocks that are sequentially linked, creating a continuously growing chain as new transactions are recorded. Each block confirms the timing and sequence of transactions, securely maintained within a network that operates under mutually agreed-upon rules. Key characteristics of blockchain include the following:

1. Decentralisation: Data are spread across a network of computers rather than being stored in a central database, reducing dependency on any single authority and enhancing security.
2. Transparency: All transactions are visible to network participants, ensuring that any data recorded are easily verifiable and auditable, fostering trust among users.
3. Immutability: Once data are entered into the blockchain, it is nearly impossible to modify them, ensuring data integrity.
4. Security: Advanced cryptographic techniques protect data, preventing unauthorised access and ensuring that each transaction is securely linked to the previous one.

2.5. Blockchain and IoMT Integration

In healthcare, blockchain's significant potential lies in its ability to revolutionise medical data management and patient care [24]. It can enhance how medical information is shared and managed among various stakeholders, including hospitals, doctors, and insurance companies. By offering a secure and transparent environment, blockchain ensures that medical records are not only protected against unauthorised access but are also immutable and traceable. This can lead to improved treatment structures and more coordinated patient care based on a globally accessible system.

Overall, blockchain can play a crucial role in healthcare, offering robust solutions for secure data storage, the efficient exchange of patient records, and a foundation for various other applications. This technology not only promises to safeguard sensitive information but also to significantly improve operational efficiencies and patient outcomes in the healthcare sector.

Cryptography is one of the main components of security in blockchain technology;' it is pivotal for ensuring the technology's reliability and trustworthiness while preserving data integrity [25]. By employing cryptographic methods, sensitive information like identity verification and financial transactions remains protected, fostering trust among users and reinforcing the security of blockchain networks. Public-key or asymmetric cryptography is essential for securely engaging with the blockchain, enabling the verification of transactions. Additionally, hash functions play a crucial role by facilitating the creation of digital signatures and linking blocks. Despite their importance, both public-key cryptosystems and hash functions are vulnerable to potential threats posed by the advancement of quantum computing technology, which could compromise the security of blockchain systems in the future [26].

3. Post-Quantum Cryptography

Quantum computing can provide a wealth of solutions for problems that have been unsolvable using traditional computing methods. One branch of that would be to use quantum physics for securing communication, known as quantum cryptography. However, the existing cryptographic techniques and algorithms were designed considering the infeasibility of finding mathematical solutions using traditional computing and these will not be effective against quantum computing. Hence, researchers have been redesigning cryptographic algorithms that can be used to protect communication and data after quantum computers are more readily available [5].

Quantum computing exploits the vulnerabilities of traditional cryptographic algorithms and makes them obsolete for securing data or communication purposes. Public-key cryptography is severely affected by quantum speed-up of factoring large numbers, which significantly reduces the time required to break these algorithms. Similarly, symmetric encryption faces the challenge of severely reduced strength due to quantum speed-up [27].

Post-quantum cryptography (PQC) focuses on developing cryptographic algorithms based on mathematical problems that are resistant to traditional and quantum computing. Some of the most prominent approaches in this field are briefly discussed below and are compared in Table 1 [28]:

- Lattice-based cryptography [29];

- Hash-based cryptography [30];
- Code-based cryptography [31];
- Multivariate polynomial cryptography [32].

Table 1. Comparison of post-quantum cryptographic approaches.

	Lattice-Based Cryptography	Hash-Based Cryptography	Code-Based Cryptography	Multivariate Polynomial Cryptography
Technique	Hard lattice problems	Hash functions	Error correction codes	Solving polynomial equations
Applications	Key exchange encryption	Digital signature	Key exchange encryption	Digital signature encryption
Strength	Hardness of lattice problems	Well-tested hash-based security	Proven security in decoding	Effective verification
Issues	Large key size, complex lattice-based maths	Large key size, large signature, non-versatile	Large key size, efficiency	Large key size, key generation
Implemented Algorithms	KYBER SABER	SPHINCS+ XMSS	McEliece BIKE	Rainbow GeMSS

A fundamental block of lattice-based cryptography is a lattice, a mathematical structure in a multi-dimensional space that is composed of a repeating grid generated by a periodic repeat of a unit cell across all dimensions. A lattice is a linear combination of basis vectors and the selection of these basis vectors defines the geometry of a lattice. The fact that a lattice is a multi-dimensional repeating structure with geometry depending upon the basis vectors provides an immense number of configurations, which is a required feature of a cryptographic technique. The geometric and computational complexity involved in solving lattice problems, which consist of certain operations in lattice structures and increase with the increase in lattice dimensions, makes these problems intractable for quantum and traditional computing. Lattice problems are quantum-resistant as there is no known algorithm capable of solving them due to their high-dimensional nature [29].

Hash-based cryptography employs a deterministic, computationally quick, pre-image and collision-resistant hash function with an avalanche effect. Some quantum algorithms may be able to find collisions in these hash functions; however, with large hash sizes, hash-based cryptography may still be quantum-resistant depending upon the size of the hash [30].

Code-based cryptography relies on Error Correcting Codes (ECCs) by adding redundant bits in data to detect and correct errors. Some of the main approaches in this area rely on difficult decoding of the randomly generated linear code. The inability of quantum algorithms to decode random linear code makes this a suitable candidate for post-quantum cryptography [31].

Multivariate polynomial cryptography is based on systems of polynomial equations involving mathematical operations and multiple variables. The computational difficulty associated with the solution of these equations in high-dimensional space makes these cryptographic schemes quantum-resistant [32].

The above-mentioned cryptographic approaches are quantum-resistant and may provide data and communication security and privacy in the post-quantum era. However, every approach has its limitations and challenges that reduce its applicability in practical scenarios. The focus of this work is to analyse the post-quantum cryptographic approaches that are necessary for the secure and optimal function of blockchain-based privacy solutions in IoMT [33].

The research, development, and standardisation of cryptographic algorithms that are quantum-resistant is an ongoing effort by the National Institute of Standards and Technology (NIST) [34]. This is a rapidly changing area of research and development where, recently, four primary quantum-resistant algorithms were selected for testing and standardisation to withstand potential threats by quantum machines.

4. Blockchains and Quantum Computing

The emergence of quantum computing in recent years requires safeguarding some of the fundamental building blocks of blockchain technology. As discussed in the previous section, in the post-quantum era cryptographic algorithms such as public key and hash-based ones may come under threat. In light of these issues, researchers in the blockchain field have started looking into finding solutions that are quantum-resistant. The first set of solutions is called post-quantum blockchain and relies on using post-quantum encryption methods for securing blockchains. The second set of approaches, known as quantum blockchain, utilises quantum computers and networks to redesign blockchain structure [35]. The focus of this paper is post-quantum blockchain techniques to secure blockchains in the quantum era.

Let us first look at the challenges faced by blockchain technology in quantum computing that can disrupt its normal working. Two quantum algorithms can play a big part in this disruption, which are briefly discussed in the following subsections:

- Shor's algorithm [36];
- Grover's algorithm [37].

4.1. Shor's Algorithm

Shor's algorithm is one of the most important developments for quantum technology as it not only factorises large numbers efficiently but also has an associated practical problem for which quantum speed-up can play a significant part. Shor's factorisation algorithm finds the prime factors of any number by using modular arithmetic. In modular arithmetic, a period is defined as the number of steps that take us back to the start of a loop. The period of $a\ modulo\ N$ can provide prime factors of N in a limited number of tries. A number of steps are involved in this process and all of those can be processed effectively on a classical computer but one, which can be extremely computationally intensive. However, quantum computers can process that step with great efficiency and hence can find the prime factors of a very large number in logarithmic time [36].

One of the most used asymmetrical cartographic algorithms is the Rivest–Shamir–Adleman (RSA) algorithm, which uses public and private key pairs for the encryption and decryption of data [38]. When any data are encrypted using a public key, then they can only be decrypted using a corresponding private key. The public and private key pair is generated using a mathematical algorithm while considering a large number and its factors. If a public key is known, then finding a corresponding private key is not a trivial task. This cryptosystem is considered safe as factorisation of a considerably large number is a very difficult problem for classical computers. However, while using quantum computing, Shor's factorisation algorithm can find these factors in a very short time and, hence, can generate the private key that makes these cryptosystems vulnerable and not usable.

4.2. Grover's Algorithm

Grover's quantum algorithm is a search algorithm that can search unstructured data with quadratic improvement as compared to classical search algorithms. One of the main ideas behind Grover's algorithm is the use of a diffusion operator that amplifies the amplitude of the value that is being searched by the searching algorithm and hence increases the probability of finding that value among all possible values in unstructured data. By performing this amplification almost $\sqrt{n}$ number of times, the amplitude of the searched value will be almost one and the result can be measured [37].

Hash-based cryptography uses a mathematical function to convert a variable length input to a fixed length output. The main properties of these algorithms that make them suitable for encrypting data are that from a given output it is difficult to guess input and, secondly, there is a very low probability that any two inputs will give the same output (hash collision). A method for finding hash collision requires searching an entire search space, which is currently computationally infeasible using classical computers. Grover's quantum algorithm can search in time of order $O(\sqrt{n})$, which may result in compromised hash when using quantum computers.

4.3. Effect of Quantum Algorithms on Blockchain Security and Functioning

In a blockchain network, an address is given to every user as a unique identifier for carrying out transactions. These blockchain addresses require public-key cryptography to generate public and private key pairs. The public key is available for everyone and the user securely holds the private key. The user uses a private key to sign transactions that can be authenticated by anyone in the network with a public key to ensure that the transaction is signed by the correct user. By using a hashing function on the user's public key, a user-friendly blockchain address is created which is used for any transactions. The guarantee of authenticity for transactions signed by the user's private key is based on the assumption that the publicly available key cannot be used to regenerate the corresponding private key due to the factorisation difficulty faced by these classical computers [35].

Some of the most popular blockchain implementations use RSA and similar approaches based on factorisation difficulty assumption. The quantum speed-up with Shor's algorithm for factorisation can break such asymmetric key algorithms even when using a 2048 bit number in only a few minutes as compared to the millions of years of computing time required for this using classical computers. A quantum user can regenerate the private keys from the public keys of all other users and can start fraudulent transactions on behalf of any other user.

Blocks in blockchain networks store transactions and hash values of the previous block in order to connect to a previous block in a chain pattern. Any change to any block in the chain will break the chain if hashes for all preceding blocks are not recalculated. SHA256 [39] is a commonly used hash function to calculate hashes for blocks that are used to point to a previous block in the chain.

As previously discussed, these hash functions exhibit the non-reversible property that from the hash the input data cannot be regenerated. However, the other important property of hash functions is collision resistance which can be exploited using Grover's quantum algorithm by recreating subsequent blocks in the blockchain compromising the main strength of the blockchain, which is immutability. Furthermore, quantum miners may have an advantage as compared to miners using classical computers when it comes to proof of work census algorithms. For this algorithm, brute force searching is used to find a hash that meets specific requirements, and with quantum speed-up that may become easier for some users.

5. Post-Quantum Cryptography in Resource-Constrained Devices

The performance comparison of post-quantum cryptographic algorithms in several IoT devices highlights the importance of evaluating both Key Encapsulation Mechanisms (KEMs) and digital signatures. IoT devices, often constrained by limited computational power and memory, require cryptographic solutions that are both secure and efficient. This discussion addresses how different algorithms perform on IoT devices, focusing on KEMs and digital signature schemes.

Table 2 highlights the performance of various post-quantum cryptographic algorithms in several IoT devices, specifically addressing the concerns of resource constraints. The following are the key takeaways:

1. **Key Encapsulation Mechanisms (KEMs):** KEM algorithms are essential for securely exchanging keys between devices in an IoT network, particularly where resources are limited. The study by Halak et al. [40] demonstrated that both Kyber and SABER are efficient in terms of code size and RAM usage on ARM Cortex-M3 and Cortex-M0 devices, making them suitable for resource-constrained environments. Tasopoulos et al. [41] found that Kyber offers the best overall performance on ARM Cortex-M4 devices, while SIKE has the smallest public key and ciphertext sizes but slower execution times. Satrya et al. [42] highlighted that NTRU outperforms SABER and RSA in CPU and memory usage on Raspberry Pi-4, with Light SABER showing the best encryption/decryption delays. Mohamed et al. [43] confirmed that Kyber512 is efficient in key encapsulation and decapsulation times on a Kubernetes-managed Raspberry Pi 4 cluster, which is suitable for time-sensitive medical applications.
2. **Digital Signatures:** Digital signatures are crucial for verifying the authenticity and integrity of messages in IoT networks. Halak et al. [40] indicated that FALCON, with its low latency, is suitable for applications requiring fast verification on ARM Cortex-M3 and Cortex-M0 devices. Tasopoulos et al. [41] observed that Dilithium offers the most balanced performance on ARM Cortex-M4, while Falcon outperforms RSA at security level 1. Vidakovic et al. [44] corroborated these findings across ARM Cortex-M4, x86/x64 processors, and FPGA, noting that Dilithium provides balanced performance and Falcon excels in all operations at security level 1.

Table 2. Performance comparison of post-quantum cryptography algorithms in several IoT devices.

Ref.	Hardware	Software Library	Type	Candidates	Criteria	Result
Halak et al. [40]	ARM Cortex-M3, ARM Cortex-M0	Mbed TLS	KEM	Kyber, SABER	Code-size/RAM	Kyber, SABER
			Signature	Dilithium, FALCON	Code-size/Latency	FALCON
Tasopoulos et al. [41]	ARM Cortex-M4	wolfSSL	KEM	SABER, NTRU, SIKE, BIKE, HQC, NTRU LPRime, FrodoKEM	Execution Speed, Memory Requirements, Communication Size	Kyber offers the best overall performance; SIKE has the smallest public key and ciphertext sizes but the slowest execution time
			Signature	Dilithium, Falcon, SPHINCS+, Picnic3	Execution Speed, Memory Requirements, Communication Size	Dilithium offers the most balanced performance; Falcon outperforms RSA in all operations at security level 1
Vidakovic et al. [44]	ARM Cortex-M4, x86/x64 processors, FPGA	Not given	KEM	Not covered	Not applicable	Not applicable
			Signature	Dilithium, Falcon, SPHINCS+, Picnic3	Execution Speed, Memory Requirements, Communication Size	Dilithium offers the most balanced performance; Falcon outperforms RSA in all operations at security level 1
Mohamed et al. [43]	Raspberry Pi-4 (RPi-4)	Custom implementations for RSA, NTRU, and SABER	KEM	RSA, NTRU, SABER (including Light SABER)	CPU Usage, RAM Usage, Encryption/Decryption Time	NTRU outperforms SABER and RSA in terms of CPU and memory usage; Encryption/Decryption Time: Light SABER is the front-runner when considering encryption and decryption delays
			Signature	Not covered	Not applicable	Not applicable
Satrya et al. [42]	Raspberry Pi 4, Cluster HAT with Raspberry Pi Zero	Kubernetes (K3S) and Docker	KEM	CRYSTAL-Kyber	CPU Usage, Memory Usage, Encryption/Decryption Time, Scalability and Performance Metrics	Kyber512 demonstrated efficient key encapsulation and decapsulation times
			Lightweight Cryptography	ASCON	CPU Usage, Memory Usage, Encryption/Decryption Time, Scalability and Performance Metrics	ASCON showed effective encryption and decryption times suitable for time-sensitive medical applications

In conclusion, for resource-constrained IoT devices, lightweight cryptographic algorithms like ASCON are crucial due to their efficient CPU and memory usage. Post-quantum algorithms like CRYSTAL-Kyber, while providing robust security, need optimization to be feasible in such environments. The studies show that optimized post-quantum algorithms can offer both security and efficiency. Nevertheless, it can generally be concluded that post-quantum cryptographic schemes can be implemented and operated on current limited-resource devices in different IoT applications [45]. Furthermore, it can be observed that lattice-based schemes perform better than other types in terms of speed, memory, and energy consumption. A hybrid approach combining lightweight and post-quantum cryptography may provide the best balance for securing IoT devices without compromising performance.

6. Existing Work

In this section, first we discuss some of the research contributions targeting post-quantum IoMT privacy preservation. After that, we discuss research efforts that specifically target post-quantum blockchain-assisted privacy preservation for IoMT. In the last subsection, we briefly describe how quantum blockchain provides security and privacy for IoMT applications.

6.1. Privacy Preservation for Post-Quantum IoMT

Current privacy preservation methods are insufficient against quantum attacks and often come with high computational overheads [46]. By incorporating lattice-based cryptography, which is believed to be resistant to quantum decryption methods, these gaps could be addressed. Chen et al. [46] propose a scheme that utilises hash operations and error reconciliation technology to provide highly secure and flexible authentication suitable for the cloud environment

Li et al. [47] proposed the Healthchain system, which focuses on the privacy-preserving sharing of electronic medical records (EMRs) using a group signature scheme (GSS). The Healthchain system employs blockchain technology to avoid data tampering and ensures the privacy and security of user data. The system uses an EMR group verification model where EMR data are verified by a creating group to form a transaction and transaction data are verified by system-maintaining nodes to achieve network consensus. The incorporation of a lattice-based GSS strengthens the quantum security of the EMR verification model, providing secure verification, anonymity, traceability, and non-frameability. This scheme supports group members with free joining and revoking, making it a robust solution for EMR management in the IoMT environment.

Yadav et al. [48] focus on designing a privacy-preserving authenticated key mechanism for IoMT systems. The proposed protocol utilises the "ring learning with errors" (RLWE) assumption and physical unclonable functions (PUFs) to achieve robust security. The RLWE-based protocol ensures that the authentication and key exchange mechanisms are secure against quantum attacks. The use of PUFs adds a layer of security by leveraging the unique physical characteristics of devices, making them difficult to clone or forge. This combination of advanced cryptographic techniques provides a high level of security and privacy for IoMT data, protecting them from classical and quantum threats.

6.2. Post-Quantum Blockchain-Assisted Privacy Preservation for IoMT

Shuaib et al. [49] discussed the significant impact of quantum computing on the security and reliability of blockchain-based EHR systems. Quantum computing poses a substantial threat to the traditional cryptographic methods used in blockchain due to its ability to solve complex mathematical problems quickly. The paper highlights the necessity of transitioning to quantum-resistant cryptographic algorithms to safeguard EHR systems against quantum attacks. It emphasizes the importance of post-quantum cryptographic solutions, such as lattice-based cryptography, to ensure the continued security and integrity of blockchain-based healthcare data management systems in the face of emerging quantum computing capabilities.

liu et al. [50] presented a lattice-based proxy-oriented public auditing scheme for electronic health records in cloud-assisted Wireless Body Area Networks (WBANs). The proposed scheme utilises identity-based cryptography to avoid complex certificate management and introduces a proxy to handle signature generation, significantly reducing the computational burden on resource-constrained mobile devices. Additionally, the scheme incorporates Ethereum blockchain technology to protect against malicious proxies. The approach ensures security against quantum attacks, proxy protection, unforgeability, and privacy preservation.

Zhao et al. [51] present an advanced approach for securing IoT in smart healthcare. The Brooks Iyengar quantum Byzantine Agreement-centred blockchain Networking (BIQBA-BCN) model ensures the sincerity and equity of health data exchange. It uses a mutual authentication system based on the Blum Blum Shub and Okamoto Uchiyana Cryptosystem (BBS-OUC) and a Key Weight Block Function-Quasi-Cyclic Moderate Density Parity Check (KWBF-QCMDPC) algorithm to safeguard the confidentiality and dependability of IoT user data. The BBS-OUC cryptosystem leverages the complexity of prime factorisation and discrete logarithms, making it resistant to quantum computing threats. The KWBF-QCMDPC algorithm further enhances security by providing error-correcting codes that are difficult for quantum computers to break. This combination of advanced cryptographic techniques ensures that the BIQBA-BCN model is highly effective against quantum attacks. It has been claimed that the model provides high security and scalability, achieving a security level of 94% and offering significant improvements in data throughput, consensus latency, and node communication time.

A consortium blockchain framework for securing electronic health records (EHRs) using post-quantum cryptography is proposed [52] that employs the CRYSTALS Kyber-768 public key cryptosystem to provide security against quantum attacks. This system ensures data security, confidentiality, and integrity while giving individuals absolute authority over their health data. By using lattice-based cryptography, which relies on the hardness of problems like Learning with Errors (LWEs), the framework offers robust resistance to quantum attacks. The use of CRYSTALS Kyber-768 ensures that even with the advent of quantum computers, the cryptographic security of the EHRs remains intact. This approach enhances privacy and provides a scalable solution for managing health records securely in a post-quantum era.

Mazumdar et al. [53] introduce a quantum-inspired heuristic algorithm combined with Krill Herd Optimisation (QKHO) for healthcare prediction. This model leverages quantum-inspired techniques to enhance the accuracy, precision, recall, and F1-score of healthcare predictions. By integrating blockchain technology, the model ensures secure data transmission to the server, surpassing the security levels of existing RSA and Diffie–Hellman algorithms. The QKHO algorithm provides a highly secure and scalable solution, making it effective against quantum computing threats and improving the overall security of healthcare data transmission.

Chen et al. [54] introduce an Anti-Quantum Attribute-based Signature (AQABS) scheme designed to resist quantum computing attacks in E-health scenarios. The AQABS scheme combines the security of attribute-based signatures with the resilience of quantum-resistant cryptographic techniques. By integrating IPFS with consortium blockchain, the AQABS scheme ensures fully distributed EMR storage, encrypted-EMR searchability, fine-grained access control, low overhead, and utmost scalability. The scheme achieves EMR unforgeability and integrity, signatories' anonymity, and resistance to collusion and quantum computing attacks. Experimental results demonstrate that the AQABS scheme is efficient and lightweight in key extraction, signature generation, and verification overhead compared to existing systems.

Bhavin et al. [55] propose a hybrid scheme that combines blockchain technology with quantum blind signatures to enhance the security of healthcare data. The proposed scheme leverages quantum blind signatures to protect traditional encryption systems from quantum attacks during block creation using Hyperledger Fabric blockchain. This hybrid approach ensures that healthcare data remain secure against various quantum computing threats. The results show that the proposed scheme improves transaction throughput, reduces resource consumption, and decreases network traffic compared to state-of-the-art schemes.

Wu et al. [56] propose a blockchain-enabled EMR storage management scheme to enhance the security and privacy of healthcare data. The scheme leverages the decentralized and immutable ledger properties of blockchain technology to ensure data integrity, transparency, and robust access control. By integrating smart contracts, the scheme ensures that data access is tightly controlled and transparently logged, mitigating the risks associated with centralized data storage. However, while the scheme effectively addresses classical cybersecurity threats, it remains potentially vulnerable to quantum computing attacks.

These studies highlight various approaches to integrating blockchain and quantum cryptography techniques to secure healthcare data. Table 3 summarizes and compares these studies, emphasising their strengths against quantum attacks, potential risks, and overall impact on healthcare data security.

Table 3. Comparison of blockchain and quantum cryptography techniques for securing healthcare data.

Paper	Approach	Strengths Against Quantum Attack	Risk	Impact
Qu et al. [19]	Quantum cryptography	Uses quantum signatures and quantum identity authentication, leverages quantum cloud computing.	Dependency on quantum cloud computing, potential complexity in implementation.	Enhances security and privacy in IoMT, ensures data integrity, and prevents unauthorized access.
Shuaib et al. [49]	Discusses the broader spectrum of post-quantum cryptography	Highlights the need for quantum-resistant cryptographic algorithms.	Need to transition to quantum-resistant cryptographic algorithms.	Highlights the importance of post-quantum cryptographic solutions.
Liu et al. [50]	Lattice-based cryptography	Employs lattice-based cryptography, introduces a proxy to handle signature generation, incorporates Ethereum blockchain.	Reliance on identity-based cryptography may present a single point of failure.	Reduces computational burden on mobile devices, ensures security and privacy.
Zhao et al. [51]	Code-based cryptography	Utilises Brooks Iyengar quantum Byzantine Agreement-centred blockchain Networking model, employs BBS-OUC cryptosystem and KWBF-QCMDPC algorithm.	Complexity of integrating multiple advanced cryptographic techniques.	Ensures confidentiality and dependability of IoT user data, improves security and scalability.
Bansal et al. [52]	Lattice-based cryptography	Employs CRYSTALS Kyber-768 public key cryptosystem, uses lattice-based cryptography.	Potential computational overhead of CRYSTALS Kyber-768.	Ensures data security, confidentiality, and integrity of EHRs.
Mazumdar et al. [53]	Quantum-inspired heuristic algorithm	Integrates quantum-inspired heuristic algorithm with blockchain technology.	Complexity of quantum-inspired heuristic algorithm.	Guarantees secure data transmission surpasses RSA and Diffie–Hellman algorithms.

Table 3. Cont.

Paper	Approach	Strengths Against Quantum Attack	Risk	Impact
Chen et al. [54]	Lattice-based cryptography	Introduces Anti-Quantum Attribute-based Signature (AQABS) scheme, integrates IPFS with consortium blockchain.	Complexity of integrating multiple advanced cryptographic techniques.	Ensures secure and scalable EMR storage and retrieval.
Bhavin et al. [55]	Multivariate polynomial cryptography	Combines blockchain technology with quantum blind signatures, uses Hyperledger Fabric blockchain.	Potential resource consumption and network traffic.	Improves transaction throughput and reduces resource consumption.
Wu et al. [56]	Lattice-based cryptography	Leverages blockchain technology, integrates smart contracts.	Vulnerability to quantum computing attacks.	Improves security, transparency, and access control.
Azzaoui et al. [57]	Hash-based cryptography	Combines Quantum Terminal Machines (QTMs) and blockchain technology, uses Quantum One-Time Pad (Q-OTP) encryption.	Potential vulnerability in converting classical data into quantum bits.	Ensures security and scalability of medical data processing.
Venkatesh et al. [58]	Hash-based cryptography	Utilises quantum cryptographic principles, integrates quantum key distribution and blockchain technology.	Potential high computational cost.	Enhances security, integrity, and privacy of EMRs.
Christo et al. [59]	Quantum cryptography	Combines blockchain technology with quantum cryptography, AES, and SHA algorithms.	Complexity of integrating multiple cryptographic techniques.	Significantly improves security, scalability, and efficiency of healthcare data management.

6.3. Quantum Blockchain-Assisted Privacy Preservation for IoMT

Azzaoui et al. [57] proposed a novel Quantum Cloud-as-a-Service (QCaaS) architecture for secure and efficient processing of medical big data. The architecture combines Quantum Terminal Machines (QTMs) and blockchain technology to ensure security and scalability. QTMs act as intermediaries, converting classical data into quantum bits for processing on quantum servers, while blockchain authenticates and secures communication between nodes. The experimental results confirm that the proposed Quantum One-Time Pad (Q-OTP) encryption-based system can effectively ensure the security of medical data.

Venkatesh and Hanumantha [58] introduced a privacy-preserving quantum blockchain technique for securing electronic medical records (EMRs). The proposed method leverages quantum cryptographic principles to resist various attacks, including intercept, intercept–resend, entangle–measure, man-in-the-middle, collective, and coherent attacks, while reducing communication and computation costs compared to traditional techniques. By integrating quantum key distribution and blockchain technology, the system enhances the security, integrity, and privacy of EMRs, making them resilient to quantum threats.

Qu et al. [19] proposed a QB-IMD system, which integrates quantum blockchain technology to enhance security and privacy in IoMT. This system features a quantum blockchain structure and the QEMR algorithm to ensure data integrity and prevent tampering. The use of quantum signatures and quantum identity authentication provides robust protection against quantum attacks, making it a pioneering solution in post-quantum privacy preservation for medical data. The system also leverages quantum cloud computing for delegated computations, ensuring that diagnostic data are processed without being exposed, thus maintaining user privacy.

Christo et al. [59] propose a hybrid blockchain scheme that combines blockchain technology with quantum cryptography, AES, and SHA algorithms to enhance the security of healthcare data. The proposed scheme employs quantum cryptography for authentication, AES for encryption, and SHA for data retrieval, ensuring robust protection against frequent attacks. The scheme is structured into three phases: authentication, encryption, and data retrieval, leveraging blockchain to provide a secure, decentralized, and transparent system for managing medical records. This approach addresses the vulnerabilities of centralized healthcare systems by ensuring data integrity, confidentiality, and access control. The results indicate that the proposed scheme significantly improves security, scalability, and efficiency in managing healthcare data compared to existing systems.

In Table 3, we compare all the research work discussed in this section on the basis of strengths, risks, and impacts.

7. Discussion and Future Research Directions

The advancements in post-quantum cryptography and blockchain technology have led to innovative solutions for securing healthcare data against quantum threats. The approaches reviewed in this work demonstrate various strategies that combine quantum-resistant cryptographic techniques with blockchain technology that brings decentralisation and immutability. These solutions address the critical need for the enhanced security, privacy, and integrity of electronic medical records and other sensitive healthcare data. The key approaches against quantum attacks discussed in this paper are the following:

1. Post-Quantum Cryptographic Techniques: Several research efforts focused on using post-quantum cryptographic approaches [50,52] employ advanced cryptographic methods, such as lattice-based cryptography, to secure healthcare data. These techniques are resistant to quantum attacks but may introduce additional computational overhead.
2. Integration of Quantum-Inspired Algorithms: Quantum-inspired heuristic algorithms are integrated with blockchain, providing enhanced security measures that surpass traditional algorithms like RSA and Diffie–Hellman [53,58].
3. Quantum Signatures and Quantum Identity Authentication: Solutions such as the QB-IMD system [19] and the advanced IoT security model [51] use quantum signatures and quantum identity authentication to ensure data integrity and prevent tampering in IoMT and IoT environments.
4. Combination of Quantum and Blockchain Technologies: The combination of quantum cryptographic principles with blockchain technology is proposed to enhance security against quantum attacks, which includes using Quantum Terminal Machines and quantum key distribution [57,58].
5. Hybrid and Multi-Phase Approaches: Hybrid schemes combining blockchain with quantum cryptography and other advanced algorithms like AES and SHA are proposed to ensure a multi-layered security approach [55,59].

The above-mentioned approaches provide security and privacy protection to sensitive healthcare data to some degree. Still, there are limitations and challenges that need to be addressed for the security of quantum-secured blockchain systems in healthcare data management. Some of the limitations are as follows:

1. **Research of Post-Quantum Cryptographic Algorithms:** Research-wise, this area has not matured yet as further efforts are required to completely understand the complexity, strengths, and weaknesses of post-quantum cryptographic algorithms like lattice-based cryptography, hash-based cryptography, and other quantum-resistant techniques.
2. **Computational Efficiency:** For both post-quantum and quantum cryptographic algorithms, computational costs can be a limiting factor for their development and adoption. Research focusing on the optimisation of these algorithms is required to ensure they can be implemented efficiently in resource-constrained environments.

3. **Scalability, Interoperability, and Real-World Testing:** Technical and procedural complexities associated with blockchain, cryptographic algorithms, and healthcare data make it difficult to test large-scale solutions. Hence, more efforts are required to test if post-quantum blockchain systems can scale effectively and interoperate with existing healthcare infrastructure. Research should explore ways to integrate these technologies seamlessly into current systems without compromising performance.
4. **Integration of Quantum Key Distribution:** Practical limitations and challenges need to be explored and addressed for integrating QKD into broader healthcare data security frameworks utilising blockchain and other cryptographic protocols.

8. Conclusions

In this work, we delved into the critical domain of privacy preservation for Internet of Medical Things (IoMT)-based healthcare applications. Our study specifically concentrated on blockchain-based solutions, given their potential to provide decentralised, transparent, and immutable frameworks for managing sensitive medical data. We conducted an exhaustive review of the latest research in post-quantum and quantum blockchain-based privacy-preservation techniques tailored to IoMT. By categorizing these research efforts based on the primary security methods employed, we aimed to comprehensively map the current landscape and understand the breadth of strategies being explored. Our findings reveal that numerous innovative approaches show significant promise in safeguarding healthcare data from the threats posed by quantum computing advancements.

Despite these promising developments, our research underscores the imperative need for continued investigation. It is crucial to ensure that these security measures evolve in tandem with new and emerging security threats and are robust enough to defend against unpredictable and unprecedented zero-day attacks. This ongoing research is vital to maintain the integrity and confidentiality of healthcare data in the face of advancing technological threats.

Author Contributions: Conceptualisation, F.S. and S.S.; methodology, F.S. and S.S.; writing—original draft preparation, F.S., S.S., and U.U.T.; writing—review and editing, F.S., S.S., and U.U.T.; supervision, F.S.; project administration, F.S. and S.S. All authors have read and agreed to the published version of the manuscript.

Funding: This research received no external funding.

Data Availability Statement: Data are contained within the article.

Conflicts of Interest: The authors declare no conflicts of interest.

Abbreviations

The following abbreviations are used in this manuscript:

EMRs	electronic medical records.
FL	federated learning.
IoMT	Internet of Medical Things.
PQC	post-quantum cryptographic.
PUF	physical unclonable functions.
QCaaS	Quantum Cloud-as-a-Service.
QKD	quantum key distribution.
Q-OTP	Quantum One-Time Pad.
QTMs	Quantum Terminal Machines.

References

1. Cherbal, S.; Zier, A.; Hebal, S.; Louail, L.; Annane, B. Security in internet of things: A review on approaches based on blockchain, machine learning, cryptography, and quantum computing. *J. Supercomput.* **2023**, *80*, 3738–3816. [CrossRef]
2. Khatiwada, P.; Yang, B. An Overview on Security and Privacy of Data in IoMT Devices: Performance Metrics, Merits, Demerits, and Challenges. *pHealth* **2022**, *2022*, 126–136.

3. Selvarajan, S.; Mouratidis, H. A quantum trust and consultative transaction-based blockchain cybersecurity model for healthcare systems. *Sci. Rep.* **2023**, *13*, 7107. [CrossRef] [PubMed]
4. Dhinakaran, D.; Srinivasan, L.; Udhaya Sankar, S.; Selvaraj, D. Quantum-based privacy-preserving techniques for secure and trustworthy internet of medical things an extensive analysis. *Quantum Inf. Comput.* **2024**, *24*, 0227–0266. [CrossRef]
5. Long, B. Classical Solutions for Quantum Challenges: An Introduction to Postquantum Cryptography. *ACM SIGCAS Comput. Soc.* **2024**, *52*, 23–25. [CrossRef]
6. Al-Turjman, F.; Nawaz, M.H.; Ulusar, U.D. Intelligence in the Internet of Medical Things era: A systematic review of current and future trends. *Comput. Commun.* **2020**, *150*, 644–660. [CrossRef]
7. Razdan, S.; Sharma, S. Internet of medical things (IoMT): Overview, emerging technologies, and case studies. *IETE Tech. Rev.* **2022**, *39*, 775–788. [CrossRef]
8. Mukhopadhyay, M.; Banerjee, S.; Mukhopadhyay, C.D. Internet of Medical Things and the Evolution of Healthcare 4.0: Exploring Recent Trends. *J. Electron. Electromed. Eng. Med Inform.* **2024**, *6*, 182–195. [CrossRef]
9. Du, J.; Jiang, C.; Gelenbe, E.; Xu, L.; Li, J.; Ren, Y. Distributed Data Privacy Preservation in IoT Applications. *IEEE Wirel. Commun.* **2018**, *25*, 68–76. [CrossRef]
10. Ghubaish, A.; Salman, T.; Zolanvari, M.; Unal, D.; Al-Ali, A.; Jain, R. Recent Advances in the Internet-of-Medical-Things (IoMT) Systems Security. *IEEE Internet Things J.* **2021**, *8*, 8707–8718. [CrossRef]
11. Salim, M.M.; Kim, I.; Doniyor, U.; Lee, C.; Park, J.H. Homomorphic encryption based privacy-preservation for iomt. *Appl. Sci.* **2021**, *11*, 8757. [CrossRef]
12. Putra, K.T.; Arrayyan, A.Z.; Hayati, N.; Damarjati, C.; Bakar, A.; Chen, H.C. A Review on the Application of Internet of Medical Things in Wearable Personal Health Monitoring: A Cloud-Edge Artificial Intelligence Approach. *IEEE Access* **2024**, *12*, 21437–21452. [CrossRef]
13. Gazi, T. Data to the rescue: How humanitarian aid NGOs should collect information based on the GDPR. *J. Int. Humanit. Action* **2020**, *5*, 9. [CrossRef]
14. Liu, B.; Ding, M.; Shaham, S.; Rahayu, W.; Farokhi, F.; Lin, Z. When machine learning meets privacy: A survey and outlook. *ACM Comput. Surv. (CSUR)* **2021**, *54*, 1–36. [CrossRef]
15. Arachchige, P.C.M.; Bertok, P.; Khalil, I.; Liu, D.; Camtepe, S.; Atiquzzaman, M. A trustworthy privacy preserving framework for machine learning in industrial IoT systems. *IEEE Trans. Ind. Inform.* **2020**, *16*, 6092–6102. [CrossRef]
16. Nair, A.K.; Sahoo, J.; Raj, E.D. Privacy preserving Federated Learning framework for IoMT based big data analysis using edge computing. *Comput. Stand. Interfaces* **2023**, *86*, 103720. [CrossRef]
17. Husnoo, M.A.; Anwar, A.; Chakrabortty, R.K.; Doss, R.; Ryan, M.J. Differential privacy for IoT-enabled critical infrastructure: A comprehensive survey. *IEEE Access* **2021**, *9*, 153276–153304. [CrossRef]
18. Li, C.; Dong, M.; Xin, X.; Li, J.; Chen, X.B.; Ota, K. Efficient privacy-preserving in IoMT with blockchain and lightweight secret sharing. *IEEE Internet Things J.* **2023**, *10*, 22051–22064. [CrossRef]
19. Qu, Z.; Meng, Y.; Liu, B.; Muhammad, G.; Tiwari, P. QB-IMD: A secure medical data processing system with privacy protection based on quantum blockchain for IoMT. *IEEE Internet Things J.* **2023**, *11*, 40–49. [CrossRef]
20. Elkhodr, M.; Gide, E.; Darwish, O.; Al-Eidi, S. BioChainReward: A Secure and Incentivised Blockchain Framework for Biomedical Data Sharing. *Int. J. Environ. Res. Public Health* **2023**, *20*, 6825. [CrossRef]
21. Rahmadika, S.; Astillo, P.V.; Choudhary, G.; Duguma, D.G.; Sharma, V.; You, I. Blockchain-based privacy preservation scheme for misbehavior detection in lightweight IoMT devices. *IEEE J. Biomed. Health Inform.* **2022**, *27*, 710–721. [CrossRef] [PubMed]
22. Ali, A.; Pasha, M.F.; Guerrieri, A.; Guzzo, A.; Sun, X.; Saeed, A.; Hussain, A.; Fortino, G. A novel homomorphic encryption and consortium blockchain-based hybrid deep learning model for industrial internet of medical things. *IEEE Trans. Netw. Sci. Eng.* **2023**, *10*, 2402–2418. [CrossRef]
23. Saraji, S. Introduction to Blockchain. In *Sustainable Oil and Gas Using Blockchain*; Springer: Cham, Switzerland, 2023; pp. 57–74.
24. Stafford, T.F.; Treiblmaier, H. Characteristics of a Blockchain Ecosystem for Secure and Sharable Electronic Medical Records. *IEEE Trans. Eng. Manag.* **2020**, *67*, 1340–1362. [CrossRef]
25. Storublevtcev, N. Cryptography in blockchain. In Proceedings of the Computational Science and Its Applications–ICCSA 2019: 19th International Conference, Saint Petersburg, Russia, 1–4 July 2019; Proceedings, Part II 19; Springer: Cham, Switzerland, 2019; pp. 495–508.
26. Fernández-Caramès, T.M.; Fraga-Lamas, P. Towards Post-Quantum Blockchain: A Review on Blockchain Cryptography Resistant to Quantum Computing Attacks. *IEEE Access* **2020**, *8*, 21091–21116. [CrossRef]
27. Aumasson, J.P. The impact of quantum computing on cryptography. *Comput. Fraud Secur.* **2017**, *2017*, 8–11. [CrossRef]
28. Kumar, M. Post-quantum cryptography Algorithm's standardization and performance analysis. *Array* **2022**, *15*, 100242. [CrossRef]
29. Nejatollahi, H.; Dutt, N.; Ray, S.; Regazzoni, F.; Banerjee, I.; Cammarota, R. Post-quantum lattice-based cryptography implementations: A survey. *ACM Comput. Surv. (CSUR)* **2019**, *51*, 1–41. [CrossRef]
30. Sundaram, B.V.; Ramnath, M.; Prasanth, M.; Sundaram, V. Encryption and hash based security in Internet of Things. In Proceedings of the 2015 3rd International Conference on Signal Processing, Communication and Networking (ICSCN), Chennai, India, 26–28 March 2015; IEEE: Piscataway, NJ, USA, 2015; pp. 1–6.
31. Sendrier, N. Code-based cryptography: State of the art and perspectives. *IEEE Secur. Priv.* **2017**, *15*, 44–50. [CrossRef]

32. Faugère, J.C.; Perret, L. An efficient algorithm for decomposing multivariate polynomials and its applications to cryptography. *J. Symb. Comput.* **2009**, *44*, 1676–1689. [CrossRef]
33. Sood, N. Cryptography in Post Quantum Computing Era. SSRN 4705470. 2024. Available online: https://papers.ssrn.com/sol3/papers.cfm?abstract_id=4705470 (accessed on 1 May 2024).
34. National Institute of Standardisation and Technology. Available online: https://www.nist.gov/ (accessed on 10 July 2024).
35. Yang, Z.; Alfauri, H.; Farkiani, B.; Jain, R.; Di Pietro, R.; Erbad, A. A survey and comparison of post-quantum and quantum blockchains. *IEEE Commun. Surv. Tutor.* **2023**, *26*, 967–1002. [CrossRef]
36. Shor, P.W. Algorithms for quantum computation: Discrete logarithms and factoring. In Proceedings of the 35th Annual Symposium on Foundations of Computer Science, Santa Fe, NM, USA, 20–22 November 1994; IEEE: Piscataway, NJ, USA, 1994; pp. 124–134.
37. Grover, L.K. A fast quantum mechanical algorithm for database search. In Proceedings of the Twenty-Eighth Annual ACM Symposium on Theory of Computing, Philadelphia, PA, USA, 22–24 May 1996; pp. 212–219.
38. Rivest, R.L.; Shamir, A.; Adleman, L. A method for obtaining digital signatures and public-key cryptosystems. *Commun. ACM* **1978**, *21*, 120–126. [CrossRef]
39. Rachmawati, D.; Tarigan, J.; Ginting, A. A comparative study of Message Digest 5 (MD5) and SHA256 algorithm. *J. Phys. Conf. Ser.* **2018**, *978*, 012116. [CrossRef]
40. Halak, B.; Gibson, T.; Henley, M.; Botea, C.B.; Heath, B.; Khan, S. Evaluation of performance, energy, and computation costs of quantum-attack resilient encryption algorithms for embedded devices. *IEEE Access* **2024**, *12*, 8791–8805. [CrossRef]
41. Tasopoulos, G.; Li, J.; Fournaris, A.P.; Zhao, R.K.; Sakzad, A.; Steinfeld, R. Performance evaluation of post-quantum TLS 1.3 on resource-constrained embedded systems. In Proceedings of the International Conference on Information Security Practice and Experience, Taipei, Taiwan, 23–25 November 2022; Springer: Cham, Switzerland, 2022; pp. 432–451.
42. Satrya, G.B.; Agus, Y.M.; Mnaouer, A.B. A comparative study of post-quantum cryptographic algorithm implementations for secure and efficient energy systems monitoring. *Electronics* **2023**, *12*, 3824. [CrossRef]
43. Mohamed, E.H.; Ankunda, P.V.; Ung, J.; Hwu, W.M. Securing the Internet of Medical Things (IoMT) with K3S and Hybrid Cryptography: Integrating Post-Quantum Approaches for Enhanced Embedded System Security. In Proceedings of the 2024 IEEE 17th Dallas Circuits and Systems Conference (DCAS), Virtual, 19–21 April 2024; IEEE: Piscataway, NJ, USA, 2024; pp. 1–6.
44. Vidaković, M.; Miličević, K. Performance and Applicability of Post-Quantum Digital Signature Algorithms in Resource-Constrained Environments. *Algorithms* **2023**, *16*, 518. [CrossRef]
45. Gharavi, H.; Granjal, J.; Monteiro, E. Post-quantum blockchain security for the Internet of Things: Survey and research directions. *IEEE Commun. Surv. Tutorials* **2024**. [CrossRef]
46. Chen, X.; Wang, B.; Li, H. A privacy-preserving multi-factor authentication scheme for cloud-assisted IoMT with post-quantum security. *J. Inf. Secur. Appl.* **2024**, *81*, 103708. [CrossRef]
47. Li, C.; Jiang, B.; Dong, M.; Xin, X.; Ota, K. Privacy preserving for electronic medical record sharing in healthchain with group signature. *IEEE Syst. J.* **2023**, *17*, 6114–6125. [CrossRef]
48. Yadav, D.K.; Yadav, D.; Pal, Y.; Chaudhary, D.; Sahu, H.; Manasa, A. Post Quantum Blockchain Assisted Privacy Preserving Protocol for Internet of Medical Things. In Proceedings of the 2023 IEEE World Conference on Applied Intelligence and Computing (AIC), Sonbhadra, India, 29–30 July 2023; IEEE: Piscataway, NJ, USA, 2023; pp. 965–970.
49. Shuaib, M.; Hassan, N.H.; Usman, S.; Alam, S.; Sam, S.M.; Samy, G.A.N. Effect of quantum computing on blockchain-based electronic health record systems. In Proceedings of the 2022 4th International Conference on Smart Sensors and Application (ICSSA), Penang, Malaysia, 10–12 September 2024; IEEE: Piscataway, NJ, USA, 2022; pp. 179–184.
50. Liu, X.; Luo, Y.; Yang, X.; Wang, L.; Zhang, X. Lattice-Based Proxy-Oriented Public Auditing Scheme for Electronic Health Record in Cloud-Assisted WBANs. *IEEE Syst. J.* **2022**, *16*, 2968–2978. [CrossRef]
51. Zhao, Z.; Li, X.; Luan, B.; Jiang, W.; Gao, W.; Neelakandan, S. Secure internet of things (IoT) using a novel brooks Iyengar quantum byzantine agreement-centered blockchain networking (BIQBA-BCN) model in smart healthcare. *Inf. Sci.* **2023**, *629*, 440–455. [CrossRef]
52. Bansal, A.; Mehra, P.S. A Post-Quantum Consortium Blockchain Based Secure EHR Framework. In Proceedings of the 2023 International Conference on IoT, Communication and Automation Technology (ICICAT), Gorakhpur, India, 23–24 June 2023; IEEE: Piscataway, NJ, USA, 2023; pp. 1–6.
53. Mazumdar, H.; Chakraborty, C.; Venkatakrishnan, S.B.; Kaushik, A.; Gohel, H.A. Quantum-inspired heuristic algorithm for secure healthcare prediction using blockchain technology. *IEEE J. Biomed. Health Inform.* **2023**, *28*, 3371–3378. [CrossRef] [PubMed]
54. Chen, X.; Xu, S.; Qin, T.; Cui, Y.; Gao, S.; Kong, W. AQ–ABS: Anti-quantum attribute-based signature for EMRs sharing with blockchain. In Proceedings of the 2022 IEEE Wireless Communications and Networking Conference (WCNC), Austin, TX, USA, 10–13 April 2022; IEEE: Piscataway, NJ, USA, 2022; pp. 1176–1181.
55. Bhavin, M.; Tanwar, S.; Sharma, N.; Tyagi, S.; Kumar, N. Blockchain and quantum blind signature-based hybrid scheme for healthcare 5.0 applications. *J. Inf. Secur. Appl.* **2021**, *56*, 102573. [CrossRef]
56. Wu, G.; Wang, Y. The security and privacy of blockchain-enabled EMR storage management scheme. In Proceedings of the 2020 16th International Conference on Computational Intelligence and Security (CIS), Guangxi, China, 27–30 November 2020; IEEE: Piscataway, NJ, USA, 2020; pp. 283–287.

57. Azzaoui, A.E.; Sharma, P.K.; Park, J.H. Blockchain-based delegated Quantum Cloud architecture for medical big data security. *J. Netw. Comput. Appl.* **2022**, *198*, 103304. [CrossRef]
58. Venkatesh, R.; Hanumantha, B.S. A Privacy-Preserving Quantum Blockchain Technique for Electronic Medical Records. *IEEE Eng. Manag. Rev.* **2023**, *51*, 137–144. [CrossRef]
59. Christo, M.S.; Sarathy, P.; Priyanka, C. An efficient data security in medical report using blockchain technology. In Proceedings of the 2019 International Conference on Communication and Signal Processing (ICCSP), Melmaruvathur, India, 4–6 April 2019; IEEE: Piscataway, NJ, USA, 2019; pp. 606–610.

Disclaimer/Publisher's Note: The statements, opinions and data contained in all publications are solely those of the individual author(s) and contributor(s) and not of MDPI and/or the editor(s). MDPI and/or the editor(s) disclaim responsibility for any injury to people or property resulting from any ideas, methods, instructions or products referred to in the content.

A Novel Static Analysis Approach Using System Calls for Linux IoT Malware Detection

Jayanthi Ramamoorthy [1,2,*,†], Khushi Gupta [1,†], Ram C. Kafle [2,†], Narasimha K. Shashidhar [1] and Cihan Varol [1]

1 Department of Computer Science, Sam Houston State University, Huntsville, TX 77340, USA; kxg095@shsu.edu (K.G.); nks001@shsu.edu (N.K.S.); cvarol@shsu.edu (C.V.)
2 Department of Mathematics and Statistics, Sam Houston State University, Huntsville, TX 77340, USA; rckafle@shsu.edu
* Correspondence: jxr153@shsu.edu
† These authors contributed equally to this work.

Abstract: The proliferation of Internet of Things (IoT) devices on Linux platforms has heightened concerns regarding vulnerability to malware attacks. This paper introduces a novel approach to investigating the behavior of Linux IoT malware by examining syscalls and library syscall wrappers extracted through static analysis of binaries, as opposed to the conventional method of using dynamic analysis for syscall extraction. We rank and categorize Linux system calls based on their security significance, focusing on understanding malware intent without execution. Feature analysis of the assigned syscall categories and risk ranking is conducted with statistical tests to validate their effectiveness and reliability in differentiating between malware and benign binaries. Our findings demonstrate that potential threats can be reliably identified with an F1 score of 96.86%, solely by analyzing syscalls and library syscall wrappers. This method can augment traditional static analysis, providing an effective preemptive measure to enhance Linux malware analysis. This research highlights the importance of static analysis in strengthening IoT systems against emerging malware threats.

Keywords: ELF static analysis; Linux system calls; machine learning; malware detection

1. Introduction

The Internet of Things (IoT) is reshaping cyber–physical systems with unprecedented impact. By the end of this decade, an estimated 25.44 billion devices will be interconnected, predominantly as IoT devices, comprising 75% of the total device count [1]. Recent reports, including the 2023 Zscaler ThreatLabz Enterprise IoT and OT Threat Report, highlight a staggering 400% surge in attacks targeting IoT devices [2]. The rapid integration of IoT technologies across various industries signals a continued escalation in such security incidents impacting various domains such as transportation, healthcare, and energy management. However, IoT devices lack security features and are inherently complex in terms of hardware and software design, making them vulnerable to cyber attacks [3].

Linux systems are the backbone of numerous IoT devices [4]. This ubiquity has led to a noticeable uptick in malware designed to exploit Linux-based environments, as evidenced by a 50% increase in new Linux malware reported by AV-ATLAS in just one year [5].

The focus of malware analysis is to examine malicious binaries (malware) to discern their behavior, purpose, and impact using a range of techniques and tools. This is typically carried out through static analysis, dynamic analysis or a hybrid of both these approaches. The extent of analysis is only limited by the objective—to quickly detect the malware or understand its functionality and behavior patterns such as network traffic, interactions with the system, evasion techniques and other features.

In this paper, we focus solely on static analysis to extract system calls (syscalls) and library syscall wrappers made by binaries to the Linux kernel for ARM architecture-based

Linux files, commonly known as ELF (Executable and Linkable Format) binaries. We analyze 1117 ARM-based Linux IoT malware samples and 1214 benign or non-malicious binaries. The syscall dataset created is evaluated using common machine learning classification models, such as Logistic Regression, Neural Networks, and Random Forest, to classify these binaries as malicious or benign.

While there are numerous research works where syscalls are used to analyze malware through dynamic analysis, to the best of our knowledge, there are no known research works that extract syscalls and library syscall wrappers from static analysis.

Static analysis has the limitation of not capturing all the system calls made by a binary, as it cannot detect indirect system calls. Static analysis is also not effective when the binaries use extensive obfuscation or encryption to pack malware. In this study, we evaluate whether the limited syscalls and library syscall wrappers that can be extracted from the disassembly of the binary are still adequate for malware detection. We limit the scope of static analysis to syscalls extracted from reverse engineering the ELF binary to determine if this approach is a viable option.

The main contributions of this paper are as follows:

- Systematic Linux syscall Categorization and Security Risk Ranking: We categorize and assign risk rankings to all extracted calls based on their potential security risks. The validity of these categories and rankings is confirmed through statistical analysis in this study.
- Static Analysis Dataset: We reverse-engineered 2331 ARM architecture-based ELF binaries to extract syscalls and library syscall wrappers using static analysis and correlated with syscall categories and security risk ranking to create a comprehensive dataset, which includes the syscalls, category and security risk ranking for each syscall, along with added statistical features for each binary.
- Malware Detection: To demonstrate the reliability of using static analysis for malware detection based on system calls extracted from the disassembled binaries, we evaluate the dataset created with standard ML classification models such as Logistic Regression, Random Forest classifier, Support Vector Classification (SVC), and Multi-Layer Perceptron (MLP) Neural Networks.

Although we focus on ARM binaries, our syscall extraction approach is architecture-agnostic due to the use of the popular reverse engineering tool radare2. Radare2 employs the ESIL (Evaluable Strings Intermediate Language) framework, which abstracts the underlying architecture details, allowing for consistent and accurate syscall extraction across different hardware platforms [6].

Binary lifting in static analysis has been successfully employed for malware variant classification by Ramamoorthy et al. [7], where the Intermediate Representation (IR) of opcode sequences was used to create a dataset. In their study, a Random Forest classification model achieved an F1 score of 97%.

The organization of this paper is as follows: Section 2 provides the background for the study, followed by Section 3, which describes relevant research work in the field of IoT malware detection. The methodology is explained in Section 4, including statistical analysis and machine learning models used for evaluation. Section 5 presents the results obtained using the proposed methods and includes an analysis of these results. Section 6 summarizes the conclusions, followed by Section 7, which discusses the limitations and future scope.

2. Preliminaries

In this section, we discuss and provide an overview of malware static analysis and system and library syscall wrappers.

2.1. System and Library Calls

In ELF ARM binaries, both library and system calls are crucial and play an important role in program functionality. They enable binaries to interact with external resources, perform operations and access system-level functionalities.

- Syscalls: In Linux, a system call (syscall) is the primary interface through which user-space programs request services and access functionalities from the operating system kernel. It acts as a bridge between the user space, where applications run, and the kernel space, where core system functions reside. System calls enable programs to perform privileged operations and access restricted system resources, such as hardware interaction, process management, file I/O operations, network communication, and memory management as shown in Figure 1. Some system calls are architecture-specific, and their implementation and availability can vary between different hardware platforms (e.g., x86, ARM, MIPS). These differences are due to the unique characteristics and requirements of each architecture, necessitating specific handling within the kernel to optimize performance and compatibility [8].
- Library syscall wrappers: When using programming languages like C++ or Java, developers often use pre-built functions from libraries such as the GNU C Library *glibc*, which contains routines for file management, memory allocation, and computational tasks. When a program calls one of these library functions, it may require system-level functionalities that reside in kernel space, like hardware interaction or process management. Consequently, library functions often make system calls and provide wrapper functions to syscalls in order to perform these tasks as shown in Figure 1.

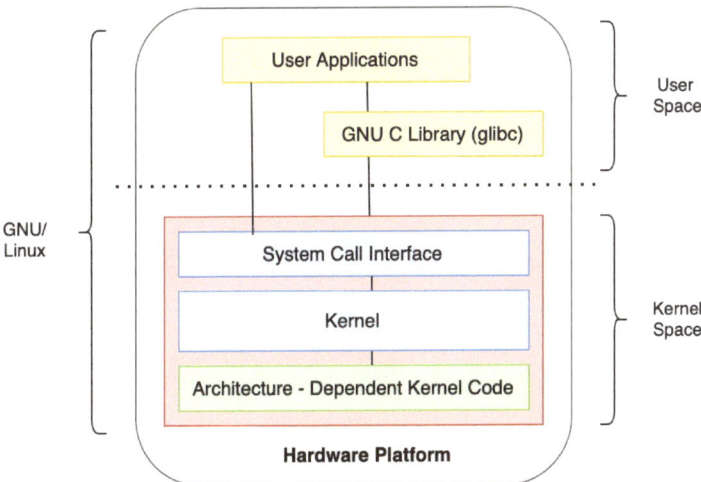

Figure 1. Architecture of the GNU/Linux operating system.

- Virtual syscalls:
Many Linux distributions also provide optimization of certain syscalls called virtual syscalls. Virtual syscalls, or vDSO (virtual dynamic shared object) calls, are a set of performance-optimized routines provided by the Linux kernel that user-space applications can use to execute certain system calls more efficiently. These virtual syscalls are mapped directly into the process address space, allowing some system call functionalities to be executed without the overhead of a traditional syscall. vDSO calls are generally captured through dynamic analysis, as they involve runtime components and optimizations. This study does not capture virtual syscalls.

2.2. Linux Malware Static Analysis

Static malware analysis examines malicious software without execution, focusing on the code and structure of the malware to identify malicious patterns, behaviors, and potential threats. This method analyzes API calls, function calls, and data structures to understand malware logic and potential behaviors, such as file modifications, network communications, and privilege escalation attempts.

Static analysis, although powerful, has limitations—it cannot detect all system calls made during runtime, which are essential for fully understanding interactions with the operating system. Dynamic analysis complements this by capturing all interactions and changes, providing a comprehensive view of malware capabilities. The approach we propose does not seek to replace dynamic analysis; rather, it offers crucial insights into malware behavior without the need for execution.

Anti-evasive malware techniques, such as obfuscation and encryption, and packed binaries can diminish the effectiveness of static analysis significantly as demonstrated by a recent study [9].

In this paper, we introduce a novel approach that studies call patterns by extracting syscalls and library syscall wrappers from binaries through static analysis, forming the basis of the dataset used in our research.

3. Literature Review

In the past decade, malware has continued to pose a significant threat to global infrastructure. As the majority of the infrastructure systems heavily rely on the Internet of Things (IoT), the importance of Linux malware analysis has become crucial, particularly because Linux is the preferred operating system for major IoT devices and servers. In this section, we present some of the previous literature on the usage of syscalls to classify malware.

Asmitha et al. [10,11] propose a novel non-parametric statistical approach using machine learning techniques for identifying previously unknown malicious Executable Linkable Files (ELF). They used a dataset of 226 Linux malware samples and 442 benign samples from which syscalls (features) were dynamically extracted. Their proposed method prioritizes features through non-parametric statistical techniques such as the Kruskal–Wallis ranking test (KW) and Deviation From Poisson (DFP). Three learning algorithms (J48, Adaboost, and Random Forest) were used to create a prediction model using a minimal feature set extracted from the system calls. This approach yielded an optimal feature vector, resulting in an overall classification accuracy of 97.30% in identifying unknown malicious files.

Additionally, Phu et al. [12] introduce a framework for analyzing MIPS ELF files based on syscall behaviors. They compiled a database of 3773 MIPS ELF malware samples from Detux, IoTPOT, and VirusShare and executed them in a QEMU-based sandbox called F-sandbox, derived from Firmadyne and Detux. For their final feature set, the authors took the 30 most common syscalls and applied the n-gram technique to construct the feature vectors. The authors then employed machine learning classifiers such as Support Vector Machine (SVM), Random Forest, and Naive Bayes to analyze the syscalls. Their experiments demonstrated that Random Forest achieves high accuracy in classifying MIPS malware, with an accuracy of 97.44%.

Taking it a level further, Tahir and Qadir [13] proposed a detection scheme for IoT malware using cross-architectural (MIPS, ARM and x86) analysis based on system calls. They had 69 system calls dynamically extracted from 1048 samples of both IoT malware and benign binaries. They then trained five popular machine learning models: Support Vector Machine, Random Forest, Logistic Regression, Bagging, and Multi-Layer Perceptron to detect the malware. The results showed that Random Forest achieved the highest detection accuracy at 99.04%. Additionally, their findings also suggested that feature sets based on system calls hold significant potential for detecting multi-architectural IoT malware.

Shobana and Poonkuzhali [14] carried out the same using deep learning techniques. In their approach, the authors identified malware by analyzing its behavior through the sequence of system calls it generates during execution. The system calls made by IoT malware are captured using the strace tool on Ubuntu. These recorded malicious system calls undergo preprocessing using n-gram techniques to extract necessary features. The resulting system calls are then categorized into two classes—normal and malicious sequences using a Recurrent Neural Network (RNN). The results reveal that using this technique produced 98% classification accuracy.

Abderrahmane et al. [15] present a solution that will classify whether an Android application is malicious or not. The application is installed and executed simulating its usage. After execution, system calls generated by the Linux kernel are collected, processed, and fed into the Neural Network model that will be used to classify the application. The authors used a convolutional Neural Network for this research and their findings reveal that their solution could detect malicious applications with an accuracy of 93.29%.

Lastly, Ramamoorthy et al. [7] focus on using binary lifting methods to perform static analysis and extract Intermediate Representation (IR) opcode sequences for analysis. The study introduces a series of statistical entropy-based characteristics derived from these opcode sequences. By concentrating solely on function metadata and opcode entropy, their approach is versatile across different architectures. It effectively identifies malware and accurately categorizes its variations, achieving an impressive F1 score of 97%.

To the best of our knowledge, existing research predominantly focuses on syscall extraction through dynamic analysis, using tools like strace to generate extensive syscall logs as shown in Table 1. This method requires executing malware within a controlled environment, which poses significant challenges in Linux due to the diversity of architectures and runtime environments. In contrast, our study uses a static analysis approach, extracting syscalls and library syscall wrappers directly from disassembled code. Although this method may not capture indirect syscalls to the kernel, we demonstrate that it remains an effective strategy for identifying malicious behavior.

Table 1. Comparative study of research works that use syscalls for malware detection.

Research	Statistical Analysis	Arch	Features Used	Dynamic/Static	Models	Accuracy	Comparison
Asmitha Vinod [10]	non-parametric statistical methods like Kruskal–Wallis ranking test (KW), Deviation From Poisson (DFP)	-	syscalls	Dynamic	J48, Adaboost, Random forest	97.30%	Dynamic analysis *
Asmitha Vinod [11]	-	-	syscalls	Dynamic	Naïve bayes, J48, Adaboost, RF, IBK-5	97%	Dynamic analysis *
Phu et al. [12]	n-gram, chi square	MIPS	syscalls	Dynamic	RF, NB, SVM	97%	Dynamic analysis *, MIPS architecture-specific
Tahir Qadir [13]	-	x86, MIPS, ARM	syscalls	Dynamic	SVM, LR, RF, MLP, Bagging	Bagging, RF (99%)	Dynamic analysis *
Shobana Poonkuzhali [14]	-	-	syscalls	Dynamic	RNN	98.7%	Dynamic analysis *
Abderrahmane et al. [15]	-	ARM	log files, syscalls	Dynamic	CNN	93.3%	Dynamic analysis *, ARM architecture-specific

Table 1. Cont.

Research	Statistical Analysis	Arch	Features Used	Dynamic/Static	Models	Accuracy	Comparison
DeMarinis et al. [16]	-	x86	syscalls extracted from assembly	Static	-	-	This research is a completely different use case and not directly related since it proposes a proactive defensive mechanism, but we reference this work since syscalls are filtered from disassembly. The authors use a binary analysis-based framework that automatically limits syscalls that attackers can (ab)use.
Our research	chi-squared test, Wilcoxon test	Multi architecture	syscalls, library syscall wrappers	Static	Linear Regression, Random Forest, SVC, Neural Network	96.86%	We could not find any recent works that use syscalls for static malware analysis. Static analysis requiring no execution or environment setup. Multi-architecture support. Can be combined with other static features for a higher accuracy. We conduct statistical analysis to validate syscall ranking and category assignment. These scores can be adjusted for various architectures.

'*' These works use dynamic analysis which requires a compatible environment to be set up and configured for the malware to execute. Moreover, dynamic analysis tends to generate a high volume of syscalls within a short period, as the kernel performs various internal operations during its execution. This rapid generation of syscalls can lead to significant noise, with numerous irrelevant and redundant syscalls that can obscure meaningful patterns.

4. Methodology

In this section, we provide a detailed explanation of the workflow that was utilized throughout our research. This workflow is structured into several distinct steps, each of which is outlined in this section and visualized in Figure 2 below.

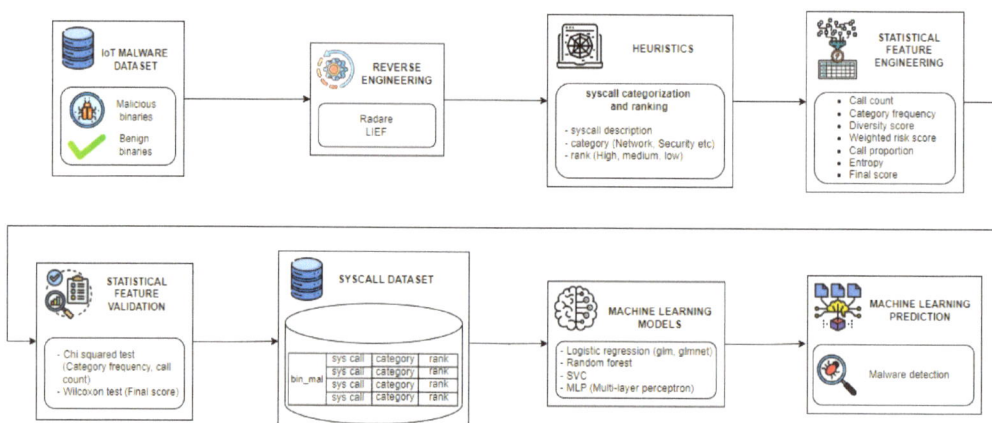

Figure 2. Workflow of the proposed approach.

4.1. Iot Malware Dataset

This paper concentrates exclusively on employing static analysis to identify system calls (syscalls) and library syscall wrappers utilized by binaries targeting the Linux kernel in ARM architecture-based Linux files, which are typically referred to as ELF (Executable and Linkable Format) binaries. Our study involves the examination of 1117 samples of ARM-based Linux IoT malware and 1214 binaries categorized as benign or non-malicious, with 251 unique syscalls as shown in Figure 3. The raw binaries are extracted from open-source IoT malware ARM architecture binary dataset from [17,18]. We selected random samples of binaries containing fewer than 2000 functions to ensure that the reverse engineering process was batched to a manageable size. Subsequent stages are illustrated as shown in Figure 2.

The proportion of syscalls by ranking and categories are shown in Figures 3 and 4 which are discussed in the Section 4.7.

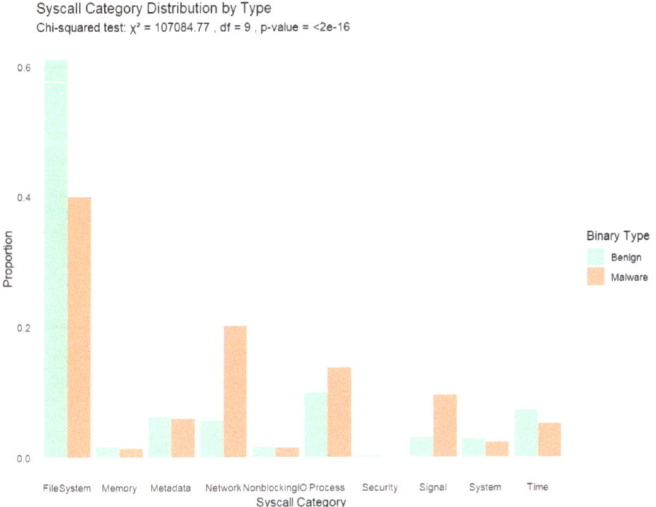

Figure 3. Proportion of syscall category in malware and benign subsets.

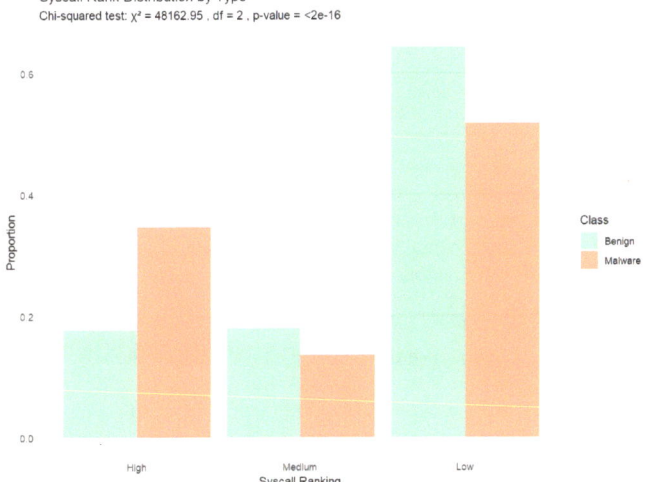

Figure 4. Proportion of security risk ranking of syscalls in malware and benign subsets.

4.2. Reverse Engineering

This study explores the identification of malicious behavior in ARM-based Linux IoT devices by analyzing syscalls and library syscall wrappers extracted solely from static analysis of ELF binaries. We employ radare2 [19], a popular reverse engineering library, to extract syscalls and library syscall wrappers from global symbols, exports, imports, and call references within the binary.

4.3. Heuristics

Additionally, each syscall is assigned a risk level—High, Medium, or Low, reflecting its potential for malicious use. The Linux syscalls are categorized as outlined in Table 2. To validate the effectiveness of our categorization and ranking system, we perform a chi-squared analysis [20] to compare the distribution of syscalls between benign and malware samples. This statistical validation confirms the significant differentiation in syscall usage patterns between the malware and benign subsets.

Our dataset distribution is outlined in Table 3. The raw binaries are reverse-engineered to extract the the following attributes: binary_id, hash, call_api, call_description, call_category, and call_rank.

Table 2. Extended categorization of Linux syscalls.

Syscall Category	Description
FileSystem	Handles file management operations such as reading, writing, and permissions.
Process	Manages process lifecycle such as creation, execution, and termination.
Memory	Controls memory allocation, deallocation, change and access critical for process management.
Network	Encompasses syscalls for network communication such as socket management, send, recv.
System	General system calls for services, system configuration and management.
Metadata	Involves retrieval and manipulation of file or system metadata.
Signal	Inter-process communication, and signal-driven interruptions.
Security	Key management, encryption, and access controls.
NonblockingIO	Non-blocking operations for input/output.
Time	Measuring time, manipulation.

Table 3. Counts of binary samples.

Class	Binary Count	Syscall Counts
Malware	1117	163,288
Benign	1214	45,284

Unlike Windows environment, Linux binaries have built-in debugging symbols with DWARF which is an integral part of the ELF format. This rich context contributes to gathering relevant symbols and their usage patterns, which is crucial for assessing the potential maliciousness of each call. We have systematically categorized all Linux system calls into the following ten categories based on their functionality:

4.4. Statistical Feature Engineering

To enhance the predictive power of our models, we enriched our dataset with several engineered features derived from the syscall and library call data. These features aim to capture the nuances of syscall behavior that distinguish malware from benign software. The engineered features include the following:

- Call Count: The total number of system calls made by each binary. This feature reflects the general activity level of the binary, which can indicate suspicious behavior.

- Distinct Call Count: The number of unique syscalls made by each binary. A high variety of calls can be indicative of complex or unusual binary behavior.
- Category Frequency: For each of the ten syscall categories, we calculated the frequency of syscalls falling into each category per binary. This helps in understanding which types of operations are predominant in a binary, aiding in profiling typical and atypical behaviors.
- High-Risk Call Proportion: The proportion of syscalls that are ranked as 'High' risk relative to the total number of syscalls. This feature specifically targets the detection of syscalls more commonly associated with malicious activities.
- Entropy of Calls: We calculated the entropy of syscall distribution within each category of a binary to measure the unpredictability and randomness of syscall usage, which can be higher in malware due to evasion techniques or diverse functionalities.
- Weighted Risk Score: By assigning weights to syscalls based on their assigned risk levels (High, Medium, and Low), we computed an overall risk score for each binary. This score provides a quantitative measure of the potential threat posed by the binary based on the observed syscalls.

Each syscall in a binary is assigned a weight based on its risk rank (High, Medium, and Low). The weighted rank score for each syscall can be calculated as:

$$\text{weighted\_rank\_score}_i = \text{weight}_i \cdot \text{count}_i \tag{1}$$

where weight_i is the risk weight for syscall i, and count_i is the count of syscall i within the binary.

The total rank score is the sum of all weighted rank scores across all syscalls in a binary:

$$\text{total\_rank\_score} = \sum_{i=1}^{n} \text{weighted\_rank\_score}_i \tag{2}$$

where n is the number of different syscalls recorded for the binary.

The diversity score, calculated using Shannon entropy, measures the unpredictability or diversity of syscall ranks in a binary. It is defined as:

$$\text{diversity\_score} = -\sum_{j=1}^{m} p_j \log_2(p_j) \tag{3}$$

where p_j is the proportion of syscalls of rank j (e.g., High, Medium, and Low), and m is the number of different ranks.

The final score is a composite metric that combines the total rank score and the diversity score, providing an overall score for each binary:

$$\text{final\_score} = \text{total\_rank\_score} + \text{diversity\_score} \tag{4}$$

These features were integrated into our analytical framework to provide a robust foundation for subsequent machine learning models to classify the samples effectively. By leveraging a combination of count-based, frequency-based, and information-theoretic metrics, we aim to capture a comprehensive profile of each binary's behavior, significantly enhancing our malware detection capabilities.

4.5. Data Organization

We index the data with a `binary_id` for organizing syscalls within each binary, to facilitate aggregation and analysis at the binary level. By grouping syscalls by their respective `binary_id`, we ensure that behavioral patterns are accurately captured for the entire binary. This structured approach enhances the reliability of our analysis and improves the model's ability to effectively learn and differentiate between distinct syscall patterns. The resulting syscall dataset created has the features as outlined in Table 4 for each `binary_id`.

Table 4. Description of the IoT Malware Syscall Analysis dataset.

Dataset Feature	Description
binary_id	Unique identifier for each binary
hash	Hash of the binary
Architecture	ARM—although our metholodolgy supports multiple architectures with binary lifting strategy
isMalware	Whether the binary is malicious or beningn
prc	32-bit/64-bit
Endian	The endianess of the binary (LSB or MSB)
Stripped	Binary attributes indicating whether it is stripped of symbol information
call_api	Name of the syscall
call_desc	Description of the function of the syscall
call_type	syscall or lib (library syscall wrapper)
call_cat	Category of the syscall, Refer Table 2
call_rank	Security risk level of the syscall based on its potential for malicious use (High, Medium, and Low)
bin_all_call_cnt	Count of all the syscalls and lib calls in the binary
bin_dist_call_cnt	Count of the distinct syscalls and libcalls in the binary
bin_dist_cat_cnt	Count of each category of syscalls and libcalls
bin_each_api_cnt	Count of unique syscalls in the binary
bin_calls_per_cat	Number of calls per category in a binary
bin_dist_call_per_cat	Number of calls in each distinct category in a binary
call_rank_n	Number of syscalls in the binary with the current syscall rank
bin_calls_per_rank	Number of syscalls and libcalls in each rank per binary
std_dev_calls_per_cat	Standard deviation of the syscalls per category in a binary
mean_calls_per_cat	Mean of calls per category in a binary
average_rank	Average rank of the syscalls and libcalls in a binary
weighted_rank_score	Weight of the syscall, Refer Equation (1)
total_rank_score	Sum of all weighted rank scores across all syscalls in a binary, Refer Equation (2)
diversity_score	Unpredictability of the syscall ranks in the binary, Refer Equation (3)
final_score	Total rank score added to the diversity score, Refer Equation (4)
cat_concentration_index	Number of syscall categories in the binary (Unused)

4.6. Handling of Outliers

Outliers highlight extreme behaviors or unusual patterns that are characteristic of malicious activities. Therefore, we retain outliers in our dataset to capture such behavior patterns.

4.7. Statistical Feature Analysis

To validate the effectiveness of the categorized and ranked syscalls in differentiating malware from benign binaries, we employ a chi-squared test. This statistical method evaluates whether there is a significant difference in the distribution of syscall ranks between malware and benign samples.

The test procedure is as follows:
- Data Collection: We calculate the frequency of each syscall rank for both malware and benign binaries.
- Normalization: The frequencies are normalized to proportions within each group.

- Chi-squared Test: We perform the chi-squared test to determine if the observed differences in the distribution of syscall ranks are statistically significant.

$$\chi^2 = \sum \frac{(O_i - E_i)^2}{E_i} \quad (5)$$

where O_i are the observed frequencies of risk rank, and E_i are the expected frequencies. We use the same methodology to assess the importance of syscall categories as well. Figure 3 shows the proportion of each syscall category for malware and benign datasets along with the chi-squared test results. Similarly, Figure 4 shows the syscall risk rank proportion along with the chi-squared test results for the malware and benign subsets.

The results show p-value $< 2.2 \times 10^{16}$ in both cases, implying that the rankings are not just arbitrary, and assigned rankings correspond to observable differences in behavior between malware and benign binaries. This reinforces the idea that higher risk ranked syscalls and certain categories of syscalls such as network, process and signal category syscalls are more critical in the context of malware activity. Also, the high-risk ranked syscalls are statistically higher in the malware subset than benign binaries.

To validate the statistical significance of the differences observed between the calculated final_score based on the entropy of the syscalls and their risk ranking, we employ the Mann–Whitney (Wilcoxon) U test. This non-parametric test is suitable for skewed data for comparison from two independent groups. Our analysis reveals a negligibly low p-value $< 2 \times 10^{-16}$, confirming that the final_score computed based on Equation (4), can reliably differentiate between malware and benign samples as shown in Figure 5.

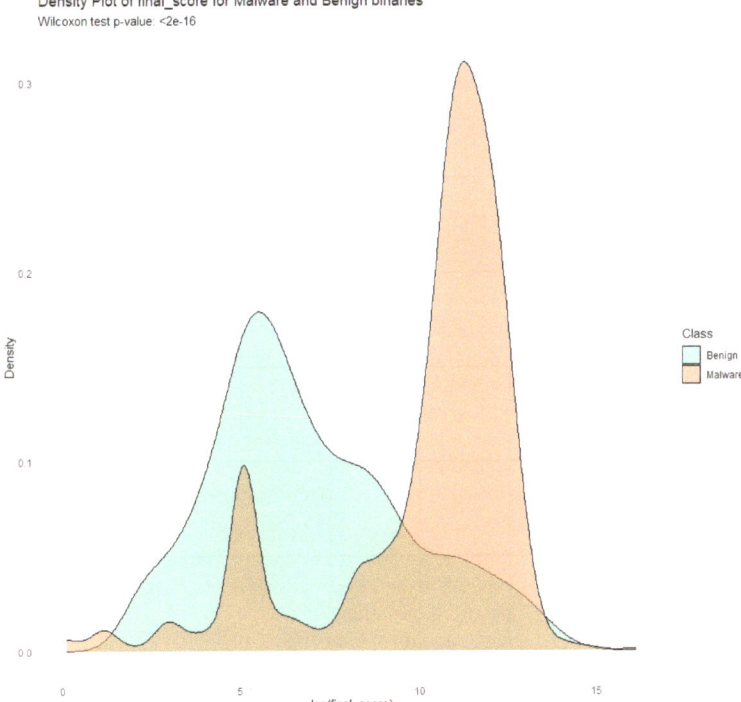

Figure 5. Differences in final score (Equation (4)) between malware and benign subsets.

4.8. Machine Learning Classification Based on Static Extraction of Syscalls

The primary focus of this study is to identify malware based on syscalls and library syscall wrappers extracted from static analysis as opposed to dynamic execution of the malware. Based on the extracted syscalls, we outlined the statistical features added to the dataset in previous sections, which are based on syscall categorization and security risk ranking of the syscalls used by the binary.

In this section, we discuss the feasibility of machine learning (ML) classification models on the created dataset. We have evaluated the approach and dataset with commonly used ML classification models: Logistic Regression, Random Forest, LinearSVC, and Multi-Layer Perceptron (MLP) Neural Networks.

- Logistic Regression: To evaluate a simple yet effective baseline for binary classification.
- Random Forest: Random Forest classifier is robust in handling outliers and anomalies typical of malware, and suitable for complex interactions between features.
- LinearSVC: LinearSVC is often effective with high-dimensional data, and used for interpretability.
- Multi-Layer Perceptron (MLP) Neural Network: MLP is effective in capturing complex patterns and interactions in the data through a layered architecture.

As discussed in Section 5, the strong performance of all the models suggests that our approach and dataset are well suited for malware detection.

To preserve the context of syscalls and library syscall wrappers within each binary, we train our models using grouped data as discussed in Section 4.5. Syscall categories and risk ranking are One-Hot-encoded to handle categorical data effectively. We then compute the Variation Inflation Factor (VIF) to identify and mitigate multicollinearity in our feature set.

The Generalized Linear Model (GLM) using the 'glmnet' statistical model package [21] is employed with Lasso Regression to train the glm model. Regularization is used to minimize overfitting and improve generalization.

The dataset is split in an 80:20 ratio, ensuring a balanced representation through shuffling of batches grouped by binary_id to preserve the frequency and group of syscalls.

4.9. Hyper Parameters

Most of the standard default values from the 'scikit-learn' library [22] were retained and minimal tuning was required for the ML models. In this section, we discuss the vectorization strategy and some of the hyperparameter settings of the individual models.

Vectorization

For text feature 'call_api' (syscall name), we use TF-IDF (Term Frequency–Inverse Document Frequency) vectorizer, with a maximum of 500 distinct tokens (unique Linux system calls) to extract, and an n-gram range of (1, 1). For numeric features, the mean value is used for imputation and numeric values are standardized (normalized) by removing the mean and scaling to unit variance. Categorical features, syscall category (call_cat) and syscall risk ranking (call_rank), are One-Hot-encoded. The pipeline for the data preprocessing and vectorization strategy is shown in Figure 6.

- Linear Regression A Linear Regression model is implemented with the preprocessing pipeline as shown in Figure 6 The syscall imbalance between malware and benign binaries is handled with class weights assigned by the Linear Regression model's balanced parameter.
- Random Forest Classifier:
  Through GridSearch, we identified the optimal number of trees as 200. The default values were retained for the maximum depth of the tree, which was set to None, and Gini impurity is used for the split criteria.
- LinearSVC:
  For LinearSVC, we used the default 'squared-hinge' loss function with an L2 regularization and set the tolerance of 1×10^{-4} for the stopping criteria.

- MLP Classifier:
  For the Neural Network model, we employed the rectified linear unit function (ReLU) for activation and the 'adam' solver for weight optimization, with an initial learning rate of 0.001.

  Our dataset is supported by statistical features and validation which is evident from the comparable results across all ML models as discussed in the next section.

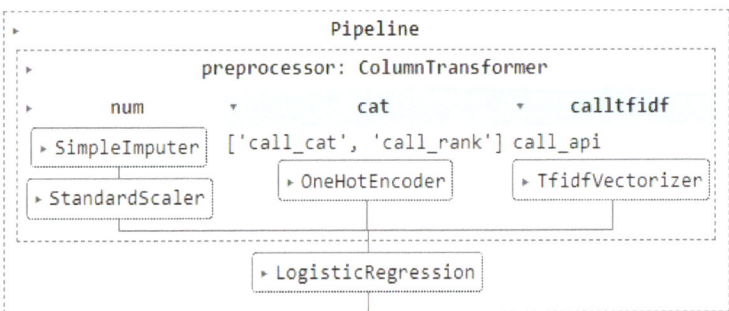

Figure 6. Preprocessing pipeline.

5. Results and Analysis

In this section, we compare and analyze the performance of various machine learning classification models on the syscall dataset comprising IoT malware and benign binaries. The primary objective is to evaluate the effectiveness of the ML models in distinguishing between malware and benign ELF binaries.

The performance metrics used to evaluate the models include accuracy, precision, and F1 score, defined as follows:

- Accuracy: The proportion of true results (both true positives and true negatives) among the total number of cases examined.

$$\text{Accuracy} = \frac{TP + TN}{TP + TN + FP + FN} \quad (6)$$

where TP, TN, FP, and FN represent the number of true positives, true negatives, false positives, and false negatives, respectively.

- Precision: The proportion of true positive results in all positive predictions.

$$\text{Precision} = \frac{TP}{TP + FP} \quad (7)$$

- F1 score: The harmonic mean of precision and recall.

$$\text{Recall} = \frac{TP}{TP + FN} \quad (8)$$

$$\text{F1 score} = 2 \times \frac{\text{Precision} \times \text{Recall}}{\text{Precision} + \text{Recall}} \quad (9)$$

The performance metrics of the ML models are shown in Table 5. The Random Forest classifier has the best performance with an F1 score of about 97%. The optimal threshold for the final prediction was found to be 0.72, and the maximum F1 score of 96.86%.

Table 5. Performance metrics of machine learning models.

Model	Accuracy	Precision	F1 Score
Random Forest	93.34%	94.71%	96.86%
Logistic Regression	92.34%	96.0%	95.06%
SVC	92.48%	96.16%	95.14%
MLP NN	92.07%	93.34%	95.03%

5.1. Discussion

As observed from the results in Table 5, batching syscalls for a single binary in the course of training likely plays a crucial role in all ML model's ability to detect malware binaries accurately since the syscall frequency and order are maintained. The Random Forest classifier performed the best, with a 96.8% F1 score likely due to the model's ability to handle outlier data which is characteristic of malware binaries. For example, there could be rarely used syscalls, or a direct syscall instead of library wrappers in malware which are anomalous behaviors.

The confusion matrices for the machine learning classifiers provide detailed insights into the performance of each model in distinguishing between malware and benign binaries. The confusion matrix is critical in understanding the classification performance, as it breaks down the results for the true positives (TP), false positives (FP), true negatives (TN), and false negatives (FN). By examining the confusion matrices, we can better understand the types of errors each model makes.

- True positives (TP): Correctly identified malware binaries.
- False positives (FP): Benign binaries incorrectly classified as malware.
- True negatives (TN): Correctly identified benign binaries.
- False negatives (FN): Malware binaries incorrectly classified as benign.

5.2. Error Analysis

The results suggest that most ML classifier models are able to accurately identify malware behavior based on static syscall analysis since the dataset includes statistical scores for the binaries based on syscall ranking and category.

The LinearSVC classifier, Logistic Regression and MLP Neural Network all show a similar number of true positives as shown in Figures 7, 8 and 9, respectively.

Analyzing Confusion Matrix of Random Forest Classifier which is the best performing model as shown in Figure 10, the labels with value 1 indicate malware and value 0 indicates benign syscalls. The Random Forest classifier accurately classified 158,223 malware syscalls with only 5065 misclassifications. The model accurately classified 36,464 benign syscalls with 8820 misclassifications. Although malware binaries are characterized by extreme outliers, the Random Forest model demonstrates a higher degree of accuracy and precision. It should be noted that all the syscalls in a single binary are set to malware or benign by the model based on the scoring and other features. The performance of the model is conducted on an individual syscall level as shown by the confusion matrix, indicating the actual performance of the models is even better.

Logistic Regression and SVC models demonstrated strong performance but had slightly higher false negative rates. This suggests that while they are effective at identifying benign binaries, they may miss certain types of malware that exhibit less common syscall patterns.

The MLP Neural Network, while effective, showed a marginally higher rate of false positives, indicating it might be more prone to overfitting on benign syscall patterns. This could be mitigated by further tuning the network architecture or using regularization techniques.

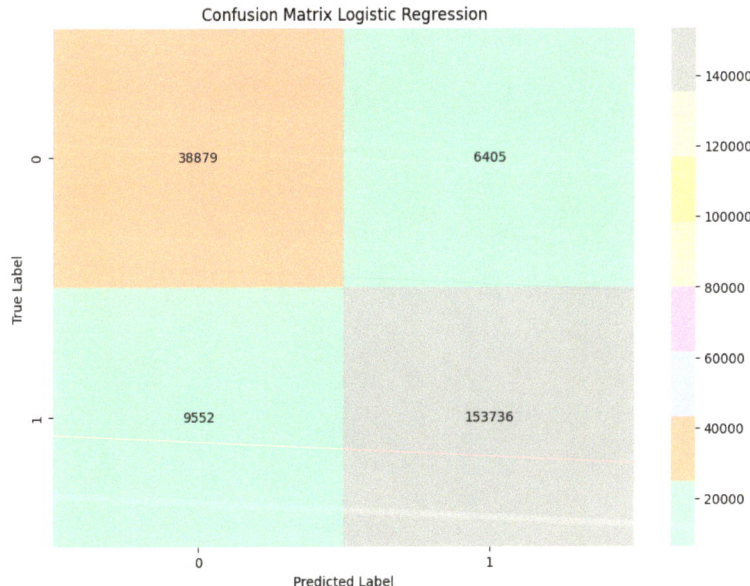

Figure 7. Logistic Regression classifier confusion matrix.

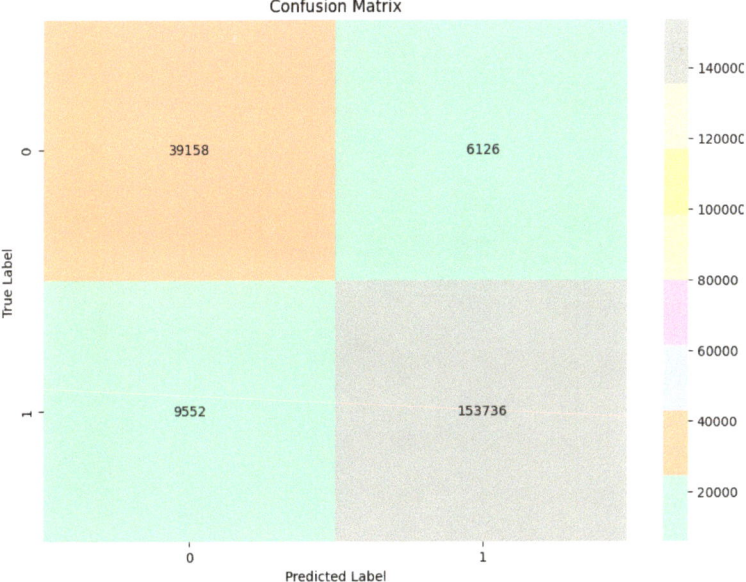

Figure 8. Linear SVC classifier confusion matrix.

Figure 9. MLP Neural Network confusion matrix.

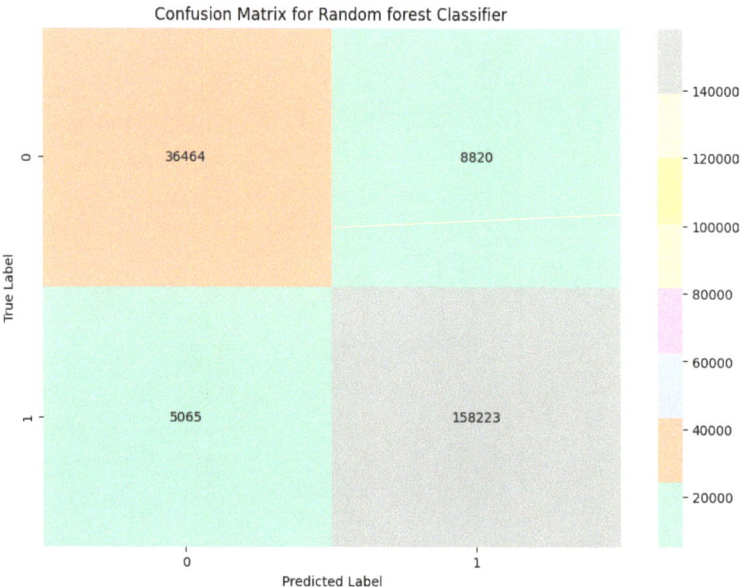

Figure 10. Random Forest classifier confusion matrix.

The calculated AUC (Area Under the Curve) score for the ROC (Receiver Operating Characteristic) curve of the Random Forest classifier is 0.964, which demonstrates robust performance under various threshold as shown in Figure 11. The optimal threshold was found to be 0.72, with a F1 score of 96.8%.

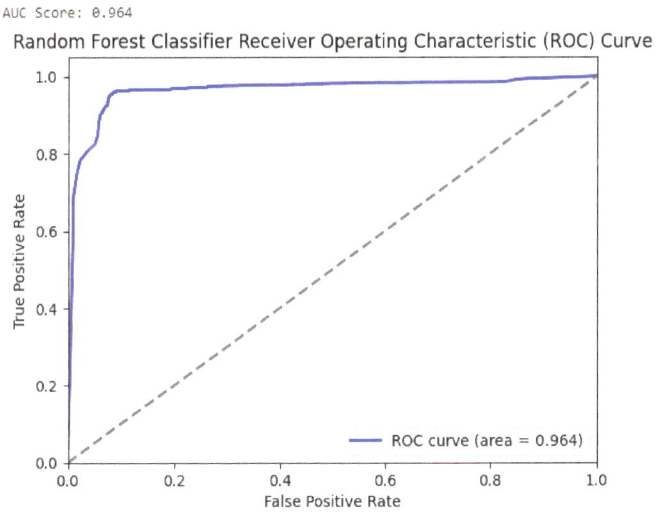

Figure 11. Random Forest classifier Receiver Operating Characteristic (ROC) curve with AUC score. Note: The dotted line indicates Random guess.

6. Conclusions

This study successfully demonstrates the viability of static analysis for detecting malware in Linux IoT systems by analyzing call patterns (syscalls and library syscall wrappers) extracted from ELF binaries. By ranking and categorizing calls based on their security implications, we have established a methodological framework that distinguishes between benign and malicious binaries effectively. The application of a chi-squared test further validates the significant differences in call patterns between these groups, reinforcing the reliability of our categorization and ranking system.

We demonstrate the efficacy of call pattern analysis from static analysis by using machine learning models such as glmnet with Lasso Regression, Random Forest classifier, SVC and MLP Neural Network. The models were able to perform malware detection with a high level of accuracy, as indicated by a Random Forest classifier F1 score of 96.86%. This is particularly significant given that this study solely uses system calls extracted using static analysis without any other significant contributing features that are important in static analysis such as sections, functions, strings and file header information. Our approach mitigates the risks associated with dynamic analysis and offers an effective alternative for malware detection.

7. Future Works

While this study demonstrates the viability of static analysis for detecting malware in Linux IoT systems with a high accuracy rate, it primarily focuses on syscalls and library syscall wrappers. However, this narrow focus may not capture all of the malware behavior patterns. Future research could consider incorporating additional static features, such as section headers, functions, and strings, for a comprehensive detection strategy.

Additionally, static analysis does not capture dynamically generated syscalls, potentially leading to incomplete data. The limitations of dynamic analysis are already discussed in this study. Augmenting static analysis with dynamic analysis in future studies could provide a more comprehensive understanding of malware behavior and improve detection rates.

Building on the findings of this study, future research will aim to extend the scope of the static analysis framework. The study could also be extended to multiple architectures for evaluation since the methodology proposed is architecture-agnostic, thereby broadening its applicability in the field of IoT malware detection.

Author Contributions: Conceptualization, J.R., R.C.K., K.G. and N.K.S.; methodology, J.R.; software, J.R.; validation, J.R., R.C.K., K.G., N.K.S. and C.V.; formal analysis, J.R.; investigation, J.R. and R.C.K.; resources, N.K.S. and C.V.; data curation, J.R.; writing—original draft preparation, J.R. and K.G.; writing—review and editing, K.G. and J.R.; visualization, J.R., K.G. and R.C.K.; supervision, R.C.K. and N.K.S.; project administration, C.V.; funding acquisition, C.V. All authors have read and agreed to the published version of the manuscript.

Funding: This research received no external funding.

Data Availability Statement: The syscall dataset created in the study is openly available at: https://github.com/jrmoorthy/Linux-Malware-Analysis, accessed on 20 July 2024.

Conflicts of Interest: The authors declare no conflicts of interest.

References

1. Howarth, J. 80+ Amazing IoT Statistics (2024–2030)—Explodingtopics.com. Available online: https://explodingtopics.com/blog/iot-stats (accessed on 9 May 2024).
2. Zscaler ThreatLabz Finds a 400 Year-over-Year. Available online: https://www.zscaler.com/press/zscaler-threatlabz-finds-400-increase-iot-and-ot-malware-attacks-year-over-year-underscoring (accessed on 9 May 2024).
3. Ngo, Q.D.; Nguyen, H.T.; Le, V.H.; Nguyen, D.H. A survey of IoT malware and detection methods based on static features. *ICT Express* **2020**, *6*, 280–286. [CrossRef]
4. Antony, A.; Sarika, S. A review on IoT operating systems. *Int. J. Comput. Appl.* **2020**, *176*, 33–40. [CrossRef]
5. AV-ATLAS Malware Portal. 2023. Available online: https://portal.av-atlas.org/malware (accessed on 10 May 2023).
6. The Official Radare2 Book: ESIL. 2024. Available online: https://book.rada.re/ (accessed on 20 July 2024).
7. Ramamoorthy, J.; Gupta, K.; Shashidhar, N.K.; Varol, C. Linux IoT Malware Variant Classification Using Binary Lifting and Opcode Entropy. *Electronics* **2024**, *13*, 2381. [CrossRef]
8. Kerrisk, M. Linux Programmer's Manual: Syscalls. 2024. Available online: https://man7.org/linux/man-pages/man2/syscalls.2.html (accessed on 18 June 2024).
9. Xu, Y.; Li, D.; Li, Q.; Xu, S. Malware Evasion Attacks against IoT and Other Devices: An Empirical Study. *Tsinghua Sci. Technol.* **2024**, *29*, 127–142. [CrossRef]
10. Asmitha, K.; Vinod, P. Linux malware detection using non-parametric statistical methods. In Proceedings of the 2014 International Conference on Advances in Computing, Communications and Informatics (ICACCI), Delhi, India, 24–27 September 2014; pp. 356–361.
11. Asmitha, K.; Vinod, P. A machine learning approach for linux malware detection. In Proceedings of the 2014 International Conference on Issues and Challenges in Intelligent Computing Techniques (ICICT), Ghaziabad, India, 7–8 February 2014; pp. 825–830.
12. Phu, T.N.; Dang, K.H.; Quoc, D.N.; Dai, N.T.; Binh, N.N. A novel framework to classify malware in mips architecture-based iot devices. *Secur. Commun. Netw.* **2019**, *2019*, 4073940. [CrossRef]
13. Tahir, I.; Qadir, S. Machine Learning-Based Detection of IoT Malware Using System Call Data. 2022. Available online: https://www.researchsquare.com/article/rs-2384013/v1 (accessed on 18 June 2024).
14. Shobana, M.; Poonkuzhali, S. A novel approach to detect IoT malware by system calls using Deep learning techniques. In Proceedings of the 2020 International Conference on Innovative Trends in Information Technology (ICITIIT), Kottayam, India, 11–12 February 2020; pp. 1–5.
15. Abderrahmane, A.; Adnane, G.; Yacine, C.; Khireddine, G. Android malware detection based on system calls analysis and CNN classification. In Proceedings of the 2019 IEEE Wireless Communications and Networking Conference Workshop (WCNCW), Marrakesh, Morocco, 15–18 April 2019; pp. 1–6.
16. DeMarinis, N.; Williams-King, K.; Jin, D.; Fonseca, R.; Kemerlis, V.P. sysfilter: Automated System Call Filtering for Commodity Software. In Proceedings of the 23rd International Symposium on Research in Attacks, Intrusions and Defenses (RAID 2020), San Sebastian, Spain, 14–16 October 2020; pp. 459–474.
17. Olsen, S.H.; OConnor, T. Toward a Labeled Dataset of IoT Malware Features. In Proceedings of the 2023 IEEE 47th Annual Computers, Software, and Applications Conference (COMPSAC), Torino, Italy, 26–30 June 2023; pp. 924–933.
18. Refade. IoT_ARM: A Collection of IoT Malware Samples for ARM Architecture. 2024. Available online: https://github.com/refade/IoT_ARM (accessed on 18 June 2024).
19. Seba, P. *Radare2 Book*; Radare Project. 2020. Available online: https://book.rada.re/credits/credits.html (accessed on 8 May 2024).
20. Tallarida, R.J.; Murray, R.B.; Tallarida, R.J.; Murray, R.B. Chi-square test. In *Manual of Pharmacologic Calculations: With Computer Programs*; Springer: New York, NY, USA, 1987; pp. 140–142.

21. Friedman, J.; Tibshirani, R.; Hastie, T. Regularization Paths for Generalized Linear Models via Coordinate Descent. *J. Stat. Softw.* **2010**, *33*, 1–22. [CrossRef]
22. Pedregosa, F.; Varoquaux, G.; Gramfort, A.; Michel, V.; Thirion, B.; Grisel, O.; Blondel, M.; Prettenhofer, P.; Weiss, R.; Dubourg, V.; et al. Scikit-learn: Machine Learning in Python. *J. Mach. Learn. Res.* **2011**, *12*, 2825–2830.

Disclaimer/Publisher's Note: The statements, opinions and data contained in all publications are solely those of the individual author(s) and contributor(s) and not of MDPI and/or the editor(s). MDPI and/or the editor(s) disclaim responsibility for any injury to people or property resulting from any ideas, methods, instructions or products referred to in the content.

Article

MSFuzz: Augmenting Protocol Fuzzing with Message Syntax Comprehension via Large Language Models

Mingjie Cheng [1,2], Kailong Zhu [1,2,*], Yuanchao Chen [1,2], Guozheng Yang [1,2], Yuliang Lu [1,2] and Canju Lu [1,2]

1. College of Electronic Engineering, National University of Defense Technology, Hefei 230037, China; chengmingjie22@nudt.edu.cn (M.C.); chenyuanchao@nudt.edu.cn (Y.C.); yangguozheng17@nudt.edu.cn (G.Y.); luyuliang@nudt.edu.cn (Y.L.); lucanju17@nudt.edu.cn (C.L.)
2. Anhui Province Key Laboratory of Cyberspace Security Situation Awareness and Evaluation, Hefei 230037, China
* Correspondence: zhukailong@nudt.edu.cn

Abstract: Network protocol implementations, as integral components of information communication, are critically important for security. Due to its efficiency and automation, fuzzing has become a popular method for protocol security detection. However, the existing protocol-fuzzing techniques face the critical problem of generating high-quality inputs. To address the problem, in this paper, we propose MSFuzz, which is a protocol-fuzzing method with message syntax comprehension. The core observation of MSFuzz is that the source code of protocol implementations contains detailed and comprehensive knowledge of the message syntax. Specifically, we leveraged the code-understanding capabilities of large language models to extract the message syntax from the source code and construct message syntax trees. Then, using these syntax trees, we expanded the initial seed corpus and designed a novel syntax-aware mutation strategy to guide the fuzzing. To evaluate the performance of MSFuzz, we compared it with the state-of-the-art (SOTA) protocol fuzzers, namely, AFLNET and CHATAFL. Experimental results showed that compared with AFLNET and CHATAFL, MSFuzz achieved average improvements of 22.53% and 10.04% in the number of states, 60.62% and 19.52% improvements in the number of state transitions, and 29.30% and 23.13% improvements in branch coverage. Additionally, MSFuzz discovered more vulnerabilities than the SOTA fuzzers.

Keywords: fuzzing; syntax aware; protocol implementations; large language models

Citation: Cheng, M.; Zhu, K.; Chen, Y.; Yang, G.; Lu, Y.; Lu, C. MSFuzz: Augmenting Protocol Fuzzing with Message Syntax Comprehension via Large Language Models. *Electronics* **2024**, *13*, 2632. https://doi.org/10.3390/electronics13132632

Academic Editors: Abdussalam Elhanashi and Pierpaolo Dini

Received: 6 June 2024
Revised: 30 June 2024
Accepted: 1 July 2024
Published: 4 July 2024

Copyright: © 2024 by the authors. Licensee MDPI, Basel, Switzerland. This article is an open access article distributed under the terms and conditions of the Creative Commons Attribution (CC BY) license (https://creativecommons.org/licenses/by/4.0/).

1. Introduction

In the digital age, network protocols serve as the foundation for information exchange, not only establishing rules and formats for data transfer but also ensuring the accuracy and efficiency of network information [1]. Although these protocols are crucial for the advancement of modern informatization and intelligentization, their complexity and diversity elevate security risks. Moreover, potential oversights or errors during the protocol development often lead to vulnerabilities. This leaves plenty of room for hackers to target network applications [2]. Therefore, effectively identifying and rectifying these implementation vulnerabilities is vital for maintaining the security and stability of cyberspace.

Fuzzing is an efficient software testing method that is widely used across various software applications. It was proven to be highly effective and powerful in discovering critical vulnerabilities [3,4]. Due to its automation and efficiency, fuzzing has become one of the popular methods for detecting security vulnerabilities in network protocols. This method involves sending large volumes of random or semi-random data to protocol implementations to trigger abnormal behavior, thereby uncovering potential vulnerabilities.

Compared with other fuzzing, network protocol fuzzing faces greater challenges in generating high-quality test cases, primarily due to the highly structured nature of its inputs. Specifically, the input consists of a series of request messages, with each divided into fields defined by strict syntactic rules with precise value constraints. If the input messages do not adhere to basic syntax requirements, the server discards them during the initial stage of processing. This strict requirement for input syntax significantly affects the quality of test cases.

In recent years, previous studies focused on acquiring the syntactic knowledge of protocol messages to generate high-quality test cases. These methods fall into four main categories: protocol specification extraction [5–10], network traffic analysis [11–15], program behavior analysis [16–19], and machine learning analysis [20,21]. Protocol specification extraction methods primarily derive message syntax formats from protocol RFC specification documents. Network traffic analysis methods learn about keywords and field boundaries within messages by collecting and analyzing network traffic. Program behavior analysis methods obtain protocol format information by analyzing how programs behave when processing message data. Machine learning analysis methods leverage natural-language-processing techniques to learn protocol syntax. Although these methods can acquire the partial knowledge of protocol syntax, they still face three key challenges in generating high-quality test cases:

Challenge 1: inadequate input knowledge. Current protocol fuzzers [5,11,16,20] lack an effective and detailed understanding of message syntax. These tools typically reveal only the basic syntactic structure or syntactic constraints of specific fields, failing to fully grasp the comprehensive syntactic knowledge of the message. This partial comprehension of message syntax limits the ability to generate high-quality inputs, thereby impacting the overall effectiveness of fuzzing.

Challenge 2: insufficient seed diversity. The effectiveness of test cases in mutation-based fuzzing largely depends on the quality of the initial seed corpus. However, the most widely used network protocol fuzzing benchmark, ProFuzzBench [22], often displays a lack of seed diversity. If the initial seeds are not comprehensive and solely rely on simplistic mutations, fuzzing may fail to adequately explore protocol implementations. This constraint limits the detection of a wider range of vulnerabilities.

Challenge 3: inefficient mutation strategies. Mutation strategies usually determine the quality of test cases when fuzzing. Most current protocol fuzzers [23–25] rely on random mutation methods that do not employ targeted or strategic approaches. These approaches fail to produce diverse and high-quality test cases that can effectively assess the robustness of protocol implementations. Such deficiencies in mutation strategies restrict the depth and breadth of fuzzing coverage, thereby significantly reducing the overall effectiveness of fuzzing.

To overcome the above challenges, in this paper, we propose MSFuzz, which is a protocol-fuzzing method with message syntax comprehension. By comparing the message syntax extracted from existing research with that from the source code, we found that the source code contained more detailed and effective message syntax knowledge. To overcome challenge 1, we present a method for leveraging large language models (LLMs) to extract the message syntax from source code to construct message syntax trees for protocol implementations. For challenge 2, we used LLMs in conjunction with the constructed message syntax trees to expand the initial seed corpus of the protocol implementation. To address challenge 3, we designed a novel syntax-aware mutation strategy that guides fuzzing mutations through the constructed message syntax trees, thereby generating high-quality test cases that satisfy syntactic constraints.

Specifically, we first filtered out code files related to message parsing from a large volume of protocol implementation source code. Then, utilizing the code comprehension abilities of LLMs, we incrementally extracted message types, and requested line parameters and their value constraints, as well as header fields and their value constraints, from the filtered source code. This process enabled us to construct the message syntax tree of the protocol implementation. Based on the constructed message syntax tree and configuration files of the protocol implementation, we expanded the initial seed corpus, generating a diverse and comprehensive set of seeds. Finally, we used the constructed message syntax tree to guide the mutation strategy. By parsing the messages to be mutated and matching them with the syntax tree, we determined the syntactic constraints of message fields and generated high-quality test cases based on these constraints. So far, we have implemented a prototype of MSFuzz.

We evaluated the performance of MSFuzz on three widely used network protocols: RTSP, FTP, and DAAP. We compared MSFuzz with two SOTA protocol fuzzers: AFLNET and CHATAFL. The experimental results showed that within 24 h, MSFuzz improved the number of states by averages of 22.53% and 10.04%, and the number of state transitions by averages of 60.62% and 19.52%, respectively, when compared with AFLNET and CHATAFL. Additionally, MSFuzz effectively explored the code space, achieving average branch coverage improvements of 29.30% and 23.13% over SOTA protocol fuzzers. In the ablation study, we found that the two key components, seed expansion, and syntax-aware mutation, significantly enhanced the fuzzing performance. Additionally, MSFuzz discovered more vulnerabilities than the SOTA fuzzers.

The main contributions of this paper are summarized as follows:

- To address the problem of generating high-quality test cases, we propose MSFuzz, which is a novel protocol-fuzzing technique. MSFuzz is built upon three core components: message syntax tree construction, seed expansion, and syntax-aware mutation.
- By employing a novel abstraction of message syntax structures, MSFuzz leverages the code-understanding capabilities of LLMs to effectively extract message syntax from the source code of protocol implementations, thereby constructing uniformly structured message syntax trees.
- MSFuzz utilizes the constructed protocol syntax trees to expand the initial seed corpus and applies syntax-aware mutation strategies to generate high-quality test cases that adhere to specified constraints.
- We evaluated MSFuzz on widely used protocol implementations. The results demonstrate that MSFuzz outperformed the SOTA protocol fuzzers in state coverage, code coverage, and vulnerability discovery.

2. Background and Motivation

In this section, we introduce protocol fuzzing, large language models, and a motivating example. First, we provide a brief overview of protocol fuzzing. Next, we offer background information on large language models and their recent advancements in vulnerability discovery. Finally, we illustrate the limitations of existing methods through a motivating example.

2.1. Protocol Fuzzing

As one of the most effective and efficient methods for discovering vulnerabilities, fuzzing has been applied in the field of network protocols. The inception of PROTOS [26], the pioneering protocol-focused fuzzing tool, marked the beginning of protocol fuzzing. Subsequently, this field has garnered extensive attention, resulting in numerous research achievements and establishing itself as a focal point in network security.

Protocol fuzzers primarily target server-side implementations by simulating client behavior. These tools continuously create and dispatch client messages to the servers. Depending on the method of message generation, protocol fuzzers can be broadly categorized into two types: generation based and mutation based.

Generation-based protocol fuzzers rely on prior knowledge of protocol formats to generate test cases [26–30]. PROTOS [26] generates erroneous inputs based on protocol specifications to trigger specific vulnerabilities. SPIKE [27] employs a block-based modeling approach, breaking down the protocol into different blocks and automatically generating valid data blocks for protocol messages based on predefined generation rules. Peach [31] defines data models of protocols by manually constructing Pit files, which are then used to generate test cases. SNOOZE [6] requires testers to manually extract protocol specifications from request for comments (RFC) documents, including protocol field characteristics, information exchange syntax, and state machines. Testers then send specific sequences of messages to reach the desired state and generate numerous random test cases based on the protocol specifications.

Mutation-based protocol fuzzers generate new test cases by mutating seeds, which consist of a set of request messages [19,24,32–34]. These mutation operators include altering field values within messages, inserting, deleting, or replacing specific sections of messages, as well as recombining or reordering messages. AFLNET [24] employs byte-level and region-level mutation strategies, capturing traffic during client-server communication as initial seeds and generating test cases during the fuzzing process. Meanwhile, SGPFuzzer [32] introduces various mutation operators to mutate selected seed files in a simple and structured manner, including sequence mutation, message mutation, binary field mutation, and variable string mutation.

Compared with other fuzzing, network protocol fuzzing faces more challenges, one of which is the highly structured nature of its input. Network protocol messages typically adhere to strict syntactic constraints, encompassing various types of messages, as well as syntactic constraints within internal fields. Deviations from these constraints may result in servers discarding received messages, thereby limiting the effectiveness and efficiency of fuzzing.

2.2. Large Language Models

Large language models (LLMs), as a form of deep learning-driven artificial intelligence technology, demonstrate powerful natural-language-processing capabilities. These models undergo extensive pre-training on large datasets, equipping them with the ability to deeply understand and generate natural language text. They possess a profound and rich understanding of linguistic knowledge, and their comprehension of context is remarkably thorough.

In recent studies, LLMs have shown great potential in the field of vulnerability detection. GPTScan [35] combines GPT with static analysis techniques for intelligent contract logic bug detection. CHATAFL [10] employs LLMs to guide protocol fuzzing, constructing protocol grammars, expanding initial seeds, and generating test cases capable of triggering state transitions through interactions with LLMs. Fuzz4All [36] utilizes LLMs as an input generation and mutation engine for fuzzing across multiple input languages and features. ChatFuzz [37] employs LLMs for seed mutation. TitanFuzz [38] utilizes LLMs for fuzzing deep learning libraries. These studies provide rich practical experience in the application of large language models for vulnerability detection, showcasing their potential and value in enhancing vulnerability detection capabilities.

Although applying LLMs to network protocol fuzzing has great potential, it faces several challenges. When using LLMs to analyze source code, directly inputting the entire source code often fails due to input size limitations. Therefore, it is essential to filter the source code and extract key snippets when using LLMs.

2.3. Motivating Example

As one of the SOTA network protocol fuzzers, CHATAFL [10] extracts message syntax structures from protocol specifications. Although protocol implementations generally adhere to RFC documents, variations exist between different implementations, and actual message syntactic constraints are often more detailed than the specifications. For example, Figure 1 illustrates the *PLAY* message syntax structure that CHATAFL extracts from the RTSP specification. Although the basic syntax of the message is determined, the specific values of key fields, like *Range*, remain unknown.

```
PLAY <Value> RTSP/1.0 \r\n
CSeq: <Value>\r\n
User-Agent: <Value>\r\n
Session: <Value>\r\n
Range: <Value>\r\n
\r\n
```

Figure 1. Syntax for the RTSP PLAY message.

Listing 1 shows a code snippet from the Live555 server based on the RTSP protocol, demonstrating the parsing process of the *Range* field. The code identifies six types of value constraints for the Range field: `npt = %lf - %lf`, `npt = %n%lf -`, `npt = now - %lf`, `npt = now - %n`, `clock = %n`, and `smpte = %n`. Live555 attempts six corresponding matches to determine the values of `rangeStart` and `rangeEnd`. If the *Range* value of the message does not adhere to any of these constraints, the function returns False, indicating a parsing failure.

Listing 1. Simplified code snippet from Live555.

When CHATAFL mutates the *Range* field in Live555 based on the message syntax shown in Figure 1, it only identifies the position of the field value without understanding the value constraints listed in Listing 1, continuing to use a random mutation strategy. The test cases generated from this coarse-grained message syntax, although adhering to the basic syntax structure, are likely to fail because the field values do not adhere to the syntactic constraints.

Therefore, obtaining the message syntax structure solely from protocol specifications is insufficient. The source code of the protocol implementation contains finer-grained client message syntax. It is necessary to analyze this source code to achieve a more comprehensive and detailed understanding of the message syntax to improve the generation of high-quality test cases. Generating high-quality test cases that adhere to protocol syntax constraints can significantly enhance the coverage of fuzzing, thereby exploring deeper code space and increasing the likelihood of discovering complex vulnerabilities.

3. Methodology

3.1. Overview

Figure 2 shows the overview of MSFuzz, which consists of four components: preprocessing, message syntax tree construction, seed expansion, and fuzzing with syntax-aware mutation. Its primary objective is to address the challenges associated with generating high-quality test cases in protocol fuzzing, thereby improving the efficiency and effectiveness of the fuzzing process.

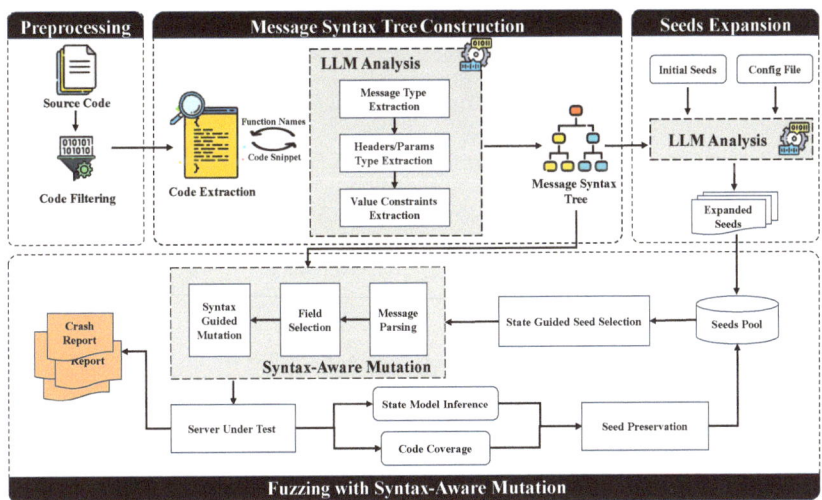

Figure 2. Overview of MSFuzz.

Preprocessing. Before extracting the message syntax from the source code, it is essential to pre-filter the source code files, as not all code is related to message parsing. By performing a preliminary analysis and filtering out irrelevant code, the search and analysis scope of LLMs during syntax tree construction is narrowed, thereby enhancing the efficiency of the LLMs.

Message Syntax Tree Construction. We investigated the structure of text-based protocol client messages and abstracted a general message syntax template. Based on the abstracted message syntax structure and heuristic rules derived from observing the source code, we designed a method to extract the message syntax from the protocol implementation source code using LLMs. This approach enables the construction of message syntax trees for the protocol implementation.

Seed Expansion. Given the critical role of initial seed diversity and quality in fuzzing, we focused on enhancing these aspects. We used the constructed message syntax tree and protocol implementation configuration files to guide the LLMs. This approach helped in expanding the initial seed corpus, thereby enhancing the seed diversity and quality.

Fuzzing with Syntax-Aware Mutation. In the fuzzing loop, we used the expanded seeds as input. For the seed mutation, MSFuzz leverages the message syntax tree to guide the mutation of the message. This ensures that the generated test cases adhered to the syntax structure and value constraints of the protocol, thereby improving the efficiency of the fuzzing. In order to ensure that MSFuzz possessed the capability to explore extreme scenarios, we also employed a random mutation strategy with a certain probability.

In the following, we present a detailed description of the core designs of MSFuzz, including *message syntax tree construction*, *seed expansion*, and *fuzzing with syntax-aware mutation*.

3.2. Message Syntax Tree Construction

MSFuzz leverages the code-understanding capabilities of LLMs to extract message syntax from preprocessed source code files. To facilitate the construction of syntax trees with a consistent structure, we first analyzed multiple text-based protocols and abstracted a general message syntax template. Then, based on this template and several heuristic rules derived from observing the source code, we designed a method for using LLMs to extract the syntax from the source code, thereby constructing the message syntax tree for the protocol implementation.

3.2.1. Message Syntax Structure

Defining a general message syntax structure is essential prior to employing LLMs for constructing message syntax trees. By analyzing multiple text-based protocols, we abstracted a general message syntax structure. This provided a standardized structure for interpreting and processing different protocol implementations. This ensured that the syntax trees generated by the LLMs were consistent and uniform.

We analyzed various text-based protocols and discovered that their client message structures could be abstracted into the general form shown in Figure 3. This general message syntax structure consisted of three parts: the request line, the request header field, and the carriage return and line feed (CRLF) characters. Not all protocols include request header fields and the ending CRLF in their client messages.

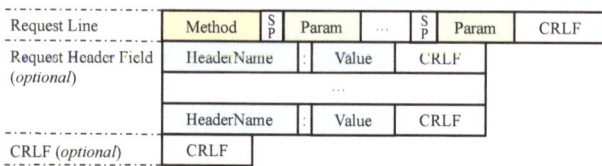

Figure 3. General message syntax structure.

Specifically, the request line contains the method name and multiple parameters, which are separated by space characters (SPs) and end with a CRLF. The request header field consists of key–value pairs in the format *HeaderName: Value* and also ends with a CRLF. The entire message typically concludes with a CRLF.

3.2.2. Extracting Syntax via LLMs

Based on the general syntax structure from Section 3.2.1, we utilized LLMs to extract the message syntax from preprocessed source code files, focusing on the request line and request header fields of the general syntax structure. Specifically, we extracted the method name (message type), parameter types, and parameter value constraints from the request line. If the request header fields exist, we extracted the header names and their value constraints. Finally, we constructed a message syntax tree for the protocol implementation based on the extracted message syntax.

Although we filtered out irrelevant files from the protocol implementation, providing all the filtered code files to LLMs often exceeds its input limit. Additionally, some irrelevant content may still remain, potentially affecting the output of LLMs. Therefore, we needed to further refine the code selection, extracting only the snippets closely related to the task at hand. Based on our analysis of network protocol implementations, we employed the following heuristic rules to extract key code snippets. This ensures that LLMs can effectively learn and extract the message syntax.

- In the source code of the protocol implementation, different types of messages or header fields have independent parsing functions. Therefore, when parsing a specific type of message or header field, LLMs can analyze only its corresponding parsing function to narrow the scope.
- Function names often follow clear naming conventions and clearly express their basic functions in the source code. Thus, providing only the function names to LLMs allows it to infer the purposes of the functions, enabling more accurate identification of the target parsing functions.

Based on the heuristic rules mentioned above, we propose an automated framework that uses LLMs to construct message syntax trees for target protocol implementations. This method employs a hierarchical, step-by-step strategy. It inputs key parsing code extracted from source code files into LLMs. From this, the framework extracts message types, request line parameters, header field types, and their value constraints, gradually constructing the message syntax tree. This method employs a hierarchical extraction strategy aligned with the syntactic structure illustrated in Figure 3. Initially, the framework identifies the message type. Subsequently, it extracts the types of request line parameters and header fields (if any). Finally, it determines the value constraints for the request line parameters and header fields. This process incrementally constructs the message syntax tree, ensuring a comprehensive representation of the protocol's message structure.

Message type extraction. MSFuzz extracts all function names from the filtered source code using the *Code Extraction* module in Figure 2. This module parses the source code to identify relevant function names, which are then utilized in prompt engineering to enable the LLM to identify message types in the target protocol implementation. The prompt includes all filtered function names and aims to identify all message types that the target protocol implementation can handle and map them to their corresponding handler functions, as shown in Figure 4a.

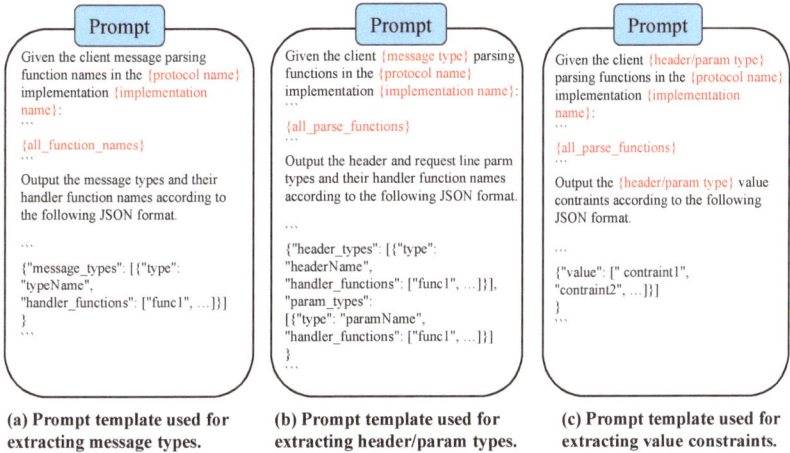

(a) Prompt template used for extracting message types.

(b) Prompt template used for extracting header/param types.

(c) Prompt template used for extracting value constraints.

Figure 4. Prompt templates used for extracting message syntax.

Header/parameter type extraction. For each message type, MSFuzz uses the *Code Extraction* module to locate and extract the corresponding function code from the source code based on the handler function names identified in the previous stage. Figure 4b illustrates the prompt used by MSFuzz to leverage the LLM for extracting message request line parameters and header fields (if any). This prompt includes the parsing function code for the message and aims to establish a mapping between the message and its request line parameters and header fields.

Value constraints extraction. To parse the values of header fields or request line parameters, MSFuzz uses the *Code Extraction* module to locate and extract the relevant function code from the source files based on the function names identified in the previous stage. Figure 4c illustrates the prompt used by MSFuzz to extract value constraints using the LLM. In this prompt, MSFuzz provides the parsing function code for the parameters or header fields, with the aim to establish a mapping between them and their corresponding value constraints.

To illustrate the construction of a message syntax tree, we utilized the MSFuzz approach in the motivating example (Figure 1) and present the results in Figure 5. Figure 5 displays only the syntactic constraints of the *Range* header field in the *PLAY* message of Live555.

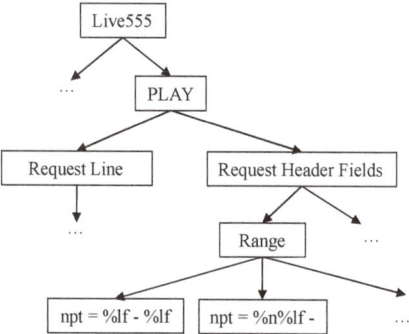

Figure 5. The message syntax tree for the motivating example.

3.3. Seed Expansion

To enhance the diversity and quality of the initial seed corpus, we employed LLMs to expand it. Although LLMs can generate new seeds for the target protocol implementation, three key problems need to be addressed: (1) How to comprehensively cover message syntactic constraints? (2) How to generate seeds specific to the target protocol implementation? (3) How to produce machine-readable seeds?

Regarding problem (1), MSFuzz constructs a message syntax tree for the target protocol implementation, as detailed in Section 3.2. This tree extensively covers messages, request line parameters, and their value constraints, as well as header field types and their value constraints. By integrating this syntax tree into our prompts, MSFuzz enables LLMs to generate more comprehensive seeds.

Regarding problem (2), configuration files for protocol implementations typically contain crucial server information, such as usernames, passwords, and network addresses. Integrating these configuration files into the LLM's prompts allows for the generation of more precise seeds that reflect the operational environment.

Regarding problem (3), an LLM demonstrates proficiency in learning from provided data and producing standardized outputs. By inputting initial seeds from the protocol implementation into the prompts and defining their format, this facilitates the production of machine-readable seeds that are immediately applicable for fuzzing.

Figure 6 illustrates the prompt used by MSFuzz to expand the seed corpus using LLMs. The prompts displayed in Figure 6 include the message syntax tree of the target protocol implementation, the configuration file, and the initial seed corpus, explicitly directing the LLM to generate outputs that adhere to the format of the initial seed corpus.

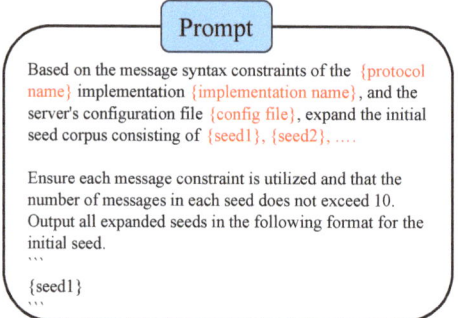

Figure 6. Prompt template used for expanding the initial seed corpus.

3.4. Fuzzing with Syntax-Aware Mutation

Although MSFuzz constructs a message syntax tree of the protocol implementation, it encounters two major problems when utilizing this syntax tree for message mutation: (1) How to select mutation locations to preserve the basic syntax structure of the message? (2) How to use the message syntax tree to precisely guide mutations? To address these problems, we designed a syntax-aware fuzzing approach, which is detailed in Algorithm 1. Specifically, MSFuzz first attempts to parse the message to obtain its basic syntax structure (line 7). After verifying the message against the syntax tree (line 9), MSFuzz randomly selects a field for mutation (line 11). Then, according to the value constraints of the field in the syntax tree, mutations are performed (lines 12–13).

Message parsing. Before mutating the messages, it is crucial to accurately identify suitable mutation locations. Randomly selecting mutation locations can disrupt the fundamental structure of the message, rendering it ineffective. Therefore, based on the general message syntax structure shown in Figure 3, MSFuzz parses the message designated for mutation. Initially, MSFuzz analyzes the request line to determine the message type and request line parameter fields and records the offsets of these fields. Subsequently, MSFuzz parses the names and values of the header fields, also noting their offsets. During mutation, MSFuzz selects positions within the request line parameter fields or header fields to preserve the integrity of the basic syntactic structure of the message.

Syntax-guided mutation. To guide message mutation using the syntax tree, it is first necessary to determine whether the message type exists within the syntax tree. If the message type does not exist, a random mutation strategy is applied. If the message type is found in the syntax tree, a request line parameter field or a header field is randomly selected as the mutation target. Next, the constraints for the selected field are identified from the syntax tree. One of these constraints is randomly chosen, and a value that adheres to this constraint is generated to replace the original field value. If the newly generated field value differs in length from the original field value, the offsets of subsequent fields are adjusted to ensure that each field in the mutated message retains its correct offset. In order to ensure that MSFuzz possesses the capability to explore extreme scenarios, we also employed a random mutation strategy with a certain probability.

For instance, consider a *PLAY* message that adheres to the syntax structure in Figure 1 and is guided to mutate by the message syntax tree in Figure 5. First, it is verified whether the *PLAY* message type exists in the Live555 syntax tree. As shown in Figure 5, this type of message syntax structure does exist. Next, a field in the *PLAY* message is randomly selected, such as the *Range* field. The value constraints of *Range* header field in the message syntax tree are looked up and one is randomly selected, such as npt = %lf - %lf. This value constraint specifies the playback range in seconds, where *%lf* represents a floating point number indicating the start and end times. MSFuzz then identifies the data type of the placeholders in the field, which, in this case, are floating point numbers. To generate a random value that matches this type, MSFuzz uses a floating point number generation function. For example, it might generate 10.5 and 20.0 for the start and end times, respectively, formatting them as npt = 10.5 - 20.0. Finally, MSFuzz replaces the original value in the *Range* field with the newly generated value, ensuring that the modified message adheres to the syntactic constraints specified in the message syntax tree. After replacing the value, MSFuzz adjusts the offset of each field within the message. This adjustment is crucial to maintain the structural integrity of the message, as altering the length of one field can affect the positions of subsequent fields. By following this process, MSFuzz can produce test cases that adhere to the syntactic constraints, thereby ensuring the validity of the generated messages.

Algorithm 1: Fuzzing with syntax-aware mutation

Input: P: protocol implementation
Input: E: expanded seeds
Input: T: message syntax tree
Output: C_x: crash reports

1: **struct** Field { type;values }
2: **struct** Message { type;params;headers }
3: StateMachine $S \leftarrow \varnothing$
4: **repeat**
5:    Seed $M \leftarrow$ StateGuidedSeedChoice(S, E)
6:    $\langle M_1, M_2, M_3 \rangle \leftarrow Split(M)$
7:    Message $m \leftarrow$ ParseMessage(T, M_2)
8:    **for** $i \leftarrow 1$ **to** AssignEnergy(M) **do**
9:      **if** $m.type \in T$ **then**
10:        Field $field$
11:        $field \leftarrow$ Choice$(m.params, m.headers)$
12:        **if** $field.type \in T$ **and** $Rand() < \varepsilon$ **then**
13:          $M_2' \leftarrow$ FieldMutate$(field, T)$
14:          $M' \leftarrow \langle M_1, M_2', M_3 \rangle$
15:        **else**
16:          $M_2' \leftarrow$ RandomMutate(M_2, T)
17:          $M' \leftarrow \langle M_1, M_2', M_3 \rangle$
18:        **end if**
19:      **else**
20:        $M_2' \leftarrow$ RandomMutate(M_2, T)
21:        $M' \leftarrow \langle M_1, M_2', M_3 \rangle$
22:      **end if**
23:      Response $R' \leftarrow$ SendToServer(P, M')
24:      **if** IsCrash(M', P) **then**
25:        $C_x \leftarrow C_x \cup \{M'\}$
26:      **end if**
27:      **if** IsInteresting(M', P, S) **then**
28:        $E \leftarrow E \cup \{(M', R')\}$
29:        $S \leftarrow$ UpdateStateMachine(S, R')
30:      **end if**
31:    **end for**
32: **until** timeout reached or abort-signal

4. Evaluation

In this section, we evaluate the performance of MSFuzz and compare it with the SOTA protocol fuzzers. We aim to answer the following research questions by evaluating MSFuzz.

- **RQ1. State coverage:** Could MSFuzz achieve a higher state space coverage than the SOTA fuzzers?
- **RQ2. Code coverage:** Could MSFuzz achieve a higher code space coverage than the SOTA fuzzers?
- **RQ3. Ablation study:** What was the impact of the two key components on the performance of MSFuzz?
- **RQ4. Vulnerability discovery:** Could MSFuzz discover more vulnerabilities than the SOTA fuzzers?

4.1. Experimental Setup

Implementation. Building upon the widely used protocol fuzzing framework AFLNET, we developed MSFuzz. The implementation of MSFuzz consisted of approximately 1.2k lines of C/C++ code and 800 lines of Python code. Specifically, we developed a Python script to interface with the LLM to acquire the message syntax of network protocol implementations and construct message syntax trees. To mitigate potential issues of incompleteness and inconsistency in the syntax extracted by the LLM, we performed three iterations and used the union of the results to construct the message syntax tree. Subsequently, leveraging the initial seed corpus and the message syntax tree, we expanded the seed corpus using the LLM. The syntax-aware mutation was predominantly implemented in C during the mutation phase of the fuzzing loop based on the constructed message syntax tree. The method of enhancing protocol fuzzing in MSFuzz does not rely on a specific LLM, and several popular LLMs on the market can be used. We selected Qwen-plus as the LLM for syntax extraction and seed expansion because it is one of the most advanced pretrained LLMs currently available. This model boasts parameters in the trillion range and was trained on a vast and diverse dataset, including software code and technical documentation. This extensive training endows the model with deep language understanding and generation capabilities, enabling it to comprehend the logical structure and semantics of source code. Additionally, it provides a substantial number of free tokens, facilitating more extensive experimentation and application. For the configuration of input parameters in the LLM, we used the default settings, such as max_token = 2000 and top_p = 0.8.

Benchmark. Table 1 provides detailed information on the benchmark network protocol implementations used in our evaluation. Our benchmark includes five network protocol implementations, encompassing three widely used protocols: RTSP, FTP, and DAAP. These protocols employ textual formats for communication and are part of the widely recognized protocol fuzzing benchmark ProFuzzBench [22]. Given their widespread usage, we consider these five target programs representative of real-world applications.

Table 1. The basic information of target protocol implementations.

Subject	Protocol	Version	Language	Size
Live555	RTSP	31284aa	C++	57K
LightFTP	FTP	139af7c	C	4.7K
Pure-FTPD	FTP	10122d9	C	29K
ProFTPD	FTP	61e621e	C	242K
Forked-daapd	DAAP	2ca10d9	C	79K

Baselines. We conducted an in-depth comparison between MSFuzz and the SOTA network protocol fuzzers (AFLNET and CHATAFL). AFLNET, the first grey-box fuzzer designed for network protocol implementations, primarily relies on mutation-based generation methods. CHATAFL, as one of the SOTA grey-box fuzzers, leverages LLMs to expand seeds, extracts message structures, and utilizes them to guide mutation.

Environment. All experiments were run on a server equipped with 64-bit Ubuntu 20.04, featuring dual Intel(R) Xeon(R) E5-2690 @ 2.90 GHz CPUs and 128 GB of RAM. Each selected protocol implementation and each fuzzer were individually set up in separate Docker containers, utilizing identical computational resources for experimental evaluation. To ensure the fairness of the experimental results, each fuzzer was subjected to 24 h of fuzzing on each protocol implementation, with the experiments repeated five times.

4.2. State Coverage

State coverage is a crucial evaluation metric in network protocol fuzzing, as it reflects the depth of coverage within the protocol state machine and the extent to which the internal logic of the protocol implementation has been explored. By measuring the number of states reached and the number of state transitions during fuzzing, it is possible to effectively assess whether the fuzzer has thoroughly explored the various states of the protocol implementation and their transitions.

Table 2 presents the average number of states and state transitions covered by different fuzzers during five times of 24 h fuzzing. To evaluate the performance of MSFuzz, we report the percentage improvement in state and state transition coverage within 24 h (*Improv*) The results indicate that compared with AFLNET and CHATAFL, MSFuzz exhibited significant advantages in discovering new states and state transitions. Specifically, MSFuzz achieved average improvements of 22.53% and 10.04% in the number of states compared with AFLNET and CHATAFL, respectively. Additionally, there were improvements of 60.62% and 19.52% in the state transitions. Compared with other protocol implementations, the state coverage improvement for LightFTP was the least significant. This was because LightFTP is a lightweight FTP protocol implementation with a simple functionality and minimal codebase. It lacks complex and deep state transitions, resulting in relatively minor improvements in the state coverage.

Table 2. The average number of states and state transitions covered by different fuzzers.

Subject	MSFuzz		Comparison with AFLNET				Comparison with CHATAFL			
	States	Transitions	States	Improv	Trans	Improv	States	Improv	Trans	Improv
Live555	15.00	159.40	11.00	36.36%	77.60	105.41%	12.00	25.00%	107.60	48.14%
LightFTP	24.00	277.60	23.00	4.35%	220.20	26.07%	23.80	0.84%	262.60	5.71%
Pure-FTPD	30.00	326.20	26.20	14.50%	202.40	61.17%	29.20	2.74%	287.20	13.58%
ProFTPD	29.40	268.40	22.20	32.43%	152.60	75.88%	26.60	10.53%	248.80	7.88%
Forked-daapd	10.00	29.60	8.00	25.00%	22.00	34.55%	9.00	11.11%	24.20	22.31%
AVG	-	-	-	**22.53%**	-	**60.62%**	-	**10.04%**	-	**19.52%**

Note: The **AVG** row is highlighted in bold to indicate the average improvement percentages across all subjects for states and state transitions.

In summary, MSFuzz could achieve higher state coverage than the SOTA fuzzers. MSFuzz not only discovered more new states but also generated more state transitions, thereby enhancing the effectiveness and comprehensiveness of the fuzzing. By exploring the state space more deeply, MSFuzz demonstrated significant advantages in the field of protocol fuzzing.

4.3. Code Coverage

Code coverage has consistently served as the standard metric for evaluating fuzzers, reflecting the amount of code executed within the protocol implementation throughout the entire fuzzing. Code coverage, as an evaluation metric, provides an effective means of assessing the performance of fuzzers.

To assess the performance of MSFuzz, we report the percentage improvement in the code branch coverage within 24 h (*Improv*) and analyzed the probability that MSFuzz outperformed baseline activities ($\hat{A}_{12}$) through random activities using the Vargha–Delaney statistic. Table 3 shows the average code branch coverage achieved by each fuzzer during five times of 24 h fuzzing.

The results demonstrate that MSFuzz achieved a higher code branch coverage than both AFLNET and CHATAFL across all five protocol implementations, validating the effectiveness of our proposed method in enhancing the code coverage. Specifically, compared with AFLNET, MSFuzz showed an average improvement of 29.30% in code branch coverage, and compared with CHATAFL, an average improvement of 23.13%. For all protocol implementations, the Vargha–Delaney effect size ($\hat{A}_{12} \geq 0.76$) indicates a significant advantage of MSFuzz in exploring the code branch coverage over baseline fuzzers.

Table 3. The average code branch coverage by different fuzzers.

Subject	MSFuzz	Comparison with AFLNET			Comparison with CHATAFL		
		Branch	Improv	$\hat{A}_{12}$	Branch	Improv	$\hat{A}_{12}$
Live555	3190.60	2796.40	14.10%	1.00	2816.40	13.29%	1.00
LightFTP	402.20	321.80	24.98%	1.00	343.00	17.26%	1.00
Pure-FTPD	1321.00	1027.60	28.55%	1.00	1147.40	15.13%	0.76
ProFTPD	5427.60	4893.20	10.92%	1.00	5123.00	5.95%	1.00
Forked-daapd	3853.20	2294.20	67.95%	1.00	2349.40	64.01%	1.00
AVG	-	-	**29.30%**	-	-	**23.13%**	-

Note: The **AVG** row is highlighted in bold to indicate the average improvement percentages across all subjects for branch coverage.

To further demonstrate the effectiveness of MSFuzz, we analyzed the average number of code branches explored by different fuzzers during five runs of 24 h and present the results in Figure 7. As illustrated in the figure, MSFuzz not only achieved the highest code coverage compared with the other fuzzers but also exhibited the fastest exploration speed. Notably, the improvements were most significant for Pure-FTPD and Forked-daapd.

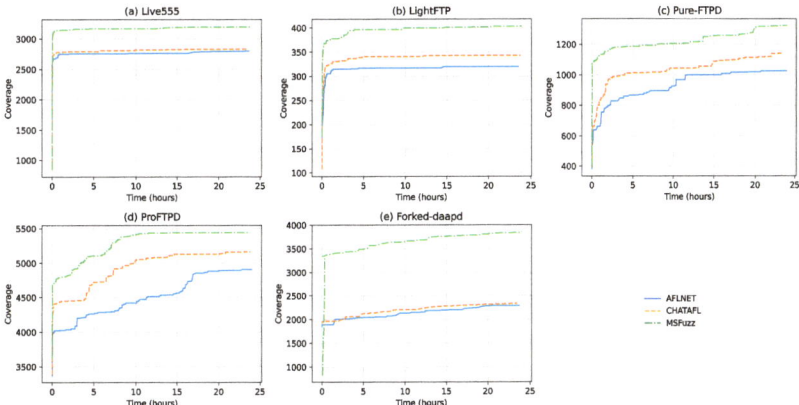

Figure 7. The average number of code branches explored by different fuzzers during 5 times of 24 h.

4.4. Ablation Study

MSFuzz employs an LLM to extract message syntax from the source code of protocol implementations, thereby constructing message syntax trees. Using these syntax trees, two strategies were employed to generate high-quality test cases that adhered to syntactic constraints, enhancing the performance of fuzzing. The first strategy involved using LLMs in conjunction with the extracted message syntax tree to expand the initial seed corpus of the protocol implementation, thereby improving the diversity and comprehensiveness of the seeds. The second strategy introduced a novel syntax-aware mutation strategy, which leveraged the constructed message syntax trees to guide the mutation process during fuzzing.

To quantitatively assess the contribution of each strategy to the overall performance of MSFuzz, we conducted an ablation study. In this study, we evaluated three tools: AFLNET (with all strategies disabled), STFuzz-E (with only the seed expansion enabled), and MSFuzz (with both the seed expansion and syntax-aware mutation strategies enabled). The experimental results are shown in Table 4. We evaluated the performance of three tools during five times of 24-h fuzzing, specifically measuring the average improvement of states, state transitions, and the percentage improvement in the code branch coverage achieved by each fuzzer. The results indicate that both strategies implemented by MSFuzz enhanced the number of states, state transitions, and code branch coverage to varying extents without negatively impacting any of these metrics.

Table 4. Improvement in the average number of states, state transitions, and branch coverage achieved by MSFuzz-E and MSFuzz.

Subject	AFLNET			MSFuzz-E			MSFuzz		
	State	Trans	Branch	State	Trans	Branch	State	Trans	Branch
Live555	11.00	77.60	2796.40	+36.36%	+88.66%	+13.00%	+36.36%	+105.41%	+14.10%
LightFTP	23.00	220.20	321.80	+0.87%	+5.36%	+17.53%	+4.35%	+26.07%	+24.98%
Pure-FTPD	26.20	202.40	1027.60	+12.98%	+55.53%	+22.99%	+14.50%	+61.17%	+28.55%
ProFTPD	22.20	152.60	4893.20	+28.83%	+50.46%	+6.87%	+32.43%	+75.88%	+10.92%
Forked-daapd	8.00	22.00	2294.20	+17.50%	+25.45%	+65.57%	+25.00%	+34.55%	+67.95%
AVG	-	-	-	**+19.31%**	**+45.09%**	**+25.19%**	**+22.53%**	**+60.62%**	**+29.30%**

Note: The **AVG** row is highlighted in bold to indicate the average improvement percentages across all subjects for states, transitions, and branch coverage.

The experimental results of AFLNET and MSFuzz-E indicate that employing the seed expansion strategy increased the number of states by 19.31%, state transitions by 45.09%, and code branch coverage by 25.19%. This demonstrated the effectiveness of the seed expansion strategy, which enhanced the quality and diversity of the seeds. By ensuring that the expanded seeds comprehensively covered the message syntax of the protocol implementation, this strategy significantly improved the fuzzing exploration of the state space and code space.

Incorporating the syntax-aware mutation strategy alongside the seed expansion strategy, as demonstrated by the results of MSFuzz-E and MSFuzz, further enhanced the three evaluation metrics. The improvement in the number of states rose from 19.31% to 22.53%, the improvement in the state transitions rose from 45.09% to 60.62%, and the improvement in the code branch coverage rose from 25.19% to 29.30%. This demonstrated the effectiveness of the syntax-aware mutation strategy, which ensured that the test cases generated after mutation adhered to the protocol syntactic constraints. This strategy prevented the server from discarding them during the initial syntax-checking phase, thereby increasing the opportunity to explore the protocol implementation.

The analysis of the time overhead and resource consumption associated with the seed expansion and syntax-aware mutation strategies demonstrated that these strategies neither slowed down the execution nor introduced significant resource consumption. The construction of message syntax trees and the expansion of the seed corpus were conducted during the preparation phase and executed only once, thus not contributing to the time overhead and resource consumption of the protocol fuzzing. Although syntax-aware mutation was performed during the fuzzing process, it could generate test cases that adhered to the protocol constraints, thereby avoiding the substantial time and inefficiency associated with traditional random mutations. This resulted in a significant improvement in the overall testing efficiency. As shown in Table 4, the experimental data for MSFuzz-E and MSFuzz substantiate this conclusion.

4.5. Vulnerability Discovery

To evaluate the vulnerability discovery performance of MSFuzz, we compared the number of unique crashes triggered by MSFuzz and the SOTA fuzzers. From the five times of 24 h fuzzing, neither AFLNET nor CHATAFL triggered any crashes across the five target protocol implementations. However, MSFuzz discovered crashes in two protocol implementations. Specifically, in LightFTP, MSFuzz detected 91 unique crashes, while in Forked-daapd, MSFuzz found 27 unique crashes. Notably, we have conducted a detailed analysis of 16 of these crashes so far, identifying two new vulnerabilities that have been reported to the Common Vulnerabilities and Exposures (CVE) database.

These experimental results validated the superior performance of MSFuzz in detecting and discovering software vulnerabilities. The differences in crash discovery capabilities among various fuzzers within the same time frame further highlight the significant advantages of MSFuzz in enhancing the fuzzing efficiency and vulnerability discovery.

5. Discussion

Although MSFuzz achieved a good performance in state coverage, code coverage, and vulnerability discovery compared with the SOTA fuzzers, it still has certain limitations.

Difficulty in applying to binary-based protocol implementations. Binary protocols have very compact data representations, with field boundaries often not clearly defined. These protocols lack explicit tags or markers to indicate the start and end of each field, which makes it challenging to discern the specific meaning and position of each field when parsing the code. This ambiguity in field delineation complicates the process of extracting and interpreting protocol messages, making it difficult to accurately construct message syntax trees. Furthermore, the variability in binary protocol structures requires a more sophisticated approach to handle the nuances of different implementations, thus posing a significant challenge for protocol-fuzzing tools like MSFuzz that rely on clear syntax demarcations.

Input capacity limitations of LLMs. Although MSFuzz provides LLMs with filtered key code functions to enhance its understanding of message syntax, the size of the function code may still exceed LLMs' input limitations in some cases. This can lead to challenges in processing large volumes of code, as LLMs may struggle to maintain context and accuracy when handling excessive input data. The limitations in input capacity can result in incomplete analysis and the potential loss of critical information needed for effective fuzzing. Consequently, the efficiency and effectiveness of MSFuzz can be compromised, as LLMs might not fully capture the nuances of the protocol implementation. Therefore, we plan to prioritize the refinement of code by removing syntax-irrelevant content within functions as a focus of our future work to optimize the processing efficiency of LLMs.

6. Related Work

6.1. Syntax-Aware Fuzzing

For protocol fuzzing, understanding the message syntax is essential [15]. Syntax-aware fuzzers strive to comprehend the detailed message syntax of protocol implementations, which allows them to generate more effective test cases. Tools such as Peach [31] and KIF [7] manually extract syntax from publicly available protocol RFC documents. This approach requires significant time from researchers and is not applicable to proprietary protocols without public documentation. AspFuzz [8] uses a specialized language to describe RFC documents for obtaining protocol syntax. PULSAR [39] and Bbuzz [40] extract the message syntax from network traffic through protocol reverse engineering methods. Polyglot [41] employs dynamic analysis techniques to extract message syntax from program behavior logs. Polar [19] combines static analysis and dynamic taint analysis to extract syntax related to protocol functionality.

In general, current syntax-aware protocol fuzzers lack a comprehensive understanding of message syntax and usually can only extract partial fields from message syntax structures. In contrast, MSFuzz is capable of extracting more detailed and comprehensive message syntax from protocol implementations. Additionally, it employs seed expansion and syntax-aware mutation methods to enhance fuzzing.

6.2. Fuzzing Based on Large Language Models

In recent years, popular large language models (LLMs) have demonstrated remarkable effectiveness in natural language processing. Researchers have attempted to integrate LLMs into fuzzing to enhance the performance of traditional fuzzing methods. CHATAFL [10] is an LLM-based protocol fuzzer that constructs the syntax structure of request messages and predicts the next message in the sequence using LLMs. Codamosa [42] uses LLMs to address the problem of stagnating code coverage in traditional fuzzing. FuzzGPT [43] leverages the CodeX and CodeGen models to automatically generate anomalous programs based on core concepts, thereby fuzzing deep learning libraries. The key insight is that historical bug-triggering programs may include rare or valuable code ingredients that are important for bug finding. KernelGPT [44] employs LLMs to automatically infer Syzkaller specifications to enhance kernel fuzzing.

All of these works rely on the knowledge acquired by LLMs during pre-training on large-scale data. In contrast, when MSFuzz extracts the message syntax of protocol implementations, it provides relevant source code snippets to LLMs. The LLM then performs syntax extraction based on these code snippets, resulting in more reliable outputs.

7. Conclusions

In this paper, we propose MSFuzz, which is a novel protocol-fuzzing method with message syntax comprehension. MSFuzz extracts key code snippets from protocol implementation source code and leverages code comprehension capabilities of LLMs to extract message syntax and construct message syntax trees. These syntax trees are then utilized to expand the seed corpus and guide seed mutation, thereby improving the effectiveness of test cases and thereby enhancing the efficiency of fuzzing. Experimental evaluation results demonstrated that MSFuzz outperformed the SOTA protocol fuzzers. Specifically, compared with AFLNET and CHATAFL, MSFuzz achieved average improvements of 22.53% and 19.52% in the number of states, 60.62% and 10.04% improvements in the number of state transitions, and 29.30% and 23.13% improvements in the branch coverage. Additionally, MSFuzz discovers more vulnerabilities than the SOTA fuzzers.

Author Contributions: M.C., K.Z. and Y.C. designed the research. M.C. performed the experiments and drafted this paper. K.Z. and Y.C. helped to organize this paper. G.Y., Y.L. and C.L. revised and finalized the paper. All authors read and agreed to the published version of this manuscript.

Funding: This research received no external funding.

Data Availability Statement: The data that support the findings of this study are available from the corresponding author upon reasonable request.

Conflicts of Interest: The authors declare no conflict of interest.

References

1. Hermann, H.; Johnson, R.; Engel, R. A framework for network protocol software. In Proceedings of the OOPSLA '95, ACM SIGPLAN Notices, Austin, TX, USA, 15–19 October 1995.
2. Yazdinejad, A.; Dehghantanha, A.; Parizi, R.M.; Srivastava, G.; Karimipour, H. Secure Intelligent Fuzzy Blockchain Framework: Effective Threat Detection in IoT Networks. *Comput. Ind.* **2023**, *144*, 103801. [CrossRef]
3. Serebryany, K. *OSS-Fuzz-Google's Continuous Fuzzing Service for Open Source Software*; USENIX: Vancouver, BC, Canada, 2017.
4. Xu, M.; Kashyap, S.; Zhao, H.; Kim, T. Krace: Data race fuzzing for kernel file systems. In Proceedings of the 2020 IEEE Symposium on Security and Privacy (SP), San Francisco, CA, USA, 18–21 May 2020; IEEE: Piscataway, NJ, USA, 2020; pp. 1643–1660.
5. Jero, S.; Pacheco, M.L.; Goldwasser, D.; Nita-Rotaru, C. Leveraging textual specifications for grammar-based fuzzing of network protocols. In Proceedings of the AAAI Conference on Artificial Intelligence, Honolulu, HI, USA, 27 January–1 February 2019; Volume 33, pp. 9478–9483.
6. Banks, G.; Cova, M.; Felmetsger, V.; Almeroth, K.; Kemmerer, R.; Vigna, G. SNOOZE: Toward a Stateful NetwOrk prOtocol fuzZEr. In Proceedings of the Information Security: 9th International Conference, ISC 2006, Samos Island, Greece, 30 August–2 September 2006; Proceedings 9; Springer: Berlin/Heidelberg, Germany, 2006; pp. 343–358.
7. Miki, H.; Setou, M.; Kaneshiro, K.; Hirokawa, N. All kinesin superfamily protein, KIF, genes in mouse and human. *Proc. Natl. Acad. Sci. USA* **2001**, *98*, 7004–7011. [CrossRef] [PubMed]
8. Kitagawa, T.; Hanaoka, M.; Kono, K. A state-aware protocol fuzzer based on application-layer protocols. *IEICE Trans. Inf. Syst.* **2011**, *94*, 1008–1017. [CrossRef]
9. Rontti, T.; Juuso, A.M.; Takanen, A. Preventing DoS attacks in NGN networks with proactive specification-based fuzzing. *IEEE Commun. Mag.* **2012**, *50*, 164–170. [CrossRef]
10. Meng, R.; Mirchev, M.; Böhme, M.; Roychoudhury, A. Large language model guided protocol fuzzing. In Proceedings of the 31st Annual Network and Distributed System Security Symposium (NDSS), San Diego, CA, USA, 26 February–1 March 2024.

11. Cui, W.; Kannan, J.; Wang, H.J. Discoverer: Automatic Protocol Reverse Engineering from Network Traces. In Proceedings of the USENIX Security Symposium, Boston, MA, USA, 6–10 August 2007; pp. 1–14.
12. Beddoe, M.A. Network protocol analysis using bioinformatics algorithms. *Toorcon* 2004, *26*, 1095–1098.
13. Cui, W.; Paxson, V.; Weaver, N.; Katz, R.H. Protocol-independent adaptive replay of application dialog. In Proceedings of the NDSS, San Diego, CA, USA, 2–3 February 2006.
14. Sun, Y.; Lv, S.; You, J.; Sun, Y.; Chen, X.; Zheng, Y.; Sun, L. IPSpex: Enabling efficient fuzzing via specification extraction on ICS protocol. In Proceedings of the International Conference on Applied Cryptography and Network Security, Rome, Italy, 20–23 June 2022; Springer: Berlin/Heidelberg, Germany, 2022; pp. 356–375.
15. Comparetti, P.M.; Wondracek, G.; Kruegel, C.; Kirda, E. Prospex: Protocol specification extraction. In Proceedings of the 2009 30th IEEE Symposium on Security and Privacy, Oakland, CA, USA, 17–20 May 2009; IEEE: Piscataway, NJ, USA, 2009; pp. 110–125.
16. Lin, Z.; Jiang, X.; Xu, D.; Zhang, X. Automatic protocol format reverse engineering through context-aware monitored execution. In Proceedings of the NDSS, San Diego, CA, USA, 10–13 February 2008; Volume 8, pp. 1–15.
17. Wondracek, G.; Comparetti, P.M.; Kruegel, C.; Kirda, E.; Anna, S.S.S. Automatic Network Protocol Analysis. In Proceedings of the NDSS, San Diego, CA, USA, 10–13 February 2008; Volume 8, pp. 1–14.
18. Cui, W.; Peinado, M.; Chen, K.; Wang, H.J.; Irun-Briz, L. Tupni: Automatic reverse engineering of input formats. In Proceedings of the 15th ACM conference on Computer and Communications Security, Alexandria, VA, USA, 27–31 October 2008; pp. 391–402.
19. Luo, Z.; Zuo, F.; Jiang, Y.; Gao, J.; Jiao, X.; Sun, J. Polar: Function code aware fuzz testing of ics protocol. *ACM Trans. Embed. Comput. Syst. (TECS)* 2019, *18*, 1–22. [CrossRef]
20. Hu, Z.; Shi, J.; Huang, Y.; Xiong, J.; Bu, X. Ganfuzz: A gan-based industrial network protocol fuzzing framework. In Proceedings of the 15th ACM International Conference on Computing Frontiers, Ischia, Italy, 8–10 May 2018; pp. 138–145.
21. Zhao, H.; Li, Z.; Wei, H.; Shi, J.; Huang, Y. SeqFuzzer: An industrial protocol fuzzing framework from a deep learning perspective. In Proceedings of the 2019 12th IEEE Conference on Software Testing, Validation and Verification (ICST), Xi'an, China, 22–27 April 2019; IEEE: Piscataway, NJ, USA, 2019; pp. 59–67.
22. Natella, R.; Pham, V.T. Profuzzbench: A benchmark for stateful protocol fuzzing. In Proceedings of the 30th ACM SIGSOFT International Symposium on Software Testing and Analysis, Virtual, 11–17 July 2021; pp. 662–665.
23. Hu, F.; Qin, S.; Ma, Z.; Zhao, B.; Yin, T.; Zhang, C. NSFuzz: Towards Efficient and State-Aware Network Service Fuzzing-RCR Report. *ACM Trans. Softw. Eng. Methodol.* 2023, *32*, 1–8. [CrossRef]
24. Pham, V.T.; Böhme, M.; Roychoudhury, A. Aflnet: A greybox fuzzer for network protocols. In Proceedings of the 2020 IEEE 13th International Conference on Software Testing, Validation and Verification (ICST), Porto, Portugal, 24–28 October 2020; IEEE: Piscataway, NJ, USA, 2020; pp. 460–465.
25. Natella, R. Stateafl: Greybox fuzzing for stateful network servers. *Empir. Softw. Eng.* 2022, *27*, 191. [CrossRef]
26. Kaksonen, R.; Laakso, M.; Takanen, A. Software security assessment through specification mutations and fault injection. In Proceedings of the Communications and Multimedia Security Issues of the New Century: IFIP TC6/TC11 Fifth Joint Working Conference on Communications and Multimedia Security (CMS'01), Darmstadt, Germany, 21–22 May 2001; Springer: Berlin/Heidelberg, Germany, 2001; pp. 173–183.
27. Aitel, D. An Introduction to SPIKE, the Fuzzer Creation Kit. 2002. Available online: https://www.blackhat.com/presen-tations/bh-usa-02/bh-us-02-aitel-spike.ppt (accessed on 6 May 2024).
28. Security, B. beSTORM Black Box Testing. 2024. Available online: https://beyondsecurity.com/solutions/bestorm.html (accessed on 6 May 2024).
29. Inc, S. Defensics Fuzz Testing. 2020. Available online: https://www.synopsys.com/software-integrity/security-testing/fuzz-testing.html (accessed on 6 May 2024).
30. Rapid7. Metasploit Vulnerability & Exploit Database. 2020. Available online: https://www.rapid7.com/db/?q=fuzzer&type=metasploit (accessed on 6 May 2024).
31. Eddington, M. Peach Fuzzing Platform. 2004. Available online: https://gitlab.com/peachtech/peach-fuzzer-community (accessed on 6 May 2024).
32. Yu, Y.; Chen, Z.; Gan, S.; Wang, X. SGPFuzzer: A state-driven smart graybox protocol fuzzer for network protocol implementations. *IEEE Access* 2020, *8*, 198668–198678. [CrossRef]
33. Schumilo, S.; Aschermann, C.; Jemmett, A.; Abbasi, A.; Holz, T. Nyx-net: Network fuzzing with incremental snapshots. In Proceedings of the Seventeenth European Conference on Computer Systems, Rennes, France, 5–8 April 2022; pp. 166–180.
34. Feng, X.; Sun, R.; Zhu, X.; Xue, M.; Wen, S.; Liu, D.; Nepal, S.; Xiang, Y. Snipuzz: Black-box fuzzing of iot firmware via message snippet inference. In Proceedings of the 2021 ACM SIGSAC Conference on Computer and Communications Security, Virtual, 15–19 November 2021; pp. 337–350.
35. Sun, Y.; Wu, D.; Xue, Y.; Liu, H.; Wang, H.; Xu, Z.; Xie, X.; Liu, Y. Gptscan: Detecting logic vulnerabilities in smart contracts by combining gpt with program analysis. In Proceedings of the IEEE/ACM 46th International Conference on Software Engineering, Lisbon, Portugal, 14–20 April 2024; pp. 1–13.
36. Xia, C.S.; Paltenghi, M.; Le Tian, J.; Pradel, M.; Zhang, L. Fuzz4all: Universal fuzzing with large language models. In Proceedings of the IEEE/ACM 46th International Conference on Software Engineering, Lisbon, Portugal, 14–20 April 2024; pp. 1–13.
37. Hu, J.; Zhang, Q.; Yin, H. Augmenting greybox fuzzing with generative ai. *arXiv* 2023, arXiv:2306.06782.

38. Deng, Y.; Xia, C.S.; Peng, H.; Yang, C.; Zhang, L. Large language models are zero-shot fuzzers: Fuzzing deep-learning libraries via large language models. In Proceedings of the 32nd ACM SIGSOFT International Symposium on Software Testing and Analysis, Seattle, WA, USA, 17–21 July 2023; pp. 423–435.
39. Gascon, H.; Wressnegger, C.; Yamaguchi, F.; Arp, D.; Rieck, K. Pulsar: Stateful black-box fuzzing of proprietary network protocols. In Proceedings of the Security and Privacy in Communication Networks: 11th EAI International Conference, SecureComm 2015, Dallas, TX, USA, 26–29 October 2015; Proceedings 11. Springer: Berlin/Heidelberg, Germany, 2015; pp. 330–347.
40. Blumbergs, B.; Vaarandi, R. Bbuzz: A bit-aware fuzzing framework for network protocol systematic reverse engineering and analysis. In Proceedings of the MILCOM 2017—2017 IEEE Military Communications Conference (MILCOM), Baltimore, MD, USA, 23–25 October 2017; IEEE: Piscataway, NJ, USA, 2017; pp. 707–712.
41. Caballero, J.; Yin, H.; Liang, Z.; Song, D. Polyglot: Automatic extraction of protocol message format using dynamic binary analysis. In Proceedings of the 14th ACM Conference on Computer and Communications Security, Alexandria, VA, USA, 31 October–2 November 2007; pp. 317–329.
42. Lemieux, C.; Inala, J.P.; Lahiri, S.K.; Sen, S. Codamosa: Escaping coverage plateaus in test generation with pre-trained large language models. In Proceedings of the 2023 IEEE/ACM 45th International Conference on Software Engineering (ICSE), Melbourne, Australia, 14–20 May 2023; IEEE: Piscataway, NJ, USA, 2023; pp. 919–931.
43. Deng, Y.; Xia, C.S.; Yang, C.; Zhang, S.D.; Yang, S.; Zhang, L. Large language models are edge-case generators: Crafting unusual programs for fuzzing deep learning libraries. In Proceedings of the 46th IEEE/ACM International Conference on Software Engineering, Lisbon, Portugal, 14–20 April 2024; pp. 1–13.
44. Yang, C.; Zhao, Z.; Zhang, L. KernelGPT: Enhanced Kernel Fuzzing via Large Language Models. *arXiv* **2023**, arXiv:2401.00563.

Disclaimer/Publisher's Note: The statements, opinions and data contained in all publications are solely those of the individual author(s) and contributor(s) and not of MDPI and/or the editor(s). MDPI and/or the editor(s) disclaim responsibility for any injury to people or property resulting from any ideas, methods, instructions or products referred to in the content.

Article

Detecting Fake Accounts on Social Media Portals—The X Portal Case Study

Weronika Dracewicz and Mariusz Sepczuk *

Faculty of Electronics and Information Technology, Warsaw University of Technology, 00-665 Warsaw, Poland; weronika.dracewicz.stud@pw.edu.pl
* Correspondence: mariusz.sepczuk@pw.edu.pl

Abstract: Today, social media are an integral part of everyone's life. In addition to their traditional uses of creating and maintaining relationships, they are also used to exchange views and all kinds of content. With the development of these media, they have become the target of various attacks. In particular, the existence of fake accounts on social networks can lead to many types of abuse, such as phishing or disinformation, which is a big challenge nowadays. In this work, we present a solution for detecting fake accounts on the X portal (formerly Twitter). The main goal behind the developed solution was to use images of X portal accounts and perform image classification using machine learning. As a result, it was possible to detect real and fake accounts and indicate the type of a particular account. The created solution was trained and tested on an adequately prepared dataset containing 15,000 generated accounts and real X portal accounts. The CNN model performing with accuracy above 92% and manual test results allow us to conclude that the proposed solution can be used to detect false accounts on the X portal.

Keywords: fake account detection; Twitter (X); machine learning; image classification

Citation: Dracewicz, W.; Sepczuk, M. Detecting Fake Accounts on Social Media Portals—The X Portal Case Study. *Electronics* **2024**, *13*, 2542. https://doi.org/10.3390/electronics13132542

Academic Editors: Abdussalam Elhanashi and Pierpaolo Dini

Received: 25 May 2024
Revised: 20 June 2024
Accepted: 24 June 2024
Published: 28 June 2024

Copyright: © 2024 by the authors. Licensee MDPI, Basel, Switzerland. This article is an open access article distributed under the terms and conditions of the Creative Commons Attribution (CC BY) license (https://creativecommons.org/licenses/by/4.0/).

1. Introduction

Social media have become an integral part of our lives. The findings in [1] stated that 59.4% of the population (4.76 billion) were active social media users as of 2023, a number that is continuously growing, with a 3% increase since 2022. Social media allow users to create and maintain relationships with others and are also a space for users to consume content or express themselves. Social media have provided us with not only entertainment for years but also, for many users, the primary source of information about the world, news, trends, or events. A social platform's role is also to provide entertainment and exciting content.

Companies also use social media as a tool to promote their products and services. A company can use social media platforms to promote its brand, products, and services by publishing photos, videos, and posts.

Among the many advantages of social media, there are also disadvantages. Particularly significant are those related to cybersecurity [2–4]. Criminals can easily create a false identity on social media and target potential victims. User activity provides a lot of valuable information, such as biographical data, material assets, needs, or personality traits. This information can be used for crimes such as the following:

- Matrimonial fraud;
- Phishing (impersonating another person or institution to obtain essential data);
- Hacking (e.g., breaking into a user's computer and taking control of it);
- Cyberstalking (online harassment).

Another disadvantage of social platforms is the speed at which people spread false information. The purpose of incorrect information (fake news), which has become extremely popular, is to mislead and harm people.

That is why it is essential to check everything from several sources and to be able to distinguish real news from fake news. However, we often do not have enough time to verify all the information we have acquired. Undoubtedly, one of the elements of such a check is to determine the authenticity of the account from which each piece of content comes. This paper focuses on a solution that detects fake accounts on the X platform based on its appearance. The main contributions of this paper are summarized as follows:

- Present a distinctive visual-based approach to account classification;
- Create an image dataset of platform X accounts;
- Validate the created dataset;
- Test the detection of the authenticity of an X portal account by using the selected machine learning model.

It should be noted that despite the many solutions related to image classification, there are few that use this approach in social networks, especially on the X platform. The text-based method is a more popular approach and performs classification based on separately extracted data, while the proposed solution treats the image as a whole. The rest of this paper is organized as follows: Section 2 contains a brief review of related work in the field of fake account detection in social networks. Section 3 describes in detail the construction of the used dataset and its fundamental characteristics. Section 4 refers to initial experiments on the created dataset and shows their results. The article ends with a discussion of the results (Section 5) and conclusions (Section 6).

2. Related Works

The statement that social media have become an integral part of our lives in recent years is not a bit exaggerated. Today, many people find it hard to imagine a day without browsing Facebook, Instagram, or Twitter. This can also be seen in the statistics of the overall use of these media between February 2023 and February 2024 [5]: Facebook (65.46%), Instagram (9.92%), Twitter (6.7%), and LinkedIn (less than 1%). Each of these portals takes a different approach to detecting fake accounts. The literature has many examples of concepts for classification using ML. In general, fake profile identification using ML is not a new idea, but new papers that show another aspect of such a detection method are still being written.

In [6], the authors focused on detecting fraudulent accounts created by humans. They used Random Forest to identify if the user's profile was authentic with TF-IDF vectorization (a numerical representation of words utterly dependent on the nature and number of documents under consideration). The paper [7] describes how to detect fake Instagram accounts based on the textual data from this OSN (online social network). ML algorithms such as K-NN, Logistic Regression (LR), and decision tree (DT) were utilized to identify these accounts. In the paper ref. [8], the authors present a supervised model for detecting fake accounts based on machine learning to deal with the spread of fake content, which could be a valuable tool for controlling the excesses of online social crime. Another text-based approach is described in [9]. Yet another text-based solution was included by Bhattacharyya and Kulkarni in the article [10]. They explained how to detect fraudulent accounts and classify them into four categories: genuine accounts, social spambots, traditional spambots, and fake followers. Goyal et al. described in [11] a new solution that combined textual, visual, and network-based features to identify the different characteristics of fake accounts. The proposed approach used a deep neural network incorporating CNNs, LSTM, and graph GCNs to analyze these features.

The paper [12] applies to a fake profile detection model that uses emotion-based features to recognize a real or fake account. The approach was trained on 12 emotion-based attributes (both positive and negative), and the experiments were conducted on Facebook posts. On the other hand, the article [13] describes a method for detecting fake accounts by using a set of 17 features which are essential to distinguishing unreal users on Facebook from real users. Another solution for detecting fake accounts on Facebook can be found in [14]. The authors describe Intergo as a scalable defense system that uses a meaningful user ranking scheme. First, the solution predicts victim accounts (real accounts connected with fake accounts) from user-level activities. Based on that, a weighted graph and a list of potential fake accounts are created. In the paper [15], considerations on mitigating the problem of stealing a real account and replacing it with a fake one are presented.

The author of the paper [16] depicted a method for identifying Instagram fake accounts. The proposed method utilizes a gathered dataset used in the bagging classifier to classify forged accounts. Moreover, the created solution was compared to several well-known machine learning classifiers in terms of classification accuracy to evaluate the method's effectiveness better. A similar approach can be found in [17]. The document [18] includes classifications of impersonator categories that can exist on Instagram. First, the authors, using crawlers, collected the activities of famous politicians during a defined period. Then, they experimented with data from these profiles and built a model that could recognize fake accounts. Another example of detecting counterfeit profiles on Instagram is presented in [19]. The paper applies to a solution of fake account identification using supervised learning machine algorithms. The detected bogus profiles were stored to help the authorities take essential action against fraudulent social media accounts.

In the context of LinkedIn, we can also find solutions for detecting fake accounts [20,21]. In [20], the authors determined the minimal set of profile data necessary to specify if a profile is fake and describe an idea of a data mining approach that can be used for such identification. The results allowed them to conclude that the approach gave roughly similar results to those based on large sets of features used for detection. In contrast, in the document [21], an approach to finding groups of fake profiles created by the same actor was depicted. The designed solution used a supervised machine learning algorithm to classify an entire profile cluster as real or forged. This can be achieved based on user-generated text, such as email address, name, or residence.

Yet another approach was used for the X portal (formerly Twitter). The author in [22] used a multi-objective hybrid feature selection idea to better classify fake profiles. The selection of features was performed by using the Minimum Redundancy–Maximum Relevance algorithm (mRMR). The paper [23] employed a Chrome extension that detects fake profiles on Twitter by analyzing the different features. This approach uses, for account identification, both Random Forest and bagging methods. An interesting case is described in [24]. The paper contains a spam recognition artificial intelligence method for Twitter social networks. The model in the proposed solution was built by using a vector support machine, a neural artificial network, and Random Forest algorithms. Finally, the document [25] considered the detection of fake accounts generated by humans, as opposed to these created by bots. For such a distinction, a dedicated set of features was selected and applied to different types of supervised machine learning models.

From a preliminary review of the literature, it can be noted that not many solutions detect fake accounts based on images (see Table 1)—they are usually datasets of specific text data associated with a particular account. In our proposed solution, detection will be based on pictures of the profiles of X accounts. A detailed description of the solution can be found in the following sections.

Table 1. Comparison of methods for detecting fake accounts on X portal.

No.	Paper	Problem	Approach
1	[10]	Fake X account detection	Text-based solution
2	[11]	Fake X, Instagram, and Facebook account detection	Combination of text, visual, and network factors
3	[22]	Fake X account detection	Text-based solution
4	[23]	Fake X account detection	Text-based solution
5	[24]	Fake X and Facebook account detection	Text-based solution
6	[25]	Fake X human-created account detection	Text-based solution
7	This paper	Fake X account detection	Image-based

3. Creating a Dataset of Twitter Accounts

In this case study, we focused on detecting fake accounts on the X portal. The X network portal was selected due to persisting areas for improvement in ensuring the privacy and authenticity of its users while accessing the portal [26]. To train the image classification model, we needed to create our dataset consisting of figures of X accounts, since the existing and publicly available datasets did not meet our needs, as none provided graphical data nor had all the necessary characteristics to define particular classes of accounts. Below, we present the process of creating our dataset, from defining features to implementing a cloned view of the original X social network and figure generation process. This procedure could be helpful for future research and similar studies for other portals and types of accounts.

3.1. Definition of Various Types of Accounts

An important consideration when dealing with the issue of detecting fake accounts is to perform an initial security analysis of social networks and characterize the types of users that appear. In this subsection, the identified types of profiles are described, along with their features and the intentions of creation.

Profiles on social networks can vary according to their features and objectives. Due to their widespread use and accessibility, social platforms are used for countless purposes. The general public can exchange messages, photos, videos, and blog posts and communicate with people. Nevertheless, being aware of the Internet's unethical side is essential to quickly recognizing and understanding its intentions. This subsection lists the types of users we have learned about and identifies their impact on online security.

According to the authors of the article [27], who studied a group of more than 100,000 users of the Twitter platform and classified their roles based on the followed/followers ratio count, Twitter users fall into three groups: (1) broadcasters, (2) acquaintances, and (3) so-called miscreants (e.g., spammers). The nature of the interconnections among users in social networks and the previously mentioned relation are presented with a scatterplot in [27].

Broadcasters. This type of user has a much larger number of followers than they are following themselves. Many of these users represent recognizable individuals, online stores, radio stations, magazines, and other large organizations that use Twitter to promote their offerings and products and interact directly with their target audiences.

Acquaintances. The second group of users, referred to as acquaintances, tend to show mutuality in their relationships, a common characteristic of online social networks. In general, these are real users interacting with the social network for its intended function to connect with friends and family or to follow people from the broadcast group.

Miscreants. The common feature of the users included in the third unique group is that they follow a much larger number of people than they have followers. This behavior is typical of spammers or people who actively promote their beliefs. While this is one of the most suspicious groups, it does not exclude the possibility that there are genuine users with many interests or preferences.

The previously referred classification of users is a fundamental knowledge base of the accounts encountered. However, we need a complete understanding of the behavior of fake accounts. Thus, this also proves that a classification based on a single account feature cannot be a sufficient basis for further research.

A deeper analysis must consider several aspects to identify the types of users correctly. The authors of the publication [28] presented the results of a similar analysis on the classification of people using Twitter into three groups: (1) human, (2) bot, and (3) cyborg—a group that includes humans assisted by bots and bots assisted by humans. This analysis included the tweets' content, the occurring URLs, the tweeting devices, the user's profile, and the number of followers and friends.

Human. A user is considered and labelled a human if their profile contains authentic, meaningful, and concrete content. In particular, real users usually write down what they do or how they feel about something that appears on Twitter. Thus, they use it as a micro-blogging platform to self-express while interacting with friends. Concreteness means that the tweet's content is presented in relatively straightforward words with awareness, e.g., the answer to the question is relevant and directly addresses its subject.

Bot. Users demonstrating a lack of human-intelligent or original content are assigned to the bot group. Examples of such behavior include endlessly forwarding (retweeting) other people's posts and posting or advertising tweets with identical content. Frequently, this results in the propagation of unethical and false information, broadly published by using fake profiles. Excessive automation and the abundant presence of spam or malicious URLs in a user's profile also expose the account's association with such a group. Attackers use malicious bots to send spam and phishing messages, spread malware, host Command and Control (C&C) channels, and launch other illegal operations. In the case of external URL links, the main warning is that there is no connection between the link and the post's content. Audited by eye-catching text, users may click on links and be redirected to harmful sites. The last and least obvious of the characteristics is aggressive behavior toward other users, aiming to attract more attention (e.g., following and massively unfollowing quickly). Bots randomly add other users as friends, aiming to reach a large audience. In such a way, spam tweets posted by the bots are displayed on other users' recommendation pages.

Cyborg. According to the definition of this group, cyborgs, upon analyzing the behavior, include users where it is possible to find indications of both human and bot roles. For example, a typical cyborg account may contain different types of tweets. Many will have content with human-like intelligence and originality, while the rest will be automatically published. That represents a usage model whereby a human uses an account occasionally while applying automated approaches most of the time.

3.2. Feature Engineering to Generate a Dataset

The essential step required to create a detection solution for fake profiles is to define a set of features, determining whether an account belongs to one of the groups described earlier. The mentioned account characteristics are discussed and described in detail in this subsection.

The paper [26] grouped the most critical studies on detecting fake Twitter accounts according to the users' profile characteristics that were taken into account. Such a list of features was considered while defining and engineering types of accounts for the database. As mentioned, the dataset of profiles' images should combine text features and account-based data. For this reason, Table 2 contains features chosen among the characteristics presented in [26] to design the profiles.

Table 2. Selected features used to identify account type.

No.	Selected Feature	Description of Feature
1	Username	Unique identifier/name of user's account
2	Biography	Short introduction written by users about themselves, their achievements, expertise, and other important information
3	Profile photo (avatar)	One of the main features of accounts; it allows one to recognize a person by their appearance more quickly and easily
4	Header photo (banner)	In addition to the previous, Twitter introduced such photos to make the user's account more attractive
5	Date of creation	The date when the user created their account and became active on the network portal
6	Website	URL link that could be the user's website or profile on other platforms
7	Number of tweets (Twitter posts)	The essential feature for fake profile detection that allows for the determination of the level of user activity
8	Number of followers	Number of other accounts that are following the user
9	Following count	Number of other accounts that are being followed by the user's profile
10	Number of likes	An important feature indicating the number of profiles that liked the content created by the user
11	Number of views	Number of profiles that have seen the content created by the user, showing how wide their audience is
12	Number of retweets	Number of how many times the user's content was shared on both Twitter and other platforms
13	Number of replies	Number of comments on the user's posts

In addition to the listed features, verification was included, which indicates account authenticity granted by Twitter.

3.3. Types and Characteristics of Generated Accounts on X Portal

The research conducted focused on four classes of accounts that were defined by applying the previously mentioned characteristics and behavioral analysis: (1) bot, (2) cyborg, (3) real, and (4) verified. Table 3 presents profile features for the selected classes of accounts.

Table 3. Attributes for different classes of accounts.

Characteristics	Classes of Accounts			
	Bot	Cyborg	Real	Verified
Profile photo	Blank or default (initials)	Blank or default (initials) or both or only profile	Blank or default (initials) + header or both or only profile photo	Yes
Header photo	No			Yes
Account description	No	No	Yes	Yes
Account website	No	No	Website URL or no website	
Number of followers	Low number of followers or no accounts following a given profile		Average	High
Number of followings	High			Average
Date of creation	Large post No. + low interactions No. (close date) or no posts (former date of account creation)		former	Former
Number of posts			Average	High
Number of interactions			Average	High
Verification	No	No	No	Standard (blue icon) or business (yellow icon) or institutional (gray icon)

The outlined definition of bots, also shown in Figure 1a, allows us to consider most bot use cases. It considers the automation processes used to publish posts and the massive creation of empty accounts with no activity. The lack of any signs of human intelligence also marks them.

The defined cyborg profiles have similarities with accounts of bot nature. However, they are unique in their combination of human and automated behavior. The features designed to create this group are also illustrated in Figure 1b.

Generated by the outlined features, accounts of actual users cover most human use cases of social networks. People will look for mutuality by following other accounts and actively sharing their experiences through posts and replies. The designed class meets the definition of both more and less active people on Twitter, the detailed scheme of which is illustrated in Figure 2a.

The social network checks the content and authenticity of the profiles belonging to the verified class when granting verification. Thus, they represent actual and well-known users who can interact with each other and have many followers. In addition, Figure 2b schematically depicts the previously discussed features of verified profiles.

The defined features of each profile type and the generating principles for a set of images were appropriately adapted to the current appearance of the X social portal. This means the proposed and designed schemes do not represent the final definition of accounts' behavior. Therefore, they should be updated continuously to keep up with the dynamically evolving social platform. It is essential to be aware that considering all the changes occurring in the social network when generating the dataset will allow us to train a more efficient model, which, on the other hand, leads to more effective predictions.

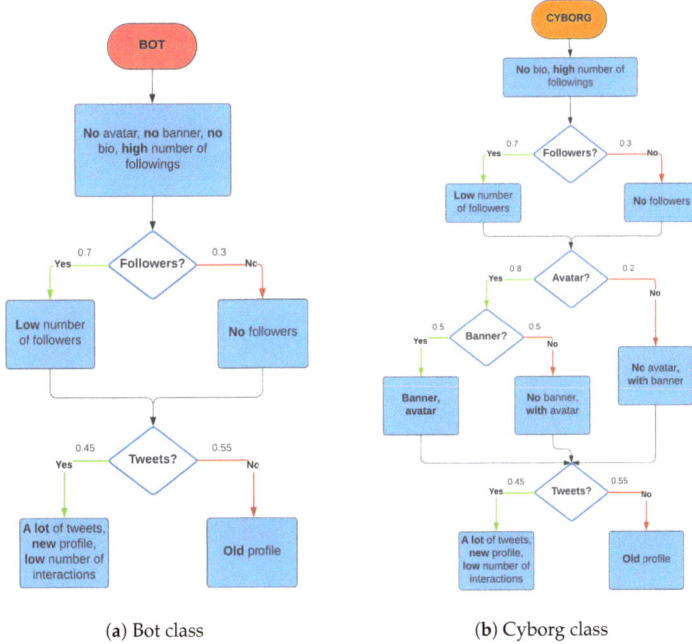

Figure 1. Feature definition scheme for profiles of (**a**) bot and (**b**) cyborg classes.

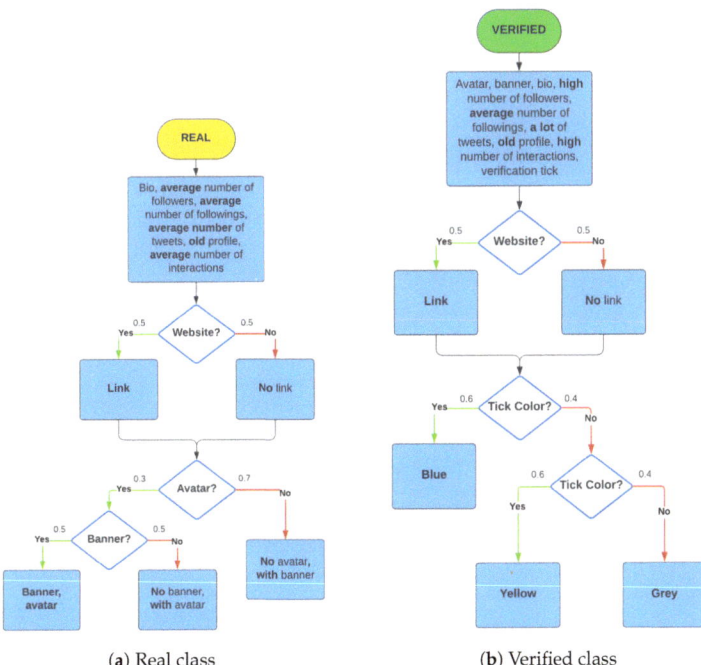

Figure 2. Feature definition scheme for profiles of (**a**) real and (**b**) verified classes.

3.4. Generation and Presentation of Accounts' Images

In order to generate a dataset, an additional website was implemented that was a mock-up view of the X social network. Following the styles of the original homepage of a user profile, we projected a primary view that allowed us to substitute data representing each class of accounts.

For this aim, standard web technologies and tools were used, such as TypeScript—to write code for a web application that is a replication of X's web view; React [29]—which was used to create graphical interfaces for a web application; and Playwright [30], which is an open-source automation library for browser testing, was used to take screenshots containing user profiles.

Additionally, the data that were used to create profiles for the dataset came from the Faker.js library [31]—used for profile pictures, first names, last names, locations, bio, links, all numeric data and text statements generation, as well as the open API Lorem picsum [32], here was used to generate profile banners.

A comparison of the web views of user profiles in the original portal and the implemented environment is shown in Figure 3a,b.

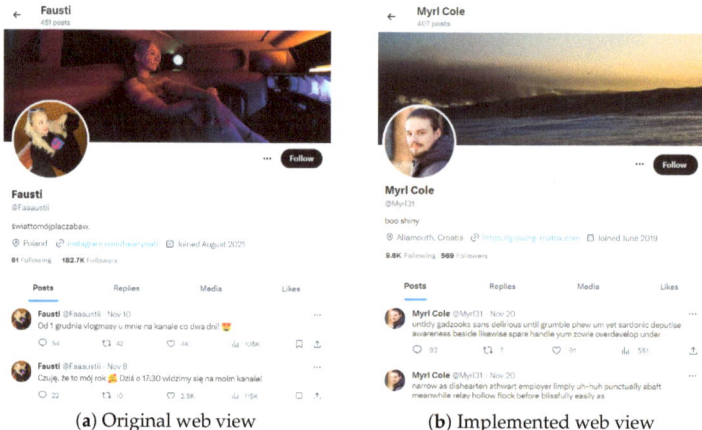

(**a**) Original web view (**b**) Implemented web view

Figure 3. Comparison of web views of (**a**) original and (**b**) implemented X profiles.

As mentioned, the website created allowed us to substitute the randomly generated classes' features defined previously dynamically. The following demonstration shows how the feature engineering performed on the dataset was applied and translated into the appearance of all types of accounts.

Figure 4 shows some examples of generated bot profiles. Their main feature is the lack of photos, additional information about the user, and a low level of activity and interaction. The first use case in Figure 4a shows an account recently created. However, the large number of posts and following profiles suggest automated control of it. In contrast, the second use case, shown in Figure 4b, presents an old account with no posts, but with a large number of following users. This behavior indicates a bot account created some time ago that has not been detected or deleted yet.

On the other hand, Figure 5 shows profile instances of the cyborg group. Their characteristics are similar to those of bots while having signs of human maintenance. For example, the use cases in Figure 5a,b present the profiles that are a combination of the behaviors we discussed above in the case of bots with profile pictures or/and header images.

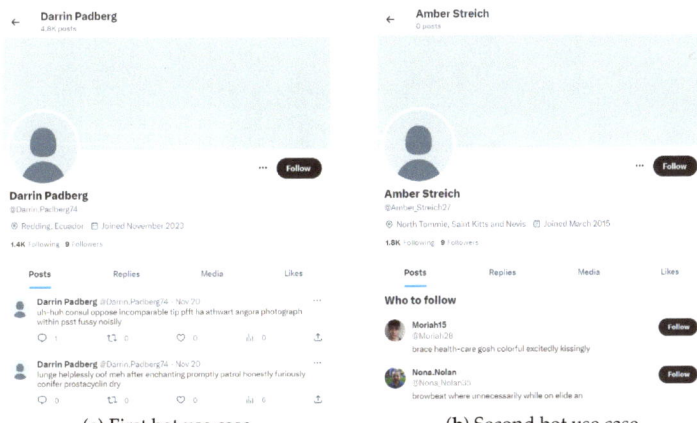

Figure 4. Web views of accounts designed as bots.

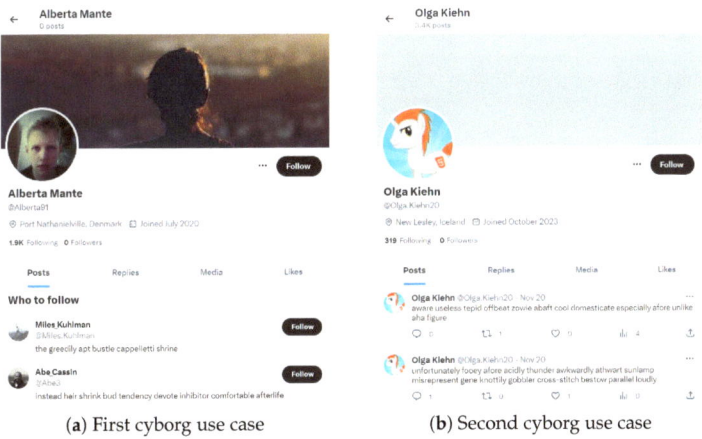

Figure 5. Web views of accounts designed as cyborgs.

The real profile class, an example of which is shown in Figure 6, represents users that tend to have profile photos and more detailed information about themselves on their social network account. Moreover, they have both a description of the profile and an adequate ratio of the number of followers/followings to the level of interaction and activity on the posted tweets. In Figure 6a, the first use case illustrates an account created some time ago but containing a banner, a URL website link, and many posts. The main reason profiles could be recognized as authentic is that an actual human using the X portal can achieve this level of characteristics. Additionally, an account with a profile picture that does not have a banner photo and a website can be classified as real due to an average user's lack of everyday use. Such an example of the profile can be seen as a second use case in Figure 6b.

Exemplary web views of accounts from the verified class, shown in Figure 7, are primarily defined as accounts of well-known and recognizable people. It is worth noting that X's verification feature significantly increases the likelihood of the account being legitimate. However, with time, this feature has become less reliable, as it is easier to gain verification without meeting all of its requirements or even purchase the verification sign. For this reason, verified accounts cannot be fully perceived as real, and analyzing them in the suggested way is essential. The main characteristics of this group remain: many followers, high-profile activity, and interaction. Despite this, truly verified users usually have profile and header photos with an extended biography and information about themselves.

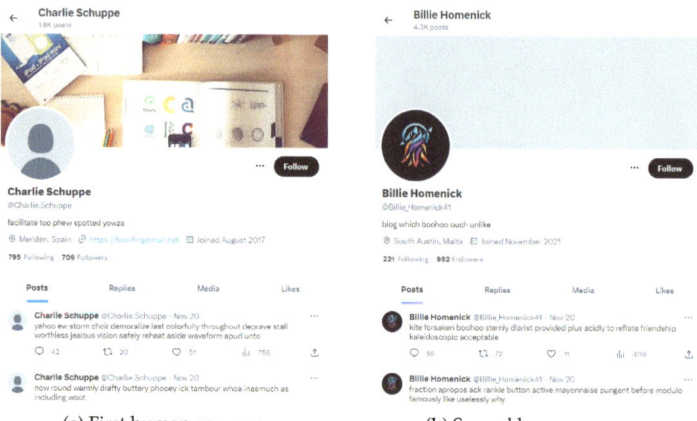

(**a**) First human use case (**b**) Second human use case

Figure 6. Web views of accounts designed as real users.

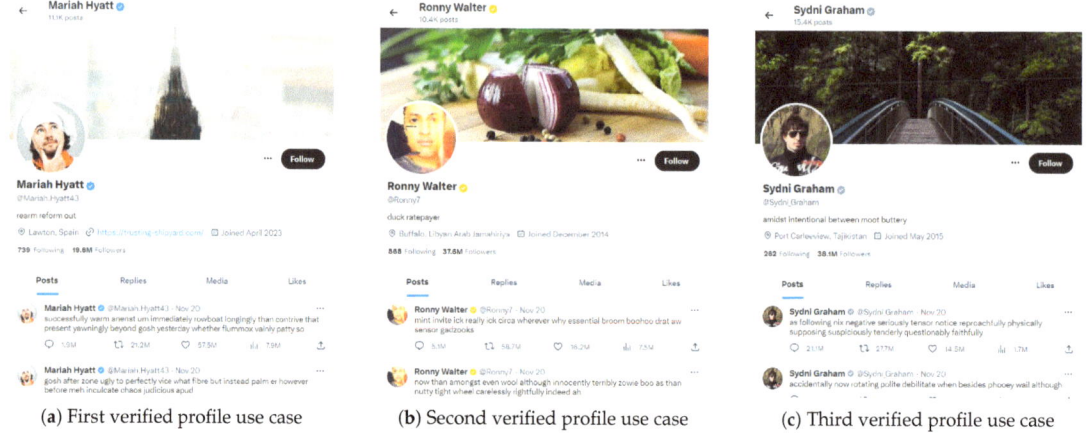

(**a**) First verified profile use case (**b**) Second verified profile use case (**c**) Third verified profile use case

Figure 7. Web views of accounts designed as verified profiles.

In the X social network, the type of verification is indicated by the color of the verification icon. Figure 7a, 7b and 7c show profiles with standard (blue icon), business (yellow/gold icon), and institutional (gray icon) types of verification, respectively.

In the manner described in this section, the substitution of data on the website was automated, and 11,000 unique profiles' images were created, where 10,000 (2500 per class) are data for training and 1000 (250 per class) for testing machine learning models. The dataset size was sufficient for machine learning, including the neural network models [33]. However, it could still be expanded for future experiments. It is worth noting that for the final evaluation of the detection tool, 4000 additional unique accounts were generated. Altogether, the developed dataset consisted of 15,000 screenshots of the accounts mentioned above with a resolution of 600 × 800 in PNG (Portable Network Graphic) format.

4. Experiments and Results

The dataset created, as discussed in the previous section, was used to train several of the most popular image classifiers. When the preferred ML model was selected and optimized, we moved on to incorporating it into the working solution, with a friendly user interface. The process of creating the specific components of the tool for detecting fake profiles on the X platform, as well as test results, are discussed in this section.

4.1. Machine Learning Model Selection and Optimization

A supervised machine-learning approach was used to detect fake accounts on the X platform. In this subsection, we present an analysis and a comparison of three image classification methods to select the most efficient detection tool for further development.

The machine learning models that were studied are (1) a model based on the Convolutional Neural Network classifier [34], (2) a model based on the Random Forest classifier [35], and (3) a model based on the Naive Bayes classifier [36]. Each of the introduced models was trained on 10,000 images of X profiles and then tested on the additional 1000 test images (10% of the training data). The tests conducted allowed us to determine metrics such as accuracy (Equation (1)), precision (Equation (2)), recall (Equation (3)), and F1-score (Equation (4)).

$$\text{Accuracy} = \frac{\text{Correct predictions}}{\text{All predictions}} \quad (1)$$

$$\text{Precision}_{class\ A} = \frac{TP_{class\ A}}{TP_{class\ A} + FP_{class\ A}} \quad (2)$$

$$\text{Recall}_{class\ A} = \frac{TP_{class\ A}}{TP_{class\ A} + FN_{class\ A}} \quad (3)$$

$$\text{F1-score}_{class\ A} = 2 \cdot \frac{\text{Precision}_{class\ A} \cdot \text{Recall}_{class\ A}}{\text{Precision}_{class\ A} + \text{Recall}_{class\ A}} \quad (4)$$

where *class A* is bot, cyborg, real, or verified; TP: true positive; TN: true negative; FP: false positive; and FN: false negative.

The metrics resulting from testing the classifiers were averaged across all classes to obtain the final macro-averaged scores shown in Table 4.

Table 4. Metric comparison of the considered models of image classifiers.

Classifier	Accuracy	Avg. Precision	Avg. Recall	Avg. F1-Score
Convolutional Neural Network	96.5	96.59	96.40	96.49
Naive Bayes	87.27	89.1	87.29	86.89
Random Forest	80.26	85.35	80.25	79.31

As the tests showed, the Convolution Neural Network turned out to be the best model of the compared classifiers. Therefore, it was decided to use the CNN model for the machine learning component to detect fake profiles.

In the next step, the optimization of the selected classification model was carried out. For this purpose, we studied the impact of the network structure and the number of neurons used in its layers on the detection process's accuracy and final loss function. The training of each model variation lasted for 25 epochs; then, the accuracy and loss function values obtained on the training (8000 images) and validation (2000 images) datasets were compared. The loss function was defined as the cross-entropy between the labels and the predictions made by the model.

The selected neural network had a sequential model consisting of four convolutional layers with ReLU activation function and 3×3 kernel size, four max-pooling layers, a flattened layer, and two dense layers. The visualization of the architecture of the addressed neural network model is presented in Figure 8.

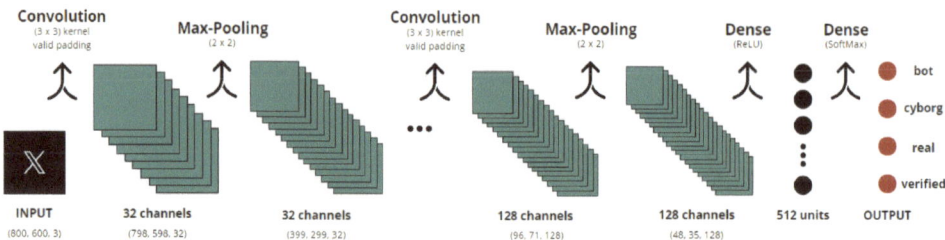

Figure 8. Design of CNN model's architecture.

Additionally, for the optimizer, we chose the Adam algorithm with a learning rate of 1×10^{-3}. The created model consisted of 110,343,876 learning parameters and weighed around 420.93 MB; the training process lasted about 2.5 h, and the final prediction time was 0.2 s.

The shape of the input data was set to (800, 600, 3), which means that the input images were 800 pixel high and 600 pixel wide and had three color channels (RGB—red, green, and blue). This corresponds to the size of screenshots taken of X's accounts, so there was no need for further data preprocessing.

Figure 9 presents graphs of the dependence of the accuracy (Figure 9a) and loss (Figure 9b) functions on the training period. The images were observed to show a correct increase in the accuracy value and a decrease in the loss value.

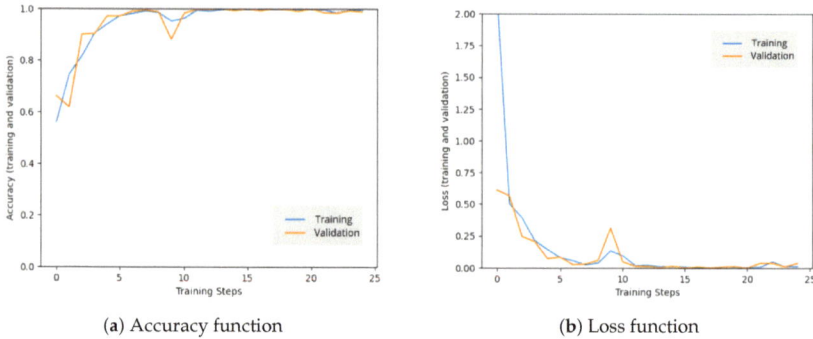

(a) Accuracy function (b) Loss function

Figure 9. Accuracy and loss function graphs for Convolutional Neural Network model.

Finally, the performance of the classification model was verified by using a test dataset (1000 images).

4.2. Detection of Fake Accounts

To check how our model will detect fake X portal accounts, a tool to ensure the most secure use of social networks was developed, enabling users to analyze any account in real time. Therefore, a web browser extension was considered the most convenient detection solution. This subsection will discuss the project assumptions and usage principles of the fake account detection tool in detail.

Among the commonly used web browsers, the Google Chrome browser based on the Chromium architecture was chosen to implement the detection solution. Moreover, an additional facade component was designed to communicate between the browser and the selected classification ML model. A pre-trained neural network classifier runs in a container environment, facilitating the developed tool's scalability, portability, and management.

We implemented WebSocket API [37] technology for communication, which made it possible to load the pre-trained model once within the first launch of the extension. In this way, obtaining the shortest and optimal waiting time for the returned detection model response was possible. Afterwards, the web socket opened a two-way communication session between the browser extension and the facade component, allowing it to load the profile's image and return the prediction result to the users. Furthermore, when forwarding to the machine learning model input, we used Base64 encoding to transfer the image from the client side and decode it on the facade side.

In terms of implementation, a web browser extension, which was given the name of "FakeDetector", making it available in the pop-up window format, made it possible to use the tool while browsing profiles on social networks. In this step, the user must navigate to the profile of interest, where a screenshot of the currently open tab is taken according to strictly set positions and sizes. Therefore, for the best performance, it is essential to correctly position the profile—ensuring all account information is visible—and set the screen scale to 100% to avoid unwanted scaling. The screenshots are taken by using the chrome.tabs API [38] and are then forwarded to the ML model input.

All of the detection components of the fake profile tool are illustrated in Figure 10.

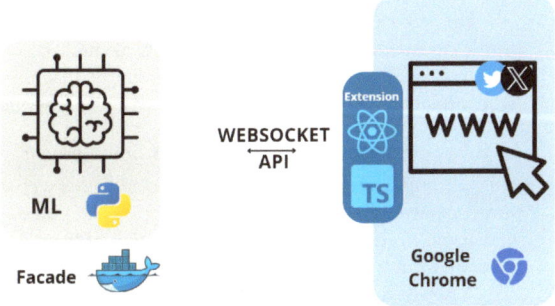

Figure 10. System implementing fake account detection approach.

To summarize the components of the tool outlined so far, we could present the process of its usage in the following steps:

1. Upon first launching the extension, the user navigates to the X social network profile of their interest (Figure 11a);
2. Within the extension, the user selects the button to detect if the account is fake (Figure 11b);
3. The extension takes a screenshot of the web page element containing the profile on the portal;
4. The image is sent to the facade component by the WebSocket API;
5. The facade forwards the image as the input to the machine learning model;
6. The model predicts whether the analyzed account is fake and returns the results;
7. Through the facade, the results are sent back to the end user (Figure 11c);
8. The probability of the account being fake is displayed to the user in the extension interface.

Figure 11 depicts the extension's interfaces in its particular states, such as ML model initialization (Figure 11a), its readiness when the ML model is already loaded (Figure 11b), and the detection result produced by the ML model (Figure 11c).

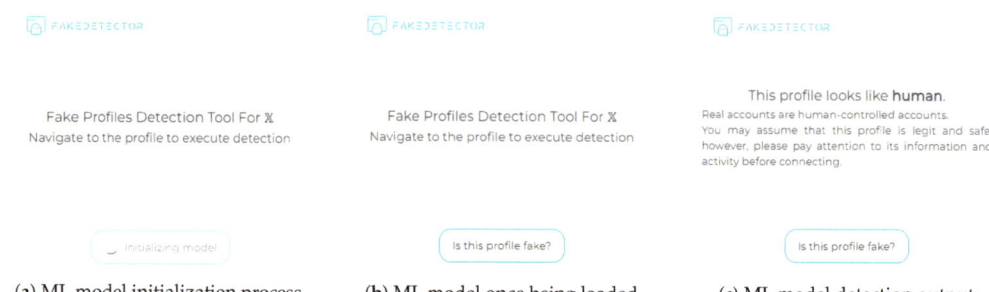

(a) ML model initialization process   (b) ML model once being loaded   (c) ML model detection output

Figure 11. Extension's interfaces in its particular states.

4.3. Tests Performed on Fake Account Detection Tool

This subsection provides the results of the tests conducted on a tool for detecting fake profiles in a social network for both the implemented copy and the original environment—the X portal. The predictions obtained from the machine learning model were placed into the confusion matrix; then, the true positive rate was calculated.

4.3.1. Testing in Implemented Copy of X Environment

The implemented environment is a copy of X's web view created for dataset generation purposes needed to train the ML classifier. The features and characteristics of the profiles were appropriately adjusted to the definition of their class. Therefore, since it most closely resembles the training dataset, high efficiency rates were expected there.

A new test set containing unique images of user profiles (1000 per class) was generated to verify the accuracy of the classification tool. The resulting confusion matrix is shown in Table 5.

Table 5. Confusion matrix of the proposed classification tested on the implemented copy of the X environment.

		Classified				Total	True Pos. %
		Bot	Cyborg	Human	Verified		
Actual	Bot	992	0	8	0	1000	99.2%
	Cyborg	24	956	20	0	1000	95.6%
	Human	2	1	997	0	1000	99.7%
	Verified	0	18	58	924	1000	92.4%

4.3.2. Testing in Original X Environment

Tests of the fake profile detection tool on the original X platform were conducted manually from the end user perspective. For this reason, the test set was significantly reduced. It is also worth noting that the initial classification of accounts could not be considered objective in the case of the bots, cyborgs, and real users classes, as they were selected on a profile appearance basis. Therefore, the result of the presented tests is only illustrative, proving the satisfactory functioning of the tool and detecting the identified reasons for incorrect model classifications.

To verify the classification tool's performance, unique X user accounts were manually selected (50 per class). Moreover, as the model was trained on a set of accounts that do not have posts with videos or photos, we preferred to choose accounts with recent text-based posts when conducting the test. The resulting confusion matrix is shown in Table 6.

Table 6. Confusion matrix of the proposed classification tested on the original X environment.

		Classified				Total	True Pos. %
		Bot	Cyborg	Human	Verified		
Actual	Bot	46	0	4	0	50	92%
	Cyborg	0	45	5	0	50	90%
	Human	0	13	37	0	50	74%
	Verified	0	6	12	32	50	64%

5. Discussion

With fake accounts exposing social networks to considerable danger and frequently being misused by attackers, more detection solutions are being developed (e.g., LinkedIn, Facebook, and X). The statistics of these tools provided by the creators of the OSN allow us to estimate their effectiveness; however, we still need to learn the details of their implementation. On the other hand, it is known that the vast majority use machine learning models for detection.

Our study significantly differs from most approaches to detecting fake accounts, as it analyzes the outlook of a user's account, thus focusing on its most essential features. Nevertheless, there is room for improvement in this tool's performance and efficiency. After analyzing the obtained test results, we noted slight misclassification in the bot class with the human class due to the overlapping percentage of accounts characterized by minimal personal information, confirmed during manual tests. The informal definition of the cyborg class led to 2.4% of profiles being mistakenly classified as bots and 2% as humans, since "cyborg" is a human-assisted "bot". The classification model performed quite differently in the case of the original X environment, showing us 10% of cyborgs being classified as humans, since the activity level of some "cyborgs" may have been similar to its extraordinarily high level of real users. The same situation also arises for the human class, with the 74% of correctly detected data, even though the classification tested on the implemented environment ended up with the best true positive rate of 99.7%. The class of verified profiles both on the copy and the original environments was the most misunderstood by the trained model. First of all, the reason for this behavior of the model is the confusing activity level of humans and cyborgs. To make class detection work as initially intended based on the presence of a verification icon, it would be necessary to increase the weight of this factor.

Following the statistics mentioned above, the main reason for the mistakes is that the class definitions did not precisely correspond to the profile activity taking place on X. Hence, it would be essential to create more precise features of the image database to obtain the most accurate match to the current state of the social network and expand user profile views to include posts with graphical content (images and videos). However, this was not the study's primary objective, as we mainly intended to introduce the applicability of image classification to the fake account detection problem by demonstrating the effectiveness of this approach.

This study is limited in a certain way, as it uses predefined account types to generate the image dataset, which may need to be revised in light of the social network's real behavior and appearance. First and foremost, this is indicated by the statistics obtained during the test analysis of the detection tool, where the model performed much better in the environment according to which it was taught. Online social networks are being actively developed, and a sudden change in any of the elements of the environment under study can have a negative impact on the accuracy of the detection method.

A further challenge may be the high sensibility of the created plugin tool to possible browser or system over-scaling, as well as the incorrect position of the profile in the browser window, causing the information of the account to be shifted. Given the entire spectrum of existing devices, modes, and fonts, supporting and correctly addressing every modification would require much effort. Although we can extend the tool to support different types of devices or include other modes and fonts, the most reliable solution would be to require users to employ the specified setup. Any exceptions to supported settings may be followed up with a message to the user about existing limitations and recommended setup to ensure detection effectiveness.

6. Conclusions

In this article, we presented the results of our work on detecting fake accounts on the X platform. In this case study, we focused on defining and classifying features of fake profiles to create a dataset with 15,000 images of unique accounts. This provides an opportunity to pave the way for further research.

Moreover, we conducted a comparative analysis of classification algorithms, selecting the most suitable one. An extension for the Google Chrome browser was created to use the trained model, allowing for the real-time interactive analysis of accounts. Finally, the initial tests were conducted, which concluded that the classes of bot and cyborg accounts were found to have the highest classification accuracy. Profiles belonging to groups of real and verified users were relatively often confused with each other, which is acceptable from the perspective of the tool's primary purpose. The unpredictable behavior of the detection model is due to the limitations encountered.

In future work, we plan to expand research on image-based account detection to other social networks (where possible). Moreover, we plan to verify whether using different types of neural networks will increase the detection level of fake accounts and their kind. Finally, due to the dynamic nature of social network web pages, the solution should be continuously adapted to the most up-to-date version of the network.

Regarding the stability of the created solution, it is worthwhile to reduce the sensitivity of the browser plugin, so the screenshot would not be taken according to fixed parameters but against a specific page element identifier that contains the necessary data for correct analysis. Moreover, despite X's default settings, it is worth supporting the most basic settings (e.g., system fonts and light and dark modes) and including support for at least two device types—desktop and mobile.

Author Contributions: Conceptualization, W.D. and M.S.; methodology, W.D.; software, W.D.; validation, W.D. and M.S.; formal analysis, W.D.; writing—original draft preparation, W.D. and M.S.; visualization, W.D. and M.S. All authors have read and agreed to the published version of the manuscript.

Funding: This research received no external funding.

Data Availability Statement: Data are contained within the article.

Conflicts of Interest: The authors declare no conflicts of interest.

References

1. Meltwater, W.A.S. Digital 2023 Global Overview Report. Available online: https://datareportal.com/reports/digital-2023-global-overview-report (accessed on 13 June 2024).
2. Almadhoor, L. Social media and cybercrimes. *Turk. J. Comput. Math. Educ. (TURCOMAT)* **2021**, *12*, 2972–2981.
3. Di Domenico, G.; Sit, J.; Ishizaka, A.; Nunan, D. Fake news, social media and marketing: A systematic review. *J. Bus. Res.* **2021**, *124*, 329–334. [CrossRef]
4. Shu, K.; Bhattacharjee, A.; Alatawi, F.; Nazer, T.H.; Ding, K.; Karami, M.; Liu, H. Combating disinformation in a social media age. *Wiley Interdiscip. Rev. Data Min. Knowl. Discov.* **2020**, *10*, 1–39. [CrossRef]
5. Social Media Use Statistics. Available online: https://gs.statcounter.com/social-media-stats (accessed on 4 March 2024).
6. Umbrani, K.; Shah, D.; Pile, A.; Jain, A. Fake Profile Detection Using Machine Learning. In Proceedings of the 2024 ASU International Conference in Emerging Technologies for Sustainability and Intelligent Systems (ICETSIS), Manama, Bahrain, 28–29 January 2024; pp. 966–973. [CrossRef]
7. Durga, P.; Sudhakar, D.T. The use of supervised machine learning classifiers for the detection of fake Instagram accounts. *J. Pharm. Negat. Results* **2023**, *14*, 267–279. [CrossRef]

8. Prakash, O.; Kumar, R. Fake Account Detection in Social Networks with Supervised Machine Learning. In *International Conference on IoT, Intelligent Computing and Security. Lecture Notes in Electrical Engineering*; Agrawal, R., Mitra, P., Pal, A., Sharma Gaur, M., Eds.; Springer: Singapore, 2023; Volume 982, pp. 287–295. [CrossRef]
9. Kanagavalli, N.; Sankaralingam, B.P. Social Networks Fake Account and Fake News Identification with Reliable Deep Learning. *Intell. Autom. Soft Comput.* **2022**, *33*, 191–205. [CrossRef]
10. Bhattacharyya, A.; Kulkarni, A. Machine Learning-Based Detection and Categorization of Malicious Accounts on Social Media. In *Social Computing and Social Media. HCII 2024. Lecture Notes in Computer Science*; Coman, A., Vasilache, S., Eds.; Springer: Cham, Switzerland, 2024; Volume 14703, pp. 328–337. [CrossRef]
11. Goyal, B.; Gill, N.S.; Gulia, P.; Prakash, O.; Priyadarshini, I.; Sharma, R.; Obaid, A.J.; Yadav, K. Detection of Fake Accounts on Social Media Using Multimodal Data With Deep Learning. *IEEE Trans. Comput. Soc. Syst.* **2023**, 1–12. [CrossRef]
12. Wani, M.A.; Agarwal, N.; Jabin, S.; Hussain, S.Z. Analyzing real and fake users in Facebook network based on emotions. In Proceedings of the 2019 11th International Conference on Communication Systems & Networks (COMSNETS), Bengaluru, India, 7–11 January 2019; pp. 110–117. [CrossRef]
13. Gupta, A.; Kaushal, R. Towards detecting fake user accounts in facebook. In Proceedings of the 2017 ISEA Asia Security and Privacy (ISEASP), Surat, India, 29 January–1 February 2017; pp. 1–6. [CrossRef]
14. Boshmaf, Y.; Logothetis, D.; Siganos, G.; Lería, J.; Lorenzo, J.; Ripeanu, M.; Beznosov, K. Integro: Leveraging victim prediction for robust fake account detection in OSNs. In Proceedings of the Network and Distributed System Security Symposium 2015 (NDSS'15), San Diego, CA, USA, 8–11 February 2015; pp. 1–15. [CrossRef]
15. Conti, M.; Poovendran, R.; Secchiero, M. Fakebook: Detecting fake profiles in on-line social networks. In Proceedings of the 2012 International Conference on Advances in Social Networks Analysis and Mining, Istanbul, Turkey, 26–29 August 2012; pp. 1071–1078. [CrossRef]
16. Sheikhi, S. An Efficient Method for Detection of Fake Accounts on the Instagram Platform. *Rev. d'Intell. Artif.* **2020**, *34*, 429–436. [CrossRef]
17. Akyon, F.C.; Esat Kalfaoglu, M. Instagram Fake and Automated Account Detection. In Proceedings of the 2019 Innovations in Intelligent Systems and Applications Conference (ASYU), Izmir, Turkey, 31 October–2 November 2019; pp. 1–7. [CrossRef]
18. Zarei, K.; Farahbakhsh, R.; Crespi, N. Deep dive on politician impersonating accounts in social media. In Proceedings of the 2019 IEEE Symposium on Computers and Communications (ISCC), Barcelona, Spain, 29 June–3 July 2019; pp. 1–6. [CrossRef]
19. Harris, P.; Gojal, J.; Chitra, R.; Anithra, S. Fake Instagram Profile Identification and Classification using Machine Learning. In Proceedings of the 2021 2nd Global Conference for Advancement in Technology (GCAT), Bangalore, India, 1–3 October 2021; pp. 1–5. [CrossRef]
20. Adikari, S.; Dutta, K. Identifying fake profiles in linkedin. In Proceedings of the Pacific Asia Conference on Information Systems (PACIS), Chengdu, China, 24–28 June 2014; pp. 1–30. [CrossRef]
21. Xiao, C.; Freeman, D.; Hwa, T. Detecting Clusters of Fake Accounts in Online Social Networks. In Proceedings of the 8th ACM Workshop on Artificial Intelligence and Security, Denver, CO, USA, 16 October 2015; pp. 91–101. [CrossRef]
22. Rostami, R.R. Detecting Fake Accounts on Twitter Social Network Using Multi-Objective Hybrid Feature Selection Approach. *Webology* **2020**, *17*, 1–18. [CrossRef]
23. Sahoo, S.R.; Gupta, B.B. Real-Time Detection of Fake Account in Twitter Using Machine-Learning Approach. In *Advances in Computational Intelligence and Communication Technology. Advances in Intelligent Systems and Computing*; Springer: Singapore, 2020; Volume 1086, pp. 149–159. [CrossRef]
24. Prabhu Kavin, B.; Karki, S.; Hemalatha, S.; Singh, D.; Vijayalakshmi, R.; Thangamani, M.; Haleem, S.L.A.; Jose, D.; Tirth, V.; Kshirsagar, P.R.; et al. Machine Learning-Based Secure Data Acquisition for Fake Accounts Detection in Future Mobile Communication Networks. *Wirel. Commun. Mob. Comput.* **2022**, *2022*, 6356152. [CrossRef]
25. Van Der Walt, E.; Eloff, J. Using Machine Learning to Detect Fake Identities: Bots vs Humans. *IEEE Access* **2018**, *6*, 6540–6549. [CrossRef]
26. Roy, P.K.; Chahar, S. Fake Profile Detection on Social Networking Websites: A Comprehensive Review. *IEEE Trans. Artif. Intell.* **2020**, *1*, 271–285. [CrossRef]
27. Krishnamurthy, B.; Gill, P.; Arlitt, M.F. A few chirps about twitter. In Proceedings of the WOSN '08: Proceedings of the First Workshop on Online Social Networks, Seattle, WA, USA, 17–22 August 2008; pp. 19–24. [CrossRef]
28. Chu, Z.; Gianvecchio, S.; Wang, H.; Jajodia, S. Who is Tweeting on Twitter: Human, Bot, or Cyborg? In Proceedings of the 26th Annual Computer Security Applications Conference, Austin, TX, USA, 6–10 December 2010; pp. 21–30. [CrossRef]
29. Meta Open Source React. The Library for Web and Native User Interfaces. 2023. Available online: https://react.dev/ (accessed on 13 June 2024).
30. Microsoft Corp. Playwright. 2023. Available online: https://playwright.dev/ (accessed on 13 June 2024).
31. Faker Open Source, Faker. 2023. Available online: https://fakerjs.dev/ (accessed on 13 June 2024).
32. Marby, D.; Yonskai, N. Lorem Picsum. Images. 2023. Available online: https://picsum.photos/images (accessed on 13 June 2024).
33. Khaled, S.; El-Tazi, N.; Mokhtar, H.M.O. Detecting Fake Accounts on Social Media. In Proceedings of the IEEE International Conference on Big Data (Big Data), Seattle, WA, USA, 10–13 December 2018; pp. 3672–3681. [CrossRef]
34. O'shea, K.; Nash, R. An introduction to convolutional neural networks. *arXiv* **2015**, arXiv:1511.08458. [CrossRef]
35. Pal, M. Random forest classifier for remote sensing classification. *Int. J. Remote. Sens.* **2005**, *26*, 217–222. [CrossRef]

36. Rish, I. An empirical study of the naive Bayes classifier. In Proceedings of the IJCAI 2001 Workshop on Empirical Methods in Artificial Intelligence, Seattle, WA, USA, 4–6 August 2001; Volume 3, pp. 41–46.
37. Lubbers, P.; Albers, B.; Salim, F. Using the WebSocket API. In *Pro HTML5 Programming*; Springer: Berlin/Heidelberg, Germany, 2011; pp. 159–191. [CrossRef]
38. Interfejs API Chrome.tabs, On-Line Documentation. Available online: https://developer.chrome.com/docs/extensions/reference/api/tabs (accessed on 4 March 2024).

Disclaimer/Publisher's Note: The statements, opinions and data contained in all publications are solely those of the individual author(s) and contributor(s) and not of MDPI and/or the editor(s). MDPI and/or the editor(s) disclaim responsibility for any injury to people or property resulting from any ideas, methods, instructions or products referred to in the content.

Article

Linux IoT Malware Variant Classification Using Binary Lifting and Opcode Entropy

Jayanthi Ramamoorthy *,†, Khushi Gupta †, Narasimha K. Shashidhar and Cihan Varol

Department of Computer Science, Sam Houston State University, Huntsville, TX 77340, USA; kxg095@shsu.edu (K.G.); nks001@shsu.edu (N.K.S.); cxv007@shsu.edu (C.V.)
* Correspondence: jxr153@shsu.edu
† These authors contributed equally to this work.

Abstract: Binary function analysis is fundamental in understanding the behavior and genealogy of malware. The detection, classification, and analysis of Linux IoT malware and its variants present significant challenges due to the wide range of architectures supported by the Linux IoT platform. This study concentrates on static analysis using binary lifting techniques to extract and analyze Intermediate Representation (IR) opcode sequences. We introduce a set of statistical entropy-based features derived from these IR opcode sequences, establishing a practical and straightforward methodology for machine learning classification models. By exclusively analyzing function metadata and opcode entropy, our architecture-agnostic approach not only efficiently detects malware but also classifies its variants with a high degree of accuracy, achieving an F1 score of 97%. The proposed approach offers a robust alternative for enhancing malware detection and variant identification frameworks for IoT devices.

Keywords: ELF static analysis; binary lifting; opcode sequence analysis; machine learning; malware detection; malware classification

Citation: Ramamoorthy, J.; Gupta, K.; Shashidhar, N.K.; Varol, C. Linux IoT Malware Variant Classification Using Binary Lifting and Opcode Entropy. *Electronics* **2024**, *13*, 2381. https://doi.org/10.3390/electronics13122381

Academic Editors: Abdussalam Elhanashi and Pierpaolo Dini

Received: 29 May 2024
Revised: 12 June 2024
Accepted: 15 June 2024
Published: 18 June 2024

Copyright: © 2024 by the authors. Licensee MDPI, Basel, Switzerland. This article is an open access article distributed under the terms and conditions of the Creative Commons Attribution (CC BY) license (https://creativecommons.org/licenses/by/4.0/).

1. Introduction

The rapid expansion of the Fourth Industrial Revolution and the Internet of Things (IoT) is transforming cyber–physical systems on an unprecedented scale. By 2030, it is projected that 75% of all devices will be IoT devices [1]. This surge will significantly influence various domains, including transportation, healthcare, and energy management. However, the complexity of hardware and software design, along with inadequate security features, makes IoT devices increasingly vulnerable to cyber-attacks [2].

According to the Zscaler ThreatLabz Enterprise IoT and OT Threat Report of 2023, there was a 400% increase in IoT malware attacks [3]. With the rapid adoption of IoT technologies in the industry, there will be an endless rise in these attacks. In September 2016, variants of the Linux Mirai malware were responsible for 1.1 Tbps DDoS attacks directed at the Dyn Domain Name System (DNS) provider [4]. In 2017, Linux/Brickerbot, a botnet similar to Mirai, infected more than 10 million IoT devices around the world [5].

Linux operating systems dominate the landscape of IoT platforms, as indicated by Antony et al. [6]. However, this widespread adoption has caused a corresponding increase in malware targeting Linux-based systems. AV-ATLAS's report [7] underscores this concern, revealing a 50% surge in new Linux malware within a single year. This alarming trend is attributed to Linux's prevalence in IoT devices, scalable cloud infrastructures, and the growing adoption of containerized applications. As Linux operating systems remain integral to numerous digital ecosystems, mitigating these security threats remains a critical challenge.

Malware analysis is the process of examining and understanding malicious binaries (malware) to determine their behavior, purpose, and potential impact. This process involves a range of techniques and tools to analyze different aspects of malware, such as its code, behavior, network traffic, and interactions with the system. Malware analysis can be broadly classified into two categories: Static analysis and Dynamic analysis. Static analysis involves analyzing the code and structure of malware without execution, which can be performed using features extracted from API calls, strings, byte n-grams, opcodes, etc., whereas dynamic analysis involves executing the malware in a controlled environment to discern its behavior and potential impact.

Opcodes (operation codes) are the fundamental components of machine language instructions in a binary file. They specify the exact operations that the CPU must perform. In the context of a malware binary, opcodes represent the low-level instructions that the malware executes to achieve its malicious objectives. Opcodes have been previously used in static malware detection [8,9]. However, when using opcodes as the features, one of the main challenges in detecting and classifying IoT malware is the heterogeneous device architectures [10,11]. Opcodes can perform equivalent functions across different architectures but due to the diversity of instruction sets of the malware binaries it is not feasible to use a consistent analysis methodology to detect and classify malware binaries across different underling architectures.

In this paper, we detect and classify Linux IoT ARM malware using the static analysis of disassembled opcode sequences extracted from the functions of the binaries. Due to the diverse range of architectures of the binaries, we use binary lifting technique to extract and analyze the Intermediate Representation (IR) of the opcode. Furthermore, we conduct statistical analysis on the IR opcode to derive a set of features such as the entropy. These features are subsequently used to train various machine learning models, such as Logistic Regression, Support Vector Classification (SVC), Random Forest, and Multi-Layer Perceptron (MLP) Neural Network.

The main contributions of this paper are as follows:

- Architecture-agnostic methodology: We propose an architecture-agnostic approach that relies on Intermediate Representation (IR) opcode instructions along with opcode entropy features to detect and categorize IoT ARM malware variants using Binary lifting.
  Binary lifting involves translating Instruction Set Architecture (ISA) assembly code which vary significantly between architectures, into a high-level Intermediate Representation. This process standardizes different opcode sets into a consistent IR format. Our method focuses specifically on opcode instructions and omits operands and register details to provide an abstract view of function behavior.
- Statistical IR opcode entropy feature set: We introduce a statistical feature set related to the entropy of Intermediate Representation (IR) opcodes. This feature set leverages the variability within opcode sequences to enhance the accuracy of malware detection and classification.
- Comprehensive function analysis dataset: We have developed a dataset that encompasses function metadata, the function IR opcode sequence, and statistical IR opcode entropy features for each ELF malware and benign binary. This dataset is derived from the raw IoT malware binary dataset and is structured to support machine learning classification models.

2. Preliminaries

In this section, we present an overview of ELF files, binary lifting techniques, and key characteristics of the binary files in our dataset, such as whether they are stripped of symbols and whether they are statically or dynamically linked. This discussion aims to provide a comprehensive understanding of the binaries in our dataset and elucidate the process of extracting the opcode Intermediate Representation (IR) from them.

2.1. ELF Files and Binary Lifting

An Executable and Linkable Format (ELF) file is a widely used standardized file format for executables, object code, shared libraries, and core dumps in Unix-like operating systems. Designed to be flexible and extensible, the ELF format supports various processor architectures and is the default binary format for many architectures. An ELF file is structured into distinct parts, including the ELF header, program headers, and section headers as shown in Figure 1.

- ELF header: Contains metadata about the file, such as its type, architecture, and entry point for execution.
- Program header: Describe segments that the operating system loads into memory, such as executable code and data segments.
- Section headers: Delineate various sections of the file used during linking and relocation, including sections for code ('.text'), initialized data ('.data'), and uninitialized data ('.bss').

An ELF file, is a standard binary file commonly generated in various architectures, including x86. ARM, MIPS, and others are composed of a sequence of bytes. When the ELF file is disassembled, an assembly code is generated. It is made up of opcode (commands of a specific architecture) and the operand (parameters used in an operation). This assembly code delineates the low-level operations that the CPU executes, detailing each instruction's specific role in the program's functionality. Each opcode corresponds to a specific operation, such as arithmetic calculations (addition, subtraction), data movement (loading and storing data), control flow changes (jumps, calls, conditional branches), and system calls (interacting with the operating system). These operations are encoded in a format that the CPU can directly execute.

Malware is a series of malicious behaviors, and opcodes have been used in prior works to identify malware and its attack behaviours. However, the analysis of ELF binaries can be complex due to the diverse instruction set architectures (ISAs) that define unique opcode specifications for each architecture. This diversity necessitates distinct analyses for each architecture.

Binary lifting is a sophisticated technique used to translate low-level ISA assembly code into high-level intermediate representations (IR). This process is essential for abstracting and standardizing assembly code that originates from different architectures and opcode sets. By converting diverse assembly instructions into a uniform IR format, binary lifting enables a more streamlined and consistent analysis across various platforms [12].

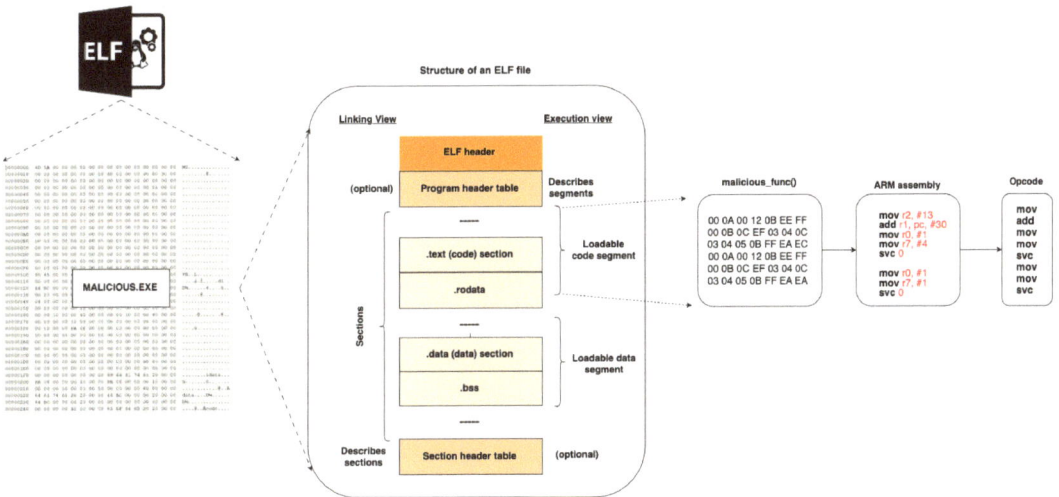

Figure 1. A dissection on the disassembly of an ELF file.

Binary lifting addresses RISC (Reduced Instruction Set Computing) architectures that have a smaller instruction set by preserving the semantics and abstracting the specifics to

Intermediate Representation (IR). For CISC (Complex Instruction Set Computing) architectures, binary lifting decomposes the instruction to simpler IR operations and ensures that all the functionality is accurately retained. In addition, differences in registers and memory access are normalized in IR, providing a consistent framework.

2.2. Characteristics of ELF Binaries

ELF binaries can vary based on whether they are stripped or non-stripped, as well as their linking characteristics, being either statically linked or dynamically linked.

2.2.1. Stripped and Non-Stripped Malware

ELF malware binaries can be stripped of all debugging information, symbol tables, and human-readable metadata. This resulting binary contains only the essential machine code required for execution. The process of stripping is commonly employed by malware authors as a means to hinder reverse engineering efforts. By eliminating function names, variable names, and other annotations, the analysis of these binaries becomes significantly more challenging. Malware Analysts will need to rely on heuristic techniques, pattern recognition, and dynamic analysis to discern stripped malware's functionality.

Conversely, non-stripped malware retains all or most of its debugging information, symbol tables, and metadata. This includes names for functions, variables, and other high-level information, providing substantial insights into the malware's operation. The presence of such information facilitates reverse engineering, enabling easy comprehension of the malware's structure, flow, and intent. We identify and label stripped binaries based on the existence of symbol tables and output from 'file' utility.

2.2.2. Static and Dynamically Linked Malware

Static linked malware refers to malicious software that incorporates all necessary libraries and dependencies directly into the executable file itself. This means that when the malware is executed, it does not rely on external libraries or shared resources from the underlying system. Instead, everything the malware needs to run is bundled within its binary. This bundling includes functions, routines, and other components required for its operation as shown in Figure 2. Static linking can increase the size of a malware's executable, as it incorporates all necessary code directly into the binary. This approach ensures that the malware operates independently on any system, eliminating potential compatibility and versioning issues with external dependencies.

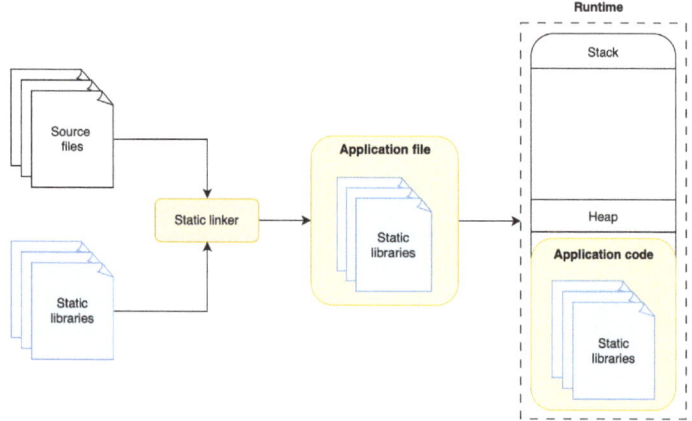

Figure 2. Inner workings of a statistically linked binary.

Dynamically linked malware, on the other hand, relies on external libraries and resources that are not included in its binary. When dynamically linked malware is executed, it accesses these libraries and resources from the system's shared libraries or external sources as shown in Figure 3. This approach reduces the size of the malware's binary since it does not need to include all dependencies within itself. However, it also means that the malware is dependent on specific library versions and system configurations. If these dependencies are not met on the target system, the malware may fail to execute or exhibit unexpected behavior.

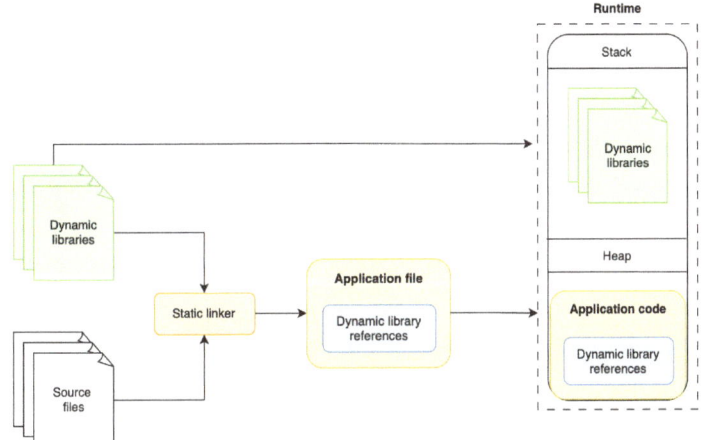

Figure 3. Inner workings on a dynamically linked binary.

3. Literature Review

Given the heavy reliance of global infrastructure systems on the Internet of Things (IoT), IoT devices are frequently exploited as entry points for cyberattacks due to their inherent security flaws. These vulnerabilities have led to the evolution of diverse IoT malware variants. In this section, we review the existing literature on malware classification using opcode sequence analysis and binary lifting.

Cozzi et al. [13] presents the largest study of IoT malware at the date of writing, reconstructing the lineage of IoT malware families using binary code similarity analysis. By tracking the relationships, evolution, and variants of these families, the study applies its technique to a dataset of over 93,000 samples submitted to VirusTotal over a period of 3.5 years. This approach facilitates the identification of various family variants and intra-family relationships due to code reuse. The paper also highlights the constant evolution of these threats by highlighting thousands of minor variations within each malware variant.

In [14], Moon et al. study the detection of IoT malware across different malware families by leveraging opcode sequence analysis. They create fixed-length training features from variable-length sequences with an entropy histogram, generating 2D visual representations that reveal intrinsic characteristics within homogeneous families while also providing robust training features. This visual differentiation aids in distinguishing between benign and malicious software, as well as correlated and uncorrelated malware. Machine learning algorithms such as 5-NN, SVM, Decision Tree, and Random Forest were then employed achieving a mean MCC of over 98.0%. Furthermore, the results also demonstrate that evolved malware can be detected with a model learned from its precedent malware.

In a similar vein, Lee et al. [15] propose a malware detection and family classification methodology. They represent IoT malware with fixed-length and low-dimensional features from opcode category information and their entropy values visualizing them as 2D images to identify patterns. The proposed features are evaluated on several ML models, including

5-NN, SVM, Decision Tree, Random Forest, and MLP yielding over 98% accuacy in malware detection and classification.

Gulmez and Sogukpinar [16] introduce a novel static analysis method for malware detection based on graph representations of opcode sequences. They disassembled PE files to obtain opcode sequences, which were then transformed into graphs. Using the histogram of node degrees within these graphs, they achieved a malware detection accuracy of 98% with machine learning algorithms such as Random Forest, KNN, Decision Tree, and SVM, with Random Forest performing the best. The study also compared the effectiveness of opcode histograms and node degree histograms, finding that the latter provided superior accuracy for malware detection.

Wang and Qian [17] introduce a classification method that leverages semantic features extracted from opcode sequences using word vectors. These sequences are treated as text sentences and fed into a text convolutional Neural network (textCNN) to classify malicious code families. The experimental results demonstrate high accuracy, with over 98% accuracy on the Microsoft Malware Challenge dataset and 91.93% accuracy on the SOREL-20M dataset. The study also optimizes model training speed by selecting key blocks containing call instructions. Overall, the proposed algorithm outperforms traditional byte n-gram representation methods in malicious code classification.

Similarly, HaddadPajouh et al. [18] explore the potential of using Recurrent Neural Networks (RNNs) to detect IoT malware. Their approach employs RNNs to analyze the operation codes (opcodes) of ARM-based IoT applications. Text mining techniques are then used to extract feature vectors from these opcodes. The authors train various machine learning models, including Random Forest, SVM, Naive Bayes, MLP, KNN, AdaBoost, and Decision Tree, using a dataset comprising of 281 malware and 270 benign samples. To evaluate the models, they tested them on 100 new IoT malware samples, using three different Long Short Term Memory (LSTM) configurations. The research findings reveal that the LSTM configuration with 10-fold cross-validation achieves the highest accuracy, reaching 98.18% in detecting new malware samples.

Furthermore, Darabian et al. [19] utilized sequential pattern mining to extract the most frequent opcode sequences in malicious IoT malware and benignware. The detected maximal frequent patterns (MFP) of opcode sequences are then used as the features to differentiate malicious applications from benignware. These features were used to train various machine learning models such as K Nearest Neighbor (KNN), Support Vector Machine (SVM), multilayer perceptron (MLP), AdaBoost, decision tree, and random forest achieving a 99% in malware classification.

Lastly, Kang et al. [20] proposed a methodology for Android malware detection and family classification using opcode n-gram features. They employed machine learning models such as Naive Bayes (NB), Support Vector Machine (SVM), partial decision tree (PART), and Random Forest. The study analyzed sequences up to 10 g, considering both binary counts (indicating the presence of specific n-opcodes in the application) and frequencies (indicating how often each n-opcode is used). The experimental results demonstrated that SVM achieved a 98% F-measure in both malware detection and family classification. Additionally, the authors concluded that binary n-opcodes provide more accurate results compared to frequency-based n-opcodes.

Addressing the challenge of classifying Linux malware across various heterogeneous architectures, Jeong et al. [21] proposes leveraging binary lifting. The core idea in this paper is to translate the binary codes of different architectures into a high-level intermediate representation (IR) using binary lifting. This creates a unified format for analyzing malware, regardless of the underlying hardware architecture. The translated IR sequences, which encapsulate malicious behavior patterns, are then fed into a deep learning model, specifically an LSTM (Long Short-Term Memory) model, for sequence learning achieving a 94% accuracy in detection and classification of various types of malware (rootkit, backdoor, worm, virus, etc.).

Our research stands out from the existing literature in several key aspects, as shown in Table 1. Unlike many studies that focus on specific architectures like ARM or PE files, our approach addresses a broader spectrum of architectures, allowing for a more comprehensive analysis. Moreover, our model achieves a competitive accuracy of 97% using Random Forest, showcasing its effectiveness. Additionally, we utilize ESIL from radare2 for Binary Lifting, offering support for a wide range of architectures and maintaining simplicity by working directly with opcode sequences without operands. What sets our research apart is the emphasis on function-wise analysis within each binary and across the dataset, coupled with rigorous statistical tests to establish the significance of differences in opcode sequences between malware and benign subsets. These unique aspects contribute to the robustness and depth of our analysis compared to the existing literature.

Table 1. Comparative study of research works that employ static analysis using opcode.

Research Work	Architecture	Features Used	Accuracy	Comparison
Moon et al. [14]	ARMv6-M	Opcode category sequences, entropy histogram	5-NN, SVM, RF (AUC: 0.99), DT (AUC: 0.97)	ARM ISA-specific. Needs opcode to be categorized.
Lee et al. [15]	ARM CPU	Sequence of opcode categories and entropy values	RF (99%)	ARM ISA-specific. Needs opcode to be categorized.
Gulmez et al. * [16]	PE files	Opcode sequences graphs	RF (98%)	PE files only. Creates graphs, subgraphs and then histograms from opcode sequences.
Wang and Qian * [17]	PE files	Vectors of opcode sequences	textCNN (98%)	PE files.
HaddadPajouh et al. [18]	ARM	Feature vector of opcodes	LSTM (94%)	ARM ISA specific.
Darabian et al. [19]	ARM	Maximal frequent patterns (MFP) of opcode sequences	Adaboost & DT (99%), MLP & KNN (96%)	ARM ISA-specific. Opcode categorization required. MFP opcode ranking based on its frequency.
Jeong and Kwak ** [21]	Multi-architecture	Opcodes IR	RNN, LSTM (94%)	Uses B2R2 for Binary Lifting and converts opcode + Operands into LowUIR representation. Results in a large amount of data (Figure 4 illustrates the LowUIR translation for one mov instruction) and therefore computationally intensive.
Our research	Multi-architecture	Opcodes IR	RF (97%)	Our research uses ESIL from radare2 for Binary Lifting and therefore supports a wide range of architectures. Use of just opcode sequence without operands or conversion to provide a high-level abstraction. Function-wise analysis—relative function analysis within each binary, and across the dataset. We conduct statistical test to establish that the opcode sequence between malware and benign subsets are significantly different.

'*' These works are for Windows binaries and do not address ELF binarie,s and therefore are not directly related to the current work. '**' The authors of this research use Binary Lifting technique but opted to generate LowUIR sequence from B2R2 (tool used for binary lifting). They also include operands in addition to opcode which is computationally intensive. Because of this, a single assembly instruction is translated to significant amount of data as illustrated in their research work. Our research uses only the opcode IR sequence to provide an abstract representation of each function, making it more efficient, less computationally intensive, and resulting in a better F1 score.

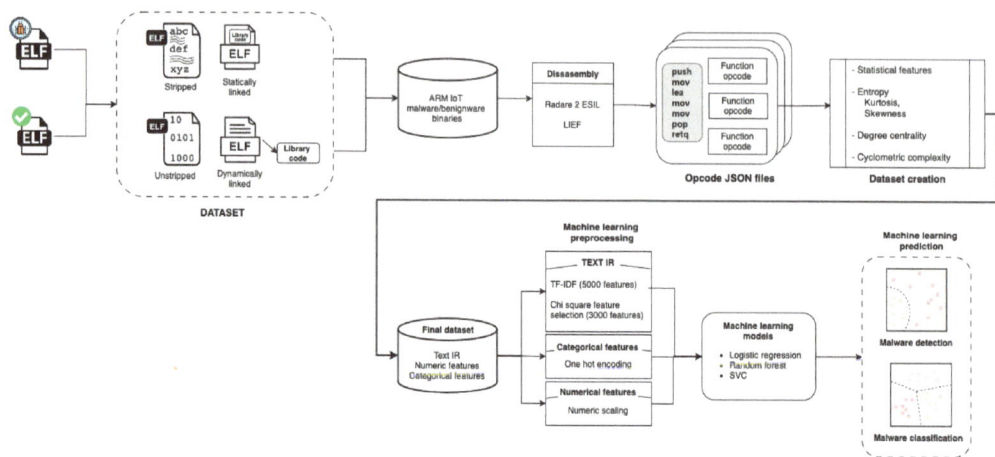

Figure 4. Workflow of the proposed approach with a dataset of raw Malware and Benign ELF binaries

4. Methodology

4.1. Dataset

This research uses ARM architecture binaries from the dataset described in [22]. It includes 65,956 open-source IoT malware binaries identified over a span of 14 years. This dataset features 1006 unique malware threat labels and is designed to encompass 15 different architectures. From this dataset, we extracted a labeled subset of ARM architecture-based ELF binaries for this study, consisting of both IoT malware variants and benign binaries, for the purpose of function analysis.

The ARM architecture binaries subset includes various malware variants such as Mirai, Gafgyt, Tsunami, Benign, Generica, Dofloo, and Jiagu. However, due to significantly less number of variants for some of these malware, we focused on the top three malware variants (Gafgyt, Mirai, and Tsunami) along with benign binary samples. Also, related works on IoT malware variants for ARM architecture have all focused on these malware. This also helped avoid extreme imbalance within our dataset. Gafgyt, Tsunami, and Mirai are all types of malware that target Internet of Things (IoT) devices, primarily using them to form botnets for various malicious activities.

Mirai, Gafgyt, and Tsunami are among the most prevalent threats targeting ARM architecture in IoT devices.

Mirai is a type of malware that primarily targets Internet of Things (IoT) devices such as routers, security cameras, and DVRs. It was first discovered in 2016 and became infamous for its role in launching massive Distributed Denial of Service (DDoS) attacks. Numerous variants have emerged since its source code was released publicly. These variants often include modifications to evade detection, add new exploits, or improve attack capabilities.

Gafgyt, also known as Bashlite, is another IoT-targeting malware that emerged around the same time as Mirai. Variants of Gafgyt have been developed to exploit different vulnerabilities and improve the efficiency of the botnet.

Tsunami, also known as Kaiten, is a malware that has been around since the early 2000s. It targets Linux-based systems and has been adapted to exploit IoT devices. While originally targeting general Linux systems, newer variants have been adapted to target ARM-based IoT devices. Variants have been developed to include additional exploits and to use more advanced command and control mechanisms.

To further refine the dataset, we selected a random sample of binaries containing fewer than 2000 functions. This step ensured that the number of functions per binary was manageable for batch analysis during reverse engineering. Following these stages, we compiled the final dataset used in this research. Table 2 presents the number of ARM architecture malware variants that were reverse engineered for function analysis.

Table 2. Counts of binary samples.

Class	Binary Count	Function Count
Gafgyt	509	146,214
Mirai	490	79,273
Tsunami	268	78,037
Benign	238	39,358

Once we extracted our dataset, we reverse engineered the binaries to retrieve its functions. Functions are discrete blocks of code within the binary that perform specific tasks. Reverse engineering and analyzing these functions is crucial for understanding malware behavior. The reverse engineering of these binaries is carried out using radare2 and LIEF libraries which are instrumental in malware analysis.

Radare2's ability to analyze multiple architectures through its Intermediate Representation (IR) instruction in ESIL (Evaluable Strings Intermediate Language) aids in architecture-agnostic analysis. The advantages of binary lifting after disassembling is highlighted in Table 3 .

Table 3. Comparison of our approach with Binary Lifting to architecture-specific disassembly.

Binary Lifting Approach	Architecture-Specific Disassembly
Allows analysis across different architectures without the need for architecture-specific tuning.	Tools and techniques can be optimized for a specific architecture, potentially providing deeper insights.
Increases scalability since we can target multiple platforms.	Fine-tuned optimizations for specific architectures.
Consistent framework and abstraction, which is efficient for comparative analysis.	Requires continuous updates and maintenance to accommodate different architectures.
Enables detection and classification of malware across a wide range of devices and environments.	Developing and maintaining multiple frameworks is tedious and resource-intensive.
Adds an additional processing step to abstract.	Multiple frameworks and toolsets can lead to fragmentation.

We assess whether an ELF binary is statically linked by examining references to libraries identified during the reverse engineering process. Additionally, we ascertain if a binary is stripped of symbols by checking for the presence of a symbol table in the binary sections, supplemented by findings from the 'file' utility. The labels for stripped and whether the binary is statically linked is determined based on best-effort, and therefore not definitive. The methodology workflow we used for this paper is outlined in Figure 4.

4.2. Static Analysis and Feature Extraction

In functional static analysis, function metadata such as the signature, number of basic blocks, cyclomatic complexity, and degree of centrality are extracted. A key aspect of our methodology is the use of radare2's Intermediate Representation (IR), specifically the ESIL (Evaluable Strings Intermediate Language), which provides an architecture-agnostic format for analyzing assembly code. This allows for a generic implementation to analyze functions.

For each function, we extract all IR opcodes maintaining the sequence and frequency. This data is encapsulated into a structured JSON format, ensuring each malware variant function is comprehensively documented. The primary dataset is then constructed with rows representing individual functions, each row tagged with identifiers like binary_id and binary hash, along with the extracted opcode sequences and additional function metadata.

4.3. Feature Engineering

In the feature engineering phase of our analysis, we include statistical metrics that attempts to capture the complexity and behavior of each function within the binaries. These metrics help in distinguishing malware and benign binaries while providing insights into the similarity of code patterns.

- IR opcode entropy: In information theory, entropy measures the randomness or unpredictability of data and mathematically represented as shown in Equation (1). We calculate the entropy of the sequence of IR opcodes for each function in the binary. Many related works have used entropy for code obfuscation and other encryption-related code.

$$H(X) = -\sum_{i=1}^{n} p(x_i) \log_2 p(x_i) \qquad (1)$$

where $p(x_i)$ is the probability of event x_i, and n is the number of different events. The logarithm is traditionally base 2.

- Skewness of opcode Entropy distribution: Skewness is a measure of the asymmetry of the probability distribution. Skewness helps us understand the distribution of opcode entropy values across all the binary functions. A positively skewed distribution suggests that most functions have lower entropy, with fewer functions exhibiting higher complexity where malicious activity s concentrated.

Skewness measures the asymmetry of distribution around the mean and is mathematically represented as shown in Equation (2).

$$\text{Skewness} = \frac{E[(X-\mu)^3]}{\sigma^3} \qquad (2)$$

where μ is the mean of the distribution, σ is the standard deviation and E is the expected value.

- Kurtosis of opcode entropy distribution: Kurtosis is a statistical measure that describes the tails of a distribution compared to a normal distribution as represented in Equation (3). In our analysis, examining the kurtosis of opcode entropy reveals whether the entropy values are heavily concentrated around the mean or if they are spread out across a wide range of values. High kurtosis in a malware binary indicates the presence of outlier functions either with extremely high complexity or extremely simple behavior such as a single instruction 'ret' or 'jmp' opcode, which are often associated with malicious payloads or complex evasion mechanisms. Low kurtosis is a suggestion of a more uniform distribution of function complexity.

$$\text{Kurtosis} = \frac{E[(X-\mu)^4]}{\sigma^4} - 3 \qquad (3)$$

where μ is the mean of the distribution, σ is the standard deviation and E is the expected value. This measures the tail of the probability distribution of the data, with the normal distribution's kurtosis adjusted to zero by subtracting 3.

4.4. Statistical Analysis

By including several statistical features with the cyclomatic complexity (ccomplexity) of the function and ratio of indegree (number of calls to the current function) and outdegree (number of outgoing function/library calls) we provide meaningful features that describe a binary functionality based on static analysis alone, for the machine learning models to train on. To validate the statistical significance of the differences observed between the opcode entropy distributions of malware and benign binaries, we employ the Mann–Whitney (Wilcoxon) U test. This non-parametric test is suitable for skewed data for comparison from two independent groups. Our analysis reveals a negligibly low p-value $< 2 \times 10^{-16}$, confirming that the entropy features can reliably differentiate between malware and benign samples as shown in Figure 5.

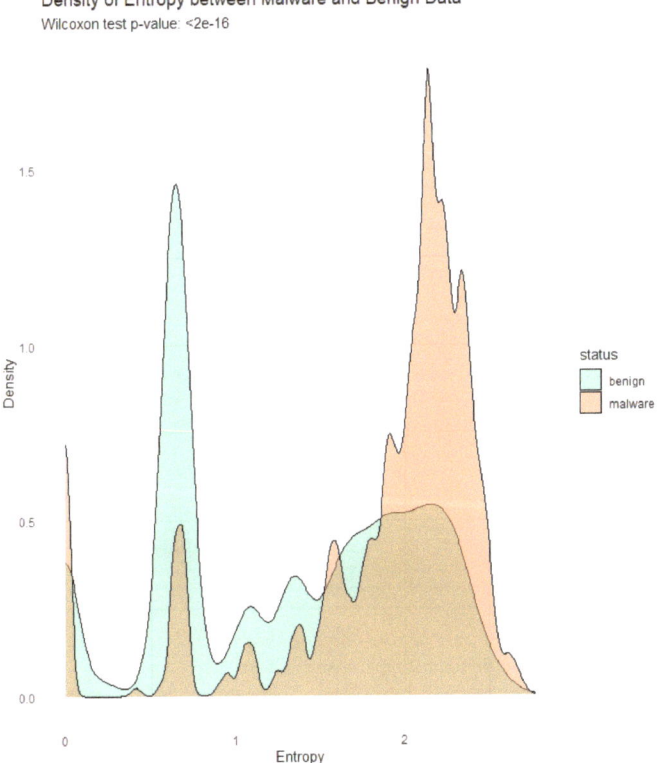

Figure 5. Opcode entropy differences between malware and benign subsets.

4.5. Data Organization

Each function within the ELF binary is indexed under a `binary_id`, facilitating aggregated analysis, and offering a comprehensive view of the overall functionality of the binary. By maintaining the functions within a `binary_id`, we capture behavioral patterns for the entire binary while preserving individual metrics such as complexity and centrality. his organization also supports the relative analysis of functions within a binary. To enable machine learning models to grasp the nuances of functions relative to the binary and across malware classes, the training data are structured in batches by `binary_id`.

The resulting dataset consists of the function and the ELF binary features as outlined in Table 4.

Table 4. Description of the IoT Malware Function Analysis dataset.

Field Name	Description
binary_id	Unique identifier for each binary
hash	Hash of the binary
endian	The endianness of the binary (LSB or MSB)
stripped	Binary attributes indicating if it is stripped of symbol information
StaticLinked	Binary attributes indicating if it is statically linked
Class	Classification of the binary (Mirai, Gafgyt, Tsunami, or benign)
Name	The name of the function within the binary

Table 4. *Cont.*

Field Name	Description
Type	The type of the function; can be function (fcn), symbol (sym), or location (loc)
Size	The size of the function in bytes
nargs	Number of arguments to the function
centrality, indegree, outdegree, indegree_ratio, outdegree_ratio	Network centrality metrics for the function
nbbs, ebbs	Number of basic blocks (nbbs) and extended basic blocks (ebbs) within the function
complexity	Cyclomatic complexity of the function, indicating the complexity of control flow
IR	Opcode sequence; binary-lifted Intermediate Representation (IR)
bin_avg_nbbs, bin_sd_nbbs, bin_avg_ebbs, bin_sd_ebbs, bin_sd_indegree, bin_sd_outdegree, bin_avg_nargs, bin_sd_nargs	Statistical features calculated across all functions in the binary, including averages and standard deviations for basic blocks, extended blocks, indegrees, outdegrees, and number of arguments
entropy	Function IR opcode entropy, measuring randomness and complexity
skewness_entropy	Skewness measure the asymmetry of entropy distribution across functions in the binary, indicating anomalies
kurtosis_entropy	Kurtosis of the entropy distribution, used to identify outliers and the "tailedness" of the distribution.

As part of our contribution, the Linux IoT Malware Function analysis dataset is available for future research efforts.

4.5.1. Data Pre-Processing for ML Based Classification

With the dataset created as outlined in the previous section, the numerical features, which include function IR opcode entropy, skewness and kurtosis scores, are standardized. The categorical features StaticLinked and Stripped columns, are One-Hot encoded. The delimited IR opcode sequences are parsed and vectorized using Term-Frequency Inverse-Document-Frequency (TF-IDF) vectorizer limited to a maximum of 5000 features. We opt to select the top 1000 features based on their highest scores, using chi-squared statistics which are suitable for multi-classification tasks. The pipeline for the data preprocessing and vectorization strategy is shown in Figure 6.

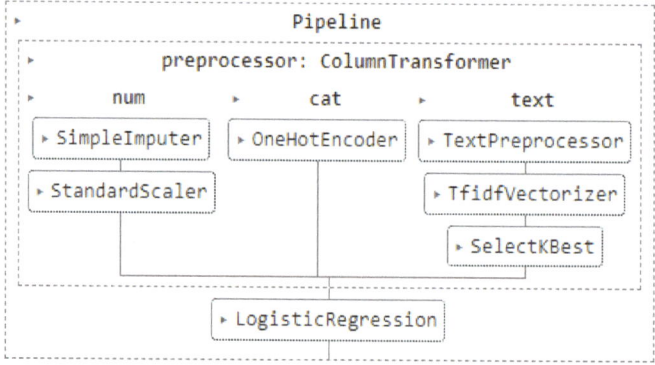

Figure 6. Data pre-processing and vectorization pipeline.

4.5.2. Machine Learning Classification of Malware Variants

The primary focus of this study is to identify and group variants based on their functionality and structure from static analysis without having to execute the malware. To this end, we employed various machine learning classifiers on the dataset we created from reverse-engineering Linux IoT malware and benign ELF binaries and augmented the dataset with additional statistical features.

The objective is to disassemble and reverse engineer malware functions to create a dataset that can be used for malware variant classification and detection irrespective of the Linux architecture with binary lifting. We have evaluated the approach and dataset with typically used ML classification models—Logistic Regression for multi-classification, Random Forest since it is robust to outliers and anomalies which is characteristic of malware, and SVC, which is effective in high-dimensional scenario where the number of dimensions exceeds the number of samples, as is often the case with detailed malware feature sets. We have also evaluated a Neural Network model, specifically MLP, which can capture complex patterns and interactions in the data through layered architecture. The strong performance of these models suggests that our dataset is well-suited for the classification task.

We compare the performance of multiple classifiers, such as Logistic Regression, Support Vector Classifier (SVC), Random Forest, and Neural networks with architectures like MultiLayer Perceptron (MLP).

4.6. Model Hyperparameters

For the dataset we generated, minimal tuning was required, and we mostly adhered to standard values from the sklearn library.

- Vectorization:
  We used a TF-IDF (Term Frequency-Inverse Document Frequency) vectorizer for function IR opcodes, with a maximum of 5000 distinct tokens to extract, and an n-gram range of (1, 1). As indicated in the methodology, we selected the top 3000 features using a chi-squared function to compute the score for each feature. The top 3000 features were chosen for computational efficiency and performance, and can be adjusted in future studies. For numeric features, we employed the mean value for imputation and standardize (normalize) by removing the mean and scaling to unit variance. For categorical features, we handled unknown categories by ignoring them, as we generated and verified the categories for incorrect of null values while extracting from the binaries.
- RandomForestClassifier:
  Through GridSearch, we identified the optimal number of trees as 200 and also adjusted initial class weights to accommodate class imbalance. The default values were retained for maximum depth of the tree, which was set to None, and Gini impurity is used for the split criteria.
- LinearSVC
  For LinearSVC, we used the default squared hinge loss function with an l2 penalty and a set the tolerance of 1×10^{-4} for the stopping criteria.
- MLP Classifier
  For the Neural network model, we employed the rectified linear unit function (ReLU) for activation and the 'adam' solver for weight optimization, with an initial learning rate of 0.001.

By following a logical approach to feature selection backed by statistical validation, the results are comparable across all the models, as discussed in the next section.

5. Results and Analysis

In this section, we compare and analyze the performance of various machine learning classification models on the function analysis dataset comprising IoT malware variants and benign binaries. The primary objective is to evaluate each model's effectiveness in distinguishing between malware functionalities and benign ELF binaries. To evaluate each model's effectiveness, we utilized the following metrics:

1. Accuracy: Accuracy measures the overall correctness of a model by calculating the ratio of correctly predicted instances to the total instances.

$$\text{Accuracy} = \frac{TP + TN}{TP + TN + FP + FN} \quad (4)$$

- TP = True Positives (correctly predicted positive instances);
- TN = True Negatives (correctly predicted negative instances);
- FP = False Positives (incorrectly predicted as positive instances);
- FN = False Negatives (incorrectly predicted as negative instances).

2. Precision: Precision measures the accuracy of positive predictions by calculating the ratio of correctly predicted positive instances to the total predicted positive instances.

$$\text{Precision} = \frac{TP}{TP + FP} \quad (5)$$

3. F1 score: F1-score is the harmonic mean of precision and recall, providing a balance between precision and recall. It considers both false positives and false negatives.

$$\text{F1-Score} = 2 \times \frac{\text{Precision} \times \text{Recall}}{\text{Precision} + \text{Recall}} \quad (6)$$

where recall is

$$\text{Recall} = \frac{TP}{TP + FN} \quad (7)$$

The models assessed include Random Forest, Logistic Regression, Support Vector Classifier (SVC), Multilayer Perceptron (MLP) Neural Network, and Long Short-Term Memory (LSTM) Networks. As outlined in Table 5, the Random Forest has the best performance with an accuracy, precision, and weighted F1 score all above 97%, indicating a high level of predictive reliability and consistency. In contrast, the LSTM model, while robust in handling sequence data, showed lower scores across all metrics, suggesting possible challenges in capturing the temporal dependencies within the static features of the dataset. The Logistic Regression and SVC models performed moderately with scores around 90%, which underscores the shortcoming of linear models in handling the nonlinear complexity of malware data. The MLP Neural Network performed better in capturing the non-linear function interactions with an F1 score of 94%.

Table 5. Performance metrics of machine learning models.

Model	Accuracy	Precision	F1 Score (Weighted)
Random Forest	97.17%	97.17%	97.176%
Logistic Regression	90.20%	90.39%	90.23%
SVC	90.09%	90.05%	90.06%
MLP NN	94.26%	94.37%	94.27%
LSTM	89.2%	89.27%	89.23%

Discussion

A closer look at the confusion matrices of the classifiers, as shown in Figures 7–10, provides accurate insights into the different models. For instance, Random Forest Classifier demonstrates the model's ability to classify different malware families and benign binaries with a higher degree of accuracy and precision score in distinguishing between the malware families, i.e., Gafgyt, Mirai, and Tsunami, indicating a strong capability to identify specific malware characteristics despite their shared functionalities. Interestingly, the false positives could also mean code-reuse or core library functionality.

Malware binaries are characterized by extreme outliers. For example, a Mirai botnet variant has a function with over 600 consecutive 'and' opcodes. Random Forest classifier performed the best, with a 97.1% F1 score likely due to the model's ability to handle outlier data. The Multi-Layer Perceptron (MLP) Neural Network followed with the next-best performance with 94.27%, possibly because of its ability to capture complex patterns among the functions within the binary. This success can be attributed to batching all the functions of a binary in a single batch, allowing the MLP to learn intricate relationships between the functions.

In contrast, the Long Short-Term Memory (LSTM) network showed the worst performance with an F1 score of 89.2%. We evaluated LSTMs, since they are suitable for sequential data and time-series prediction. Unlike Dynamic analysis, the lack of temporal data in our dataset due to the nature of static analysis might have limited its effectiveness. Moreover, LSTMs may require more extensive tuning and larger datasets to fully capture and learn the underlying patterns.

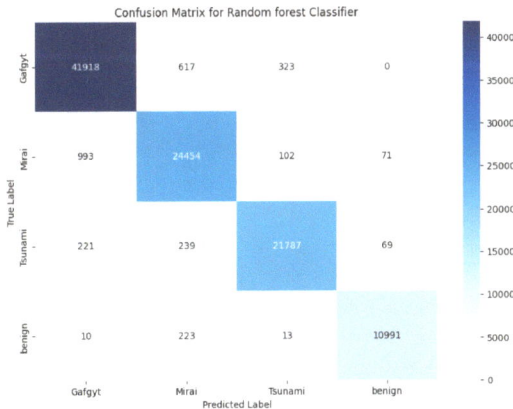

Figure 7. Random Forest classifier confusion matrix

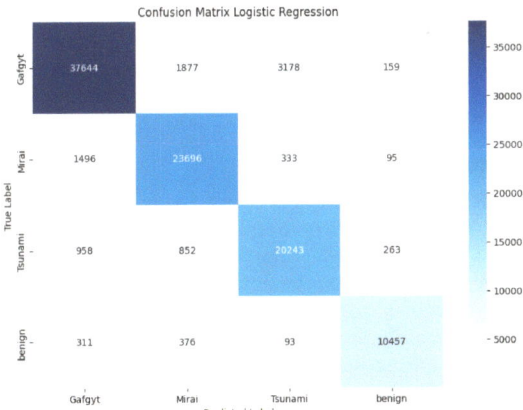

Figure 8. Logistic Regression classifier confusion matrix.

Figure 9. SVC classifier confusion matrix.

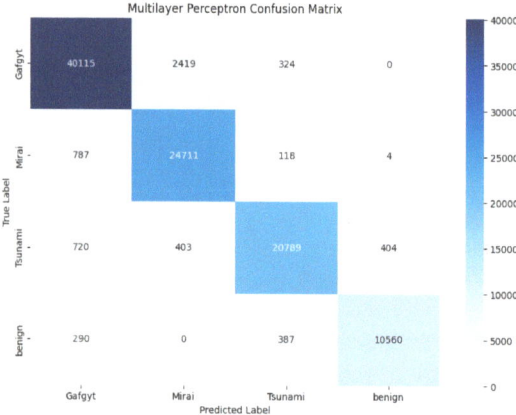

Figure 10. MLP Neural network classifier confusion matrix.

The results suggest that while there are common functionalities within the malware families, for example, use of networking libraries, encryption and other common functionality inherent to malware and benign binaries, the Classifier models are still able to differentiate these from benign behaviors.

6. Conclusions

This research uses an architecture-agnostic approach for detection and classification of Linux IoT malware variants and benign binaries. Our methodology involves the binary lifting of function opcode sequences and applying statistical entropy-based feature set based on Intermediate Representation (IR) function opcodes. We create a dataset with these features, and successfully detect and classify Linux IoT malware variants and benign binaries with multiple machine learning classifiers. The Random Forest classifier model resulted in an F1 score of 97% in detecting and classifying malware variants, including benign ELF binaries.

The proposed approach is computationally efficient as it focuses on analyzing opcode instruction sequences, enabling the extraction of statistically relevant features that abstract malware behavior from functions. Additionally, we validated the viability of this approach to detect and classify malware variants using statistical methods, confirming its effectiveness in practical applications.

Significantly, unlike related works that categorize opcodes based on different versions of the architecture, we propose a generic approach across various architectures, which is practical and reduces computational overhead. The comprehensive dataset created for this study includes function metadata alongside IR opcode sequences and entropy features, providing a solid foundation for systematic ELF binary analysis.

In conclusion, the methodologies and findings from this study provide a foundation for future detection systems that are efficient and capable of adapting to the evolving landscape of IoT malware.

7. Future Works

A limitation of the architecture-agnostic binary lifting approach is its effectiveness in handling proprietary architectures that have unique instruction sets and behaviors. These may not be supported by ESIL or other Intermediate Representation (IR) frameworks, which may result in incomplete or inaccurate results. To address this, it is essential to closely examine the IR from binary lifting to see if the assembly instructions are accurately represented and choose the right toolset for the architecture. However, these are edge cases, as most malware strive to maximize impact by supporting major architectures. Future studies can incorporate dynamic analysis to either complement or compare with the static analysis method and features used in this study, explore deep learning techniques for enhanced feature extraction, and test the models' efficacy across a broader range of IoT device architectures. Given that our approach is architecture-agnostic, we plan to apply this methodology to a diverse dataset of malware ELF binaries from various architectures. This study demonstrates that our computational and statistical approach provides a scalable and effective solution for addressing current and future challenges in IoT security.

Author Contributions: Conceptualization, J.R., K.G. and N.K.S.; methodology, J.R.; software, J.R.; validation, J.R., K.G., N.K.S. and C.V.; formal analysis, J.R.; investigation, J.R. and K.G.; resources, N.K.S. and C.V.; data curation, J.R.; writing—original draft preparation, K.G. and J.R.; writing—review and editing, K.G. and J.R.; visualization, K.G. and J.R.; supervision, N.K.S.; project administration, C.V.; funding acquisition, C.V. All authors have read and agreed to the published version of the manuscript.

Funding: This research received no external funding.

Data Availability Statement: The original dataset presented in the study is openly available at: https://github.com/jrmoorthy/Linux-Malware-Analysis (accessed on 14 June 2024).

Conflicts of Interest: The authors declare no conflicts of interest.

References

1. Howarth, J. 80+ Amazing IoT Statistics (2024–2030)—explodingtopics.com. Available online: https://explodingtopics.com/blog/iot-stats (accessed on 9 May 2024).
2. Ngo, Q.D.; Nguyen, H.T.; Le, V.H.; Nguyen, D.H. A survey of IoT malware and detection methods based on static features. *ICT Express* **2020**, *6*, 280–286. [CrossRef]
3. Zscaler ThreatLabz Finds a 400% Increase in IoT and OT Malware Attacks Year-over-Year. Available online: https://www.zscaler.com/press/zscaler-threatlabz-finds-400-increase-iot-and-ot-malware-attacks-year-over-year-underscoring (accessed on 9 May 2024).
4. Angrishi, K. Turning internet of things (iot) into internet of vulnerabilities (iov): Iot botnets. *arXiv* **2017**, arXiv:1702.03681.
5. Costin, A.; Zaddach, J. Iot malware: Comprehensive survey, analysis framework and case studies. *BlackHat USA* **2018**, *1*, 1–9.
6. Antony, A.; Sarika, S. A review on IoT operating systems. *Int. J. Comput. Appl.* **2020**, *176*, 33–40. [CrossRef]
7. AV-ATLAS Malware Portal. 2023. Available online: https://portal.av-atlas.org/malware (accessed on 10 May 2023).
8. Shabtai, A.; Moskovitch, R.; Feher, C.; Dolev, S.; Elovici, Y. Detecting unknown malicious code by applying classification techniques on opcode patterns. *Secur. Inform.* **2012**, *1*, 1. [CrossRef]
9. Santos, I.; Brezo, F.; Nieves, J.; Penya, Y.K.; Sanz, B.; Laorden, C.; Bringas, P.G. Idea: Opcode-sequence-based malware detection. In Proceedings of the Engineering Secure Software and Systems: Second International Symposium, ESSoS 2010, Pisa, Italy, 3–4 February 2010; Proceedings 2; Springer: Berlin/Heidelberg, Germany, 2010; pp. 35–43.
10. Hossain, M.M.; Fotouhi, M.; Hasan, R. Towards an analysis of security issues, challenges, and open problems in the internet of things. In Proceedings of the 2015 IEEE World Congress on Services, New York, NY, USA, 27 June–2 July 2015; IEEE: Piscataway, NJ, USA, 2015; pp. 21–28.

11. Lee, Y.T.; Ban, T.; Wan, T.L.; Cheng, S.M.; Isawa, R.; Takahashi, T.; Inoue, D. Cross platform IoT-malware family classification based on printable strings. In Proceedings of the 2020 IEEE 19th International Conference on Trust, Security and Privacy in Computing and Communications (TrustCom), Guangzhou, China, 29 December–1 January 2021; IEEE: Piscataway, NJ, USA, 2020; pp. 775–784.
12. Liu, Z.; Yuan, Y.; Wang, S.; Bao, Y. Sok: Demystifying binary lifters through the lens of downstream applications. In Proceedings of the 2022 IEEE Symposium on Security and Privacy (SP), San Francisco, CA, USA, 22–26 May 2022; IEEE: Piscataway, NJ, USA, 2022; pp. 1100–1119.
13. Cozzi, E.; Graziano, M.; Fratantonio, Y.; Balzarotti, D. Understanding linux malware. In Proceedings of the 2018 IEEE Symposium on Security and Privacy (SP), San Francisco, CA, USA, 20–24 May 2018; IEEE: Piscataway, NJ, USA, 2018; pp. 161–175.
14. Moon, S.; Kim, Y.; Lee, H.; Kim, D.; Hwang, D. Evolved IoT malware detection using opcode category sequence through machine learning. In Proceedings of the 2022 International Conference on Computer Communications and Networks (ICCCN), Honolulu, HI, USA, 25–28 July 2022; IEEE: Piscataway, NJ, USA, 2022; pp. 1–7.
15. Lee, H.; Kim, S.; Baek, D.; Kim, D.; Hwang, D. Robust IoT Malware Detection and Classification Using Opcode Category Features on Machine Learning. *IEEE Access* **2023**, *11*, 18855–18867. [CrossRef]
16. Gülmez, S.; Sogukpinar, I. Graph-based malware detection using opcode sequences. In Proceedings of the 2021 9th International Symposium on Digital Forensics and Security (ISDFS), Elazig, Turkey, 28–29 June 2021; IEEE: Piscataway, NJ, USA, 2021; pp. 1–5.
17. Wang, Q.; Qian, Q. Malicious code classification based on opcode sequences and textCNN network. *J. Inf. Secur. Appl.* **2022**, *67*, 103151. [CrossRef]
18. HaddadPajouh, H.; Dehghantanha, A.; Khayami, R.; Choo, K.K.R. A deep recurrent neural network based approach for internet of things malware threat hunting. *Future Gener. Comput. Syst.* **2018**, *85*, 88–96. [CrossRef]
19. Darabian, H.; Dehghantanha, A.; Hashemi, S.; Homayoun, S.; Choo, K.K.R. An opcode-based technique for polymorphic Internet of Things malware detection. *Concurr. Comput. Pract. Exp.* **2020**, *32*, e5173. [CrossRef]
20. Kang, B.; Yerima, S.Y.; McLaughlin, K.; Sezer, S. N-opcode analysis for android malware classification and categorization. In Proceedings of the 2016 International Conference on Cyber Security and Protection of Digital Services (Cyber Security), London, UK, 13–14 June 2016; IEEE: Piscataway, NJ, USA, 2016; pp. 1–7.
21. Jeong, H.S.; Kwak, J. Massive IoT Malware Classification Method Using Binary Lifting. *Intell. Autom. Soft Comput.* **2022**, *32*, 467–481. [CrossRef]
22. Olsen, S.H.; OConnor, T. Toward a Labeled Dataset of IoT Malware Features. In Proceedings of the 2023 IEEE 47th Annual Computers, Software, and Applications Conference (COMPSAC), Torino, Italy, 26–30 June 2023; IEEE: Piscataway, NJ, USA, 2023; pp. 924–933.

Disclaimer/Publisher's Note: The statements, opinions and data contained in all publications are solely those of the individual author(s) and contributor(s) and not of MDPI and/or the editor(s). MDPI and/or the editor(s) disclaim responsibility for any injury to people or property resulting from any ideas, methods, instructions or products referred to in the content.

Review

Research Trends in Artificial Intelligence and Security—Bibliometric Analysis

Luka Ilić [1], Aleksandar Šijan [1], Bratislav Predić [1], Dejan Viduka [2] and Darjan Karabašević [2,3,*]

[1] Faculty of Electronic Engineering, University of Niš, Aleksandra Medvedeva 14, 18000 Niš, Serbia; luka.ilic@mef.edu.rs (L.I.); aleksandar@mef.edu.rs (A.Š.); bratislav.predic@elfak.ni.ac.rs (B.P.)
[2] Faculty of Applied Management, Economics and Finance, University Business Academy in Novi Sad, Jevrejska 24, 11000 Belgrade, Serbia; dejan.viduka@mef.edu.rs
[3] College of Global Business, Korea University, Sejong 30019, Republic of Korea
* Correspondence: darjan.karabasevic@mef.edu.rs

Citation: Ilić, L.; Šijan, A.; Predić, B.; Viduka, D.; Karabašević, D. Research Trends in Artificial Intelligence and Security—Bibliometric Analysis. *Electronics* **2024**, *13*, 2288. https://doi.org/10.3390/electronics13122288

Academic Editors: Abdussalam Elhanashi and Pierpaolo Dini

Received: 7 May 2024
Revised: 6 June 2024
Accepted: 8 June 2024
Published: 11 June 2024

Copyright: © 2024 by the authors. Licensee MDPI, Basel, Switzerland. This article is an open access article distributed under the terms and conditions of the Creative Commons Attribution (CC BY) license (https:// creativecommons.org/licenses/by/ 4.0/).

Abstract: This paper provides a bibliometric analysis of current research trends in the field of artificial intelligence (AI), focusing on key topics such as deep learning, machine learning, and security in AI. Through the lens of bibliometric analysis, we explore publications published from 2020 to 2024, using primary data from the Clarivate Analytics Web of Science Core Collection. The analysis includes the distribution of studies by year, the number of studies and citation rankings in journals, and the identification of leading countries, institutions, and authors in the field of AI research. Additionally, we investigate the distribution of studies by Web of Science categories, authors, affiliations, publication years, countries/regions, publishers, research areas, and citations per year. Key findings indicate a continued growth of interest in topics such as deep learning, machine learning, and security in AI over the past few years. We also identify leading countries and institutions active in researching this area. Awareness of data security is essential for the responsible application of AI technologies. Robust security frameworks are important to mitigate risks associated with AI integration into critical infrastructure such as healthcare and finance. Ensuring the integrity and confidentiality of data managed by AI systems is not only a technical challenge but also a societal necessity, demanding interdisciplinary collaboration and policy development. This analysis provides a deeper understanding of the current state of research in the field of AI and identifies key areas for further research and innovation. Furthermore, these findings may be valuable to practitioners and decision-makers seeking to understand current trends and innovations in AI to enhance their business processes and practices.

Keywords: artificial intelligence; deep learning; machine learning; security; blockchain

1. Introduction

Artificial intelligence (AI) has become an integral part of the modern technological landscape, defining new horizons and providing extraordinary capabilities across various domains of life [1,2]. This fascinating subdomain of computer science continues to evolve [3], particularly thanks to innovations in areas such as deep learning, machine learning [4], AI security, and the application of blockchain technology [5]. Each of these segments plays a crucial role in the broader AI ecosystem, bringing new depth and diversity to our ability to create and understand intelligent systems.

Deep learning represents one of the most exciting aspects of AI, enabling computers to autonomously learn complex patterns from vast amounts of data. Inspired by the organization of the human brain [6], this technology has revolutionized how computers perceive the world around them, contributing to advancements in fields such as image recognition, natural language processing, and data analytics [7]. Machine learning, on the other hand, explores algorithms and techniques [8] that allow computers to learn from experience without explicit programming. This field is essential for developing predictive models, recommendations, and optimizations in various domains.

AI security is becoming an increasingly critical topic [9], especially with the growing number of connected devices and systems. With the proliferation of the Internet of Things (IoT) and the digitization of various industries [10], the need to preserve the security and privacy of data becomes inevitable. In this context, the development of security-aware AI becomes imperative [11], focusing on creating algorithms and systems resilient to various types of attacks and misuse while maintaining high levels of performance and efficiency. Security concerns are especially pertinent as AI systems increasingly handle sensitive data, raising questions about data protection and ethical use. The awareness of data security is essential, and it must be heightened to ensure the responsible application of AI technologies.

The importance of AI security is further underscored by recent incidents where AI systems have been exploited, leading to significant data breaches and misuse of information. As AI continues to be integrated into critical infrastructure, from healthcare to finance, the potential risks associated with inadequate security measures become more pronounced. Researchers and practitioners must, therefore, prioritize developing robust security frameworks that can anticipate and mitigate these threats. Ensuring the integrity and confidentiality of data managed by AI systems is not only a technical challenge but also a societal necessity, demanding interdisciplinary collaboration and policy development.

Blockchain technology, with its fundamental characteristics of decentralization and security, is gaining importance in the context of AI [12]. This technology provides transparency and data integrity, eliminating the need for centralized intermediaries and ensuring secure transactions and data storage. In combination with AI, blockchain opens up new possibilities for applications in various fields [13].

In line with the ubiquity of AI and its increasing impact on various spheres of society, research in this area is becoming increasingly important and comprehensive. This paper aims to analyze key trends in AI research, focusing on topics such as deep learning, machine learning, AI security, and blockchain technology. Through bibliometric analysis of publications in the Web of Science database over the past few years, we will explore the distribution of studies by year, the number of studies, and citation rankings in journals, as well as the leading countries, institutions, and authors in the field of AI. We will also analyze the distribution of studies across publication domains and the frequency of the occurrence of key author keywords to gain a deeper understanding of the thematic structure of research in this field.

It is expected that the results of this analysis will provide deeper insights into the current state and dynamics of AI research, identify key areas for further research and innovation, and inform the academic community, industry, and policymakers about current trends and potential research directions in the future.

2. Research Methodology

In this section, the methodology of reviewing papers on artificial intelligence in tertiary education is presented. This research study used the WoS database (Clarivate Analytics Web of Science Core Collection: Science Citation Index, Social Sciences Citation Index, Arts & Humanities Citation Index, Conference Proceedings Citation Index, Book Citation Index, Emerging Sources Citation Index, Index Chemicus, Current Chemical Reactions, Preprint Citation Index) [14] and a review was conducted following the PRISMA framework shown in Figure 1 [15].

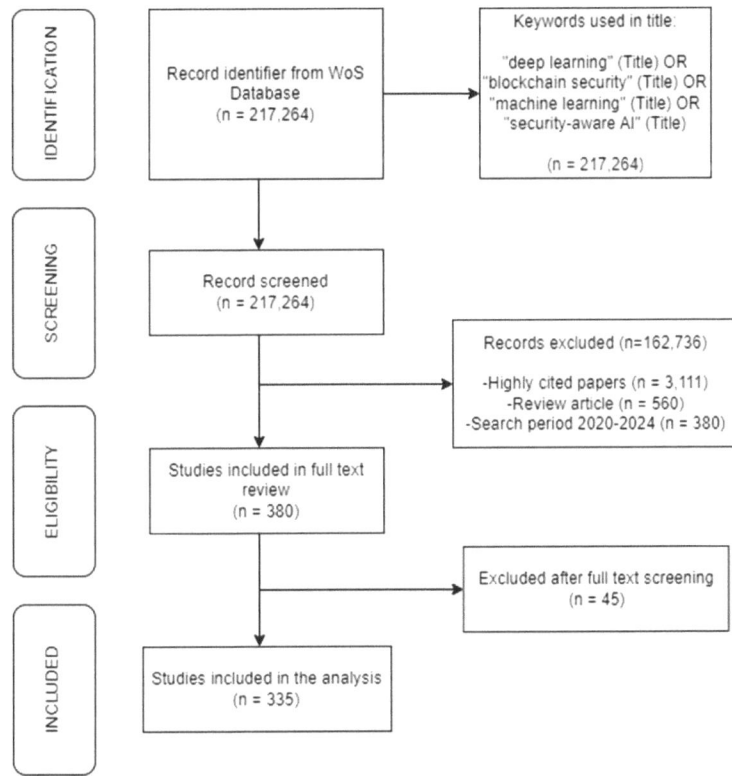

Figure 1. PRISMA flowchart (adapted from Page et al. [16]).

2.1. Defining the Research Question

In this step, the research question is defined. According to Arksey and O'Malley [17] this research aims to examine the available scientific papers and determine the extent of representation of research on the topic of artificial intelligence and its impact. Our research question is: What has been researched in the last few years based on the given keywords on the subject of artificial intelligence and what are the trends in this field?

2.2. Defining Search Sources

In order to review several AI papers, the Clarivate Analytics Web of Science (WoS) database was selected, as an initial survey (217,264) of sources showed that it contains a significant number of top scientific papers relevant to the research field. This justifies the use of WoS as a data collection source.

2.3. Defining a Search

Keywords and their meaningful combinations are defined here. Based on the research findings of Ahmed et al. [18], we chose keywords (such as machine learning, deep learning, etc.) which are strongly associated with AI as supplementary subject retrieval words to increase the comprehensiveness of the retrieved data. Search queries were performed using keywords "deep learning" (title) OR "blockchain security" (title) OR "machine learning" (title) OR "security-aware AI" (title). This was the first step in reviewing the total literature that mentions this topic. In later steps, the number of results obtained from the initial search was reduced by using filters that helped to reduce all results to those completely relevant to the research topic.

2.4. Conducting a Search

The search process was conducted according to the defined query in Section 2.3. We used Preferred Reporting Items for Systematic Reviews and Meta-Analyses (PRISMA) to comprehensively summarize previously published studies. The PRISMA guidelines include three phases: identification, screening, and inclusion [16]. Figure 1 shows the selection process. Article searches and data collection were performed in February 2024.

In the initial search, a set of 217,264 papers was obtained, but the keywords were searched only in the title to select only relevant papers, and not some that only mention those words. In addition, the authors of this paper refined the initial search results by including only papers that were selected as "highly cited papers" and the number of results dropped to 3111. The next filter we used was "document types: review article", and, in that step, only 546 papers were selected. All the papers that we selected were written in English so that we could easily analyze the quality of the papers. The search period covered the last couple of years in the WoS database, i.e., from 2020 to 2024. We received a total of 380 papers for our analysis.

2.5. Evaluation of the Quality of Results

The next step was to assess the quality of the data found and their significance was assessed according to Kitchenham, Mendes, and Travassos [19]. After a detailed review of the selected papers, the initial set of publications was reduced to 335 papers.

2.6. Primary Analysis of Scientific Papers

The necessary data were extracted according to the research question, and the entire search and selection process is visible in Figure 1. Papers were reviewed according to Web of Science categories, authors, affiliations, publication years, countries/regions, publishers, research areas, and citations by year. Bibliometric data analysis was performed with visualization of the results. During this step, several tools, such as VOSviewer version 1.6.20 [20,21], Microsoft Excel v2407, RStudio 2023.12.1 Build 402, and Biblimetrix (Boblioshiny) 4.1.4 [22,23], were used to analyze, map, and visualize the bibliometric data.

2.7. Detailed Analysis of Scientific Papers

In this step, a detailed analysis of the selected papers was performed by reading the full text of the selected papers.

2.8. Writing a Review Report

A review report was written and a discussion was held.

3. Overview of the Results

Figure 2 shows the data that we will analyze in this paper. These data were extracted from the Web of Science platform and analyzed with the help of RStudio.

Figure 2. Main info.

By analyzing the data for the period from 2020 to 2023, a significant decrease in the number of documents was observed, resulting in a negative annual growth rate of −21.53%. This rate represents the average annual decrease in the number of documents during the analyzed period. For example, taking the year 2020 as a base year with a certain number of documents, the annual growth rate of −21.53% indicates an average decrease of 21.53% in the number of documents each year compared to the previous year. This decrease may be due to various factors, such as changes in the data collection methodology, shifts in research trends, or other factors affecting document production. It is important to note that understanding this decrease has implications for interpreting the results and tracking trends in the analyzed research.

3.1. Keywords

In the center of the diagram in Figure 3, three main keywords that actually describe artificial intelligence (and their mutual connections with other keywords) stand out, namely:

- Machine learning;
- Classification;
- Deep learning.

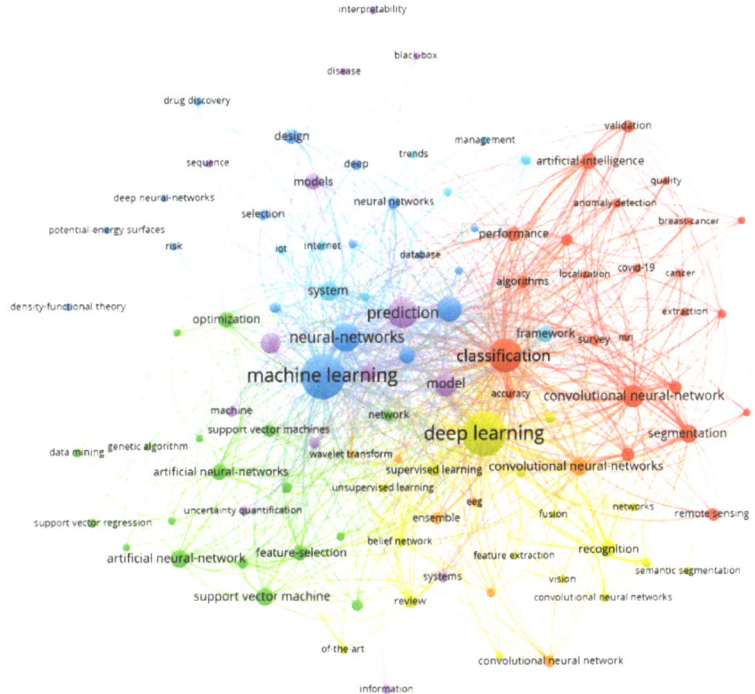

Figure 3. Displaying keywords.

These three main keywords that we can see in the diagram represent the basis of understanding and applying artificial intelligence in today's world. To understand these keywords, we need to describe or define them to continue with this paper.

3.1.1. Machine Learning

Machine learning represents the basic mechanism of learning systems to recognize patterns and make decisions based on data [24]. That is, the ability of computers to acquire and accept new knowledge and skills through experience. Machine learning is a subspecies of artificial intelligence whose task is to create a system that learns [25], where it finds conclusions and is prone to change without being explicitly (explicitly) programmed. Some of the definitions of machine learning say that it represents the design of computer algorithms that use experience when making future decisions [26].

3.1.2. Classification

The importance of "classification" is emphasized as one of the key segments in the application of artificial intelligence [27]. Through the application of deep learning algorithms, a system can accurately classify data based on its characteristics, enabling the identification of hidden patterns and connections that might be missed by the human eye. This classification process ensures improvement in various domains, from image recognition to text analysis [28], contributing to more efficient decision-making and real-time process optimization.

3.1.3. Deep Learning

Deep learning allows computer models to learn about complex data through multi-layered processing [29]. Deep learning revolutionized the technology of speech recognition, visual object recognition, and object detection. It has also made significant contributions in domains such as drug discovery. Using back-propagation algorithms, deep learning discovers complex structures in large datasets and adjusts the machine's internal parameters for better representations [30]. Convolutional networks are particularly useful for image, video, speech, and audio processing [31], while recurrent networks are effective in working with sequential data such as text and speech [30,32].

We must also mention that artificial intelligence is a term that is often mentioned in media and we need to explain what it means. Artificial intelligence is a concept that includes a large number of concepts from computer science, generally speaking, of mechanisms that use stochastic methods, that is, methods of randomness [33]. A subset of artificial intelligence is machine learning [34], which has gained great popularity among scientists and engineers, with a large number of free tools to work with. The function of machine learning is to enable a computer to perform tasks without additional programming [35]. Another branch of artificial intelligence is deep learning, whose main characteristics are multi-layer neural networks [36]. Neural networks represent a system consisting of a certain number of interconnected nodes where each node has its local memory in which it remembers the data it processes [37].

In short, the artificial intelligence ecosystem includes interrelated concepts like learning, intelligence, and deep learning, so it is no surprise why these words appear most often when it comes to artificial intelligence. This interaction shapes the modern paradigm of technological development [38], the application of which is becoming increasingly important in various areas of human life, from health to transportation, and from education to industry [39].

3.2. Categories

Analyzing the categories of scientific journals, we can see that various scientific journals appear. In addition to those shown in Figure 4, we had more categories but selected the following top 10 categories.

Figure 4. Categories of scientific journals.

1. Computer Science—Artificial Intelligence (44 papers): This category includes research that focuses on the development and application of artificial intelligence in computer science. Papers written in this area explore algorithms, machine learning, and deep learning techniques to develop intelligent systems capable of self-learning, inference, and decision-making.
2. Computer Science—Interdisciplinary Applications (29 papers): This category includes research that combines computer science with other disciplines to solve complex problems. Scientific papers in this field explore the application of computer techniques and algorithms in various fields such as health, economics, and sociology, among others.
3. Computer Science—Information Systems (19 papers): This category includes research that focuses on the development, implementation, and management of information systems. Scientific papers in this area explore how information technologies can be used to efficiently collect, process, store, and distribute information in organizations.
4. Energy Fields (19 papers): This category includes research dealing with various aspects of energy, including renewable energy sources, energy efficiency, energy distribution, and the sustainability of energy systems.
5. Engineering—Electrical Electronic (35 papers): This category includes research that focuses on electrical and electronic engineering, including the development of electrical systems, electronic components, telecommunications, control systems, and other related fields.
6. Geoscience—Multidisciplinary (22 papers): This category includes research dealing with various aspects of geoscience, including geology, geophysics, geochemistry, and other disciplines that study the structure, evolution, and processes on Earth and other planets.
7. Environmental Science (20 papers): This category includes research that focuses on the study of the environment, including ecology, environmental protection, natural resource management, and sustainable development.
8. Chemistry—Multidisciplinary (18 papers): This category includes research dealing with various aspects of chemistry, including the synthesis and characterization of chemical compounds, physical chemistry, organic and inorganic chemistry, analytical chemistry, and other disciplines.
9. Materials Science—Multidisciplinary (17 papers): This category includes research that focuses on the study of materials and their properties, including the synthesis, characterization, processing, and application of various materials in various industries.

10. Engineering—Biomedical (14 papers): This category includes research that deals with the application of engineering principles and technologies in medicine and biology. Scientific papers in this field explore the development of medical devices, diagnostic techniques, therapeutic methods, and other innovations that contribute to medical practice and health care.

Based on the number of scientific papers presented in different categories, it can be concluded that there is diversity in the research and interest from the scientific community and public in different fields. This leads to the confirmation that artificial intelligence is present in all fields and can be applied in various ways. An increased number of papers in certain categories may indicate active research and development in those areas, while, on the other hand, a lower number of papers may indicate a less researched or less popular field. In addition, certain areas may overlap or complement each other, which can affect the total number of papers. Through bibliometric analysis, the dynamics of the development of scientific disciplines can be seen and areas that require more research and attention from the scientific community in the future can be identified. The conclusion that can be drawn is that the first three categories, which include computer science, have the most scientific papers due to the active research and widespread application of artificial intelligence, information technology, and interdisciplinary applications in various fields. These areas are currently in the spotlight due to their high relevance and impact on various industries and disciplines. On the other hand, we have the category Engineering—Biomedicine, which has the smallest number of papers due to its specificity and potential limitations in research. The reason for this may be the high costs of research and the need for specialized knowledge and equipment. We must also mention the strict regulation in health sciences. This may limit the number of researchers and institutions involved in this field and the result would be a lower number of scientific papers compared to other categories.

3.3. Publications

The number of publications by year in Figure 5 shows the dynamics of research activity during the last four years. The increase in the number of publications from 2020 to 2021 can be interpreted as a result of intensive research and investment in scientific projects, which may have been stimulated by global events such as the COVID-19 pandemic [40] or other factors that may have increased interest in certain topics. However, we also have a decline in the number of publications in 2022 and 2023, which may indicate a possible trend of decreasing research activity or changing priorities in research. These numbers may be due to a variety of factors, including changes in research funding, institutional policies, resource diversion, or changes in the interest of researchers and the scientific community. Also, the lower number of publications in 2022 and 2023 may be the result of the time lag between conducting research and publishing papers. All these factors should be taken into account when analyzing the dynamics of research activity and planning future research and projects.

A decrease in the number of publications during a certain period can have a significant impact on the scientific community and the further development of research. A smaller number of published papers can contribute to slower progress in certain scientific disciplines, because the reduced volume of research results limits the availability of new research results, knowledge, and ideas in those fields. In addition, the reduction in the number of publications can have the effect of creating a gap in the literature and limit the diversity and access to certain fields. The lack of new research can slow down innovation and make it difficult to make new discoveries or develop new technologies, and also to present the achieved results. Also, fewer publications can have consequences for researchers' careers, especially for young researchers who rely on publications to advance their careers. The lack of opportunities to publish papers can limit the opportunities for obtaining financial support, creating scientific connections, and building the advancement of knowledge in one's field. However, reducing the number of publications may also encourage researchers to focus on quality instead of quantity, which may result in deeper and more thorough research. Also, it can encourage researchers to consider new approaches and methodologies in their scientific papers, which can lead to creative solutions and innovations.

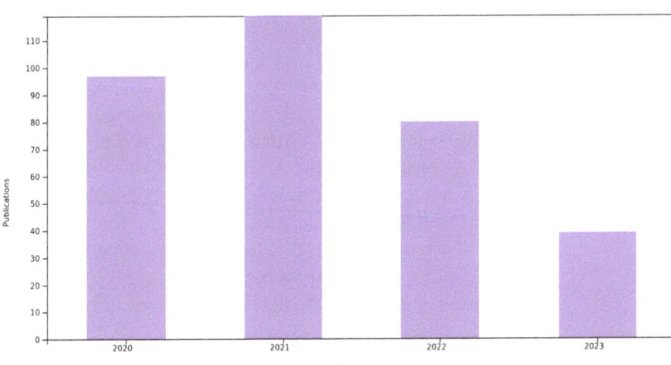

Figure 5. Number of publications.

In conclusion, it is important that the scientific community recognizes the challenges that the reduction in the number of publications can bring and works together to find solutions to support the continued progress and development of science.

3.4. Areas of Research

Analyzing Table 1, the number of papers in different fields provides insight into the diversity of the interests and focuses of the scientific community. The high number of papers in fields such as computer science can be explained by the wide application of these disciplines in various industries and areas of life, and also by the constant technological progress that requires constant development, research, and innovation. On the other hand, the smaller number of papers in areas such as biochemistry and molecular biology may be the result of the specificity of these disciplines, or the complexity of the research, as well as the smaller number of researchers dealing with these topics. Also, fields such as geology, physics, and energy may have a smaller number of AI-related papers compared to other disciplines of lesser interest. However, it is important to note that the number of researchers in a particular field may vary from year to year, depending on current trends, financial support, institutions, and other factors.

Table 1. Areas of research.

Areas of Research	Number of Papers	%
Engineering	94	28.060
Computer Science	88	26.269
Chemistry	31	9.254
Environmental Sciences Ecology	23	6.866
Geology	22	6.567
Physics	21	6.269
Science Technology Other Topics	21	6.269
Energy Fuels	19	5.672
Materials Science	17	5.075
Biochemistry Molecular Biology	15	4.478

Source: Author's own calculations.

3.5. Countries

The number of papers by country provides an insight into the activities of the research community around the world. China and the United States of America stand out as leaders in the number of papers [41], which is probably the result of their large population, wealth of resources, high investment in science and technology, and developed research infrastructure. Great Britain also has a significant amount of research, which may be the result of a rich scientific tradition, with renowned universities and institutions.

Countries like India, Australia, Canada, and Germany are also significant in the research community, with a solid amount of research. The numbers found may be the result of the population size, level of economic development, political priorities, or specific scientific interests of these countries. Iran, Singapore, and South Korea have a lower number of papers compared to the previously mentioned countries but are still significant in the global research community. These numbers may be a result of the relatively smaller populations of these countries or limited research resources.

The number of papers by country reflects the diversity of research papers around the world, with different countries contributing to scientific discoveries in different areas, according to their resources, priorities, expertise, specific characteristics, interests and capabilities.

In addition to the graph shown in Figures 6 and 7, we have also attached a map where you can see all the countries included in our filtering. We repeat, China and the United States of America stand out with great dominance, marked in darker blue.

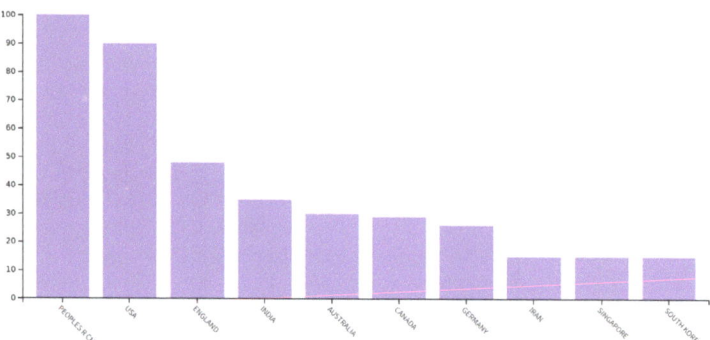

Figure 6. Countries.

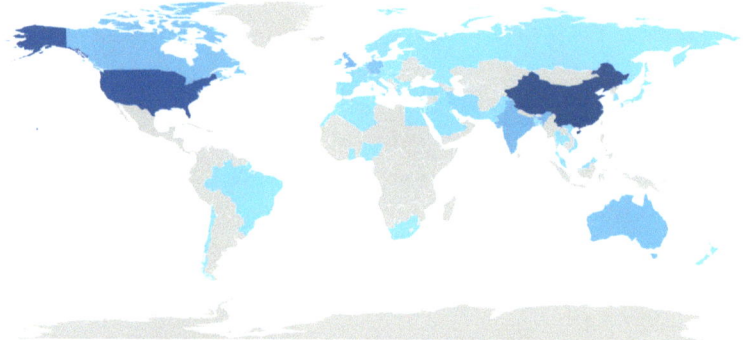

Figure 7. Country map.

In Figure 8, we can see that the United States of America was ahead of China from 2020 to mid-2022, while, from the middle of 2022, China took the lead [42]. Other selected countries had a slight increase in the number of scientific papers. It should also be noted that bilateral international visitation and academic exchange between the US and China have both significantly reduced, leading to a decline in Sino–American scientific collaboration. The number of American international students from China decreased by nearly 22%, while the number of American students studying in China dropped to only 1.8% of the 2018–2019 level. Despite this decline, influential research collaborations between the two nations remain consistently high [43].

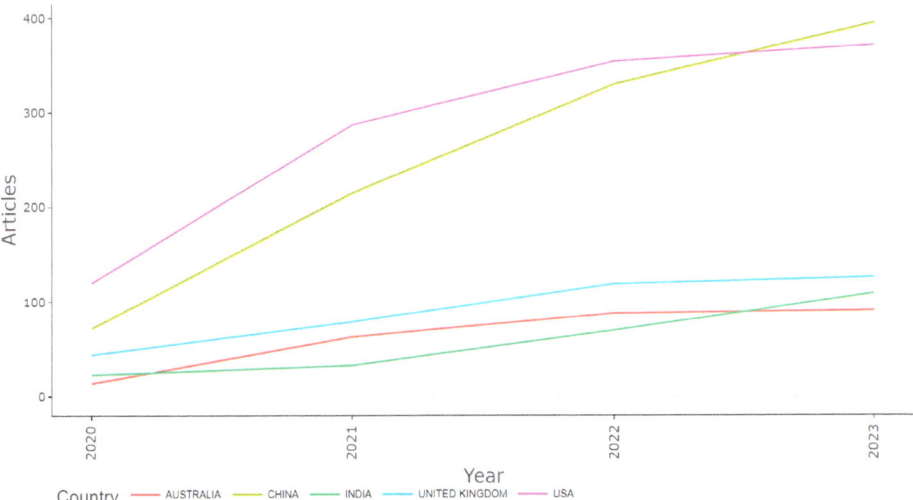

Figure 8. Overview of the number of papers by country and year.

The growth line for India is interesting, where continuous growth can be observed in the period from 2021 to 2023. On the other hand, Great Britain had an increase in the number of scientific papers in the period from 2020 to 2022, and, in the period from 2022 to 2023, the number of scientific papers was constant. The growth of the number of scientific papers from year to year is increasing, which indicates continuous progress in scientific research and activities to us. Growth can be the result of various factors, including technological progress, access to data, and the collaboration of scientific workers from all over the world.

3.6. Citations

Based on the citation trends of the last few years, 2024 represents a challenging and dynamic time in the world of scientific research. Although the number of citations in 2024 shown in Figure 9 is currently at the level of 1,000, it should be noted that this year has just begun and further growth in citations is expected as researchers publish new papers and their results become available to the general public. The previous trend of citation growth, especially the sharp increase from 2021 to 2023, suggests continued interest and activity in the scientific community. This increase may be the result of intensive research activities, increased cooperation between researchers and institutions, and the development of new technologies and methodologies that have stimulated progress in various fields of science. The period between 2020 and 2023 was marked by significant events and changes in the world, some of which directly affected the dynamics of scientific activity.

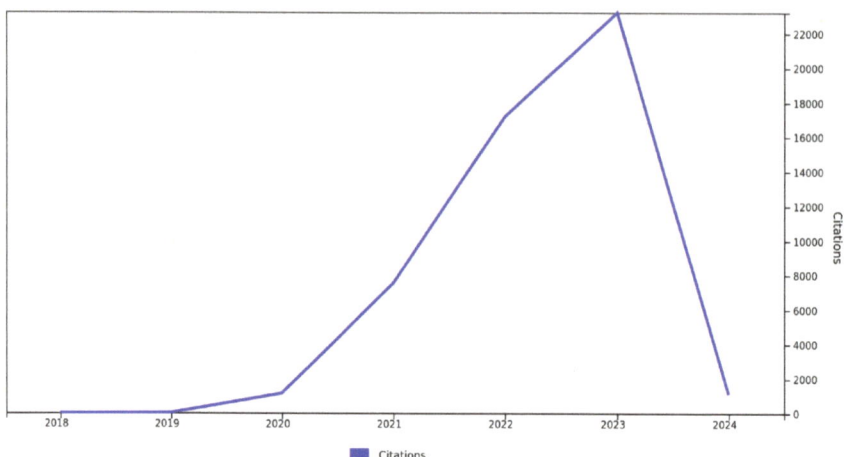

Figure 9. Number of citations.

RStudio displayed the years 2018 and 2019 with a value of 0 because these years were not included in our analysis. Alternatively, it is possible that there was an error in the Web of Science (WoS) in displaying the publication period.

Considering that the number of citations in 2024 is currently at the level of 1000, it can be assumed that this number will increase as researchers publish new papers and as these papers are integrated into existing scientific discourses. It is expected that the further development of research topics, technologies, and methodologies will contribute to the further growth of citations and the continuation of scientific progress in the future.

The Figure 10 shows the data as interpreted by RStudio. We assume that full surnames are displayed in combination with the first letter of the name. For example, "ZHANG Y" would refer to a person with the surname Zhang and the first letter of the name being Y. Based on this format, the data were organized and displayed in a graph.

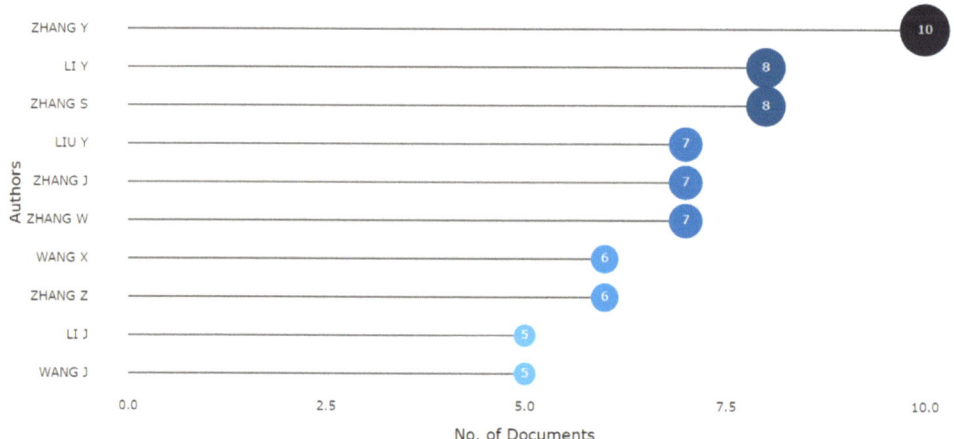

Figure 10. The most relevant authors.

3.7. Top Five Scientific Papers by Citations

In the period from 2020 to 2024, five scientific papers dealing with different aspects of machine learning were presented as. Shown in Table 2 The first article, "Review of deep learning: concepts, CNN architectures, challenges, applications, future directions" [44], recorded a constant growth of citations during the period, with a peak in 2023. Another article, "Physics-informed machine learning" [45], showed a significant increase in citations between 2020 and 2023. The third article, "Applications of machine learning to machine fault diagnosis: A review and roadmap" [46], recorded a gradual increase in citations throughout the period. The fourth article, "Machine Learning for Fluid Mechanics" [47], showed a steady level of citations over the years. The fifth article "Deep Learning for Anomaly Detection: A Review" [48] showed a significant increase in citations since 2021. The increase in the citation of these articles since 2021 can be attributed to the increasing popularity of artificial intelligence. From the end of 2021, artificial intelligence has become increasingly significant and present in various spheres of life, from technological innovations to everyday applications and changes, both in science and in practice. The increasing popularity of artificial intelligence has probably led to an increased interest in research and the application of machine learning, which is reflected by the increase in the number of citations for the analyzed articles. These articles, covering various aspects of machine learning, are becoming more relevant and in demand as artificial intelligence is increasingly integrated into our daily lives. The rise in the popularity of artificial intelligence is certainly due to OpenAI's ChatGPT, which appeared at the end of November 2022 [49].

Table 2. Papers and their citations.

Article	2020	2021	2022	2023	2024	Average per Year	Total
Review of deep learning: concepts, CNN architectures, challenges, applications, future directions	1	66	439	845	45	349	1.396
Physics-informed machine learning	0	35	485	777	33	332.5	1.330
Applications of machine learning to machine fault diagnosis: A review and roadmap	64	254	370	401	16	221	1.105
Machine Learning for Fluid Mechanics	88	247	347	361	17	213	1.065
Deep Learning for Anomaly Detection: A Review	1	112	266	311	13	175.75	703

Source: Author's own calculations.

4. Discussion

The importance of published papers and their impact on the scientific community can be viewed from several perspectives. First, published papers represent research results that contribute to the spread of knowledge and understanding in a particular field. By sharing new knowledge, researchers enable other scientists to expand their understanding, develop new theories or methods, and further explore topics covered in published papers. In addition, published papers are crucial for the academic reputation and advancement of scientists. Through the citation and recognition of their scientific papers by peers, scientists build their reputation in the scientific community, which can have a direct impact on their ability to obtain financial support for further research, access better resources, or advance their careers. Published papers have a significant impact on the improvement of educational institutions. By integrating the latest research results into teaching, educational institutions enable students to follow current trends and develop competencies that are relevant to the modern labor market. Published papers can serve as a basis for the development of new curricula or program areas, adapted to rapid changes in technology, scientific methodology, or social needs. In short, published papers are a pillar of the scientific community and a key element in the process of knowledge exchange, the development of scientists, and the improvement of educational institutions. Their importance lies in their ability to stimulate further research, enhance academic reputation, and enrich educational programs with the latest knowledge and perspectives.

5. Conclusions

The analysis of the dynamics of research activity, the number of publications by year, research area, countries, and citations provides deeper insight into the global picture of the scientific community and its current trends. The COVID-19 pandemic had a significant impact on research activity, resulting in a sharp increase in published papers in 2020 and 2021. However, the decline in the number of publications in 2022 and 2023 suggests a possible trend of reduced activity or a change in research priorities. The diversity of the interests and focuses of the scientific community is reflected in the number of papers in different fields. The high number of papers in fields such as engineering and computer science shows the wide application of these disciplines, while the lower number of papers in fields such as biochemistry may be the result of the specificity and complexity of the research in said field.

5.1. Theoretical Implications

From our findings, we concluded several important things that can help researchers understand how popular a paper is and how many times it has been cited. These results can be useful for researchers to explore new areas and find faster solutions to problems related to scientific papers popularity.

5.2. Practical Implications

Our results may be useful to researchers who want to expand their knowledge. For practitioners such as computer scientists seeking technological advancements, the research findings provide a deeper understanding of the challenges associated with scientific papers in specialized fields such as artificial intelligence.

5.3. Limitations and Future Research

This study is limited by some methodological obstacles that can be overcome by future research. We used one database for bibliometric analysis, which essentially reduced the number of papers that could be analyzed. The research period was limited to the last few years, and it would be interesting to see how much research has been carried out on this topic in the last few decades and how much of an influence these scientific papers have had on the development of this topic in scientific and professional circles. In this research, we used the WoS core collection, and for more data and future analysis, it would be useful to include other databases that could display more works and therefore give other results. This can potentially be a big challenge due to the inconsistency of data in different databases, and, for this reason, we have currently opted for the WoS database, which indexes the highest quality works. In the future, we will pay attention to this case and try to extract the maximum amount of data from other databases that are available to us at that moment. Future research should explore the impact of emerging technologies on research dynamics, such as artificial intelligence and machine learning, to see how they might change the landscape of bibliometric analysis. Additionally, investigating the effects of funding policies on publication rates could provide valuable insights into how financial support influences research output.

Overall, this analysis provides useful insights for researchers and institutions to better understand current research trends and identify areas that require additional attention and support in the future. In future research, we plan to use various databases and methods to better understand the complexity of the research environment.

Author Contributions: Conceptualization, L.I. and D.V.; methodology, D.V. and D.K.; software, A.Š. and L.I.; investigation, A.Š. and L.I.; validation, B.P. and D.K.; bibliometric analysis, L.I., D.V. and A.Š.; writing—original draft preparation, L.I. and D.V.; writing—review and editing, L.I., A.Š. and B.P.; visualization. L.I.; supervision; B.P. and D.K. All authors have read and agreed to the published version of the manuscript.

Funding: This work was partially supported by the Ministry of Science, Technological Development and Innovation of the Republic of Serbia [grant number 451-03-65/2024-03/200102].

Data Availability Statement: This manuscript is a Review, thus the data are temporary and cannot be replicated in further studies.

Conflicts of Interest: The authors declare no conflicts of interest.

References

1. Golenkov, V.; Guliakina, N.; Golovko, V.; Krasnoproshin, V. Artificial Intelligence Standardization Is a Key Challenge for the Technologies of the Future. In *Open Semantic Technologies for Intelligent System, Proceedings of the 10th International Conference, OSTIS 2020, Minsk, Belarus, 19–22 February 2020*; Golenkov, V., Krasnoproshin, V., Golovko, V., Azarov, E., Eds.; Communications in Computer and Information Science; Springer: Cham, Switzerland, 2020; Volume 1282. [CrossRef]
2. Ramírez, J.G.C.; Islam, M.M. Utilizing Artificial Intelligence in Real-World Applications. *J. Artif. Intell. Gen. Sci.* **2024**, *2*, 14–19.
3. Harikandeh, S.R.T.; Aliakbary, S.; Taheri, S. Towards Study of Research Topics Evolution in Artificial Intelligence based on Topic Embedding. In Proceedings of the 2021 11th International Conference on Computer Engineering and Knowledge (ICCKE), Mashhad, Iran, 28–29 October 2021; pp. 406–411. [CrossRef]
4. Dushyant, K.; Muskan, G.; Annu; Gupta, A.; Pramanik, S. Utilizing Machine Learning and Deep Learning in Cybersecurity: An Innovative Approach. In *Cyber Security and Digital Forensics*; Ghonge, M.M., Pramanik, S., Mangrulkar, R., Le, D.-N., Eds.; Scrivener Publishing LLC: Beverly, MA, USA, 2022. [CrossRef]
5. Wu, J.; Tran, N.K. Application of Blockchain Technology in Sustainable Energy Systems: An Overview. *Sustainability* **2018**, *10*, 3067. [CrossRef]
6. Zhao, L.; Zhang, L.; Wu, Z.; Chen, Y.; Dai, H.; Yu, X.; Liu, Z.; Zhang, T.; Hu, X.; Jiang, X.; et al. When brain-inspired AI meets AGI. *Meta-Radiol.* **2023**, *1*, 100005. [CrossRef]
7. Najafabadi, M.M.; Villanustre, F.; Khoshgoftaar, T.M.; Seliya, N.; Wald, R.; Muharemagic, E. Deep learning applications and challenges in big data analytics. *J. Big Data* **2015**, *2*, 1. [CrossRef]
8. Ascari, L.C.; Araki, L.Y.; Pozo, A.R.T.; Vergilio, S.R. Exploring machine learning techniques for fault localization. In Proceedings of the 2009 10th Latin American Test Workshop, Rio de Janeiro, Brazil, 2–5 March 2009; pp. 1–6. [CrossRef]
9. Sakhnini, J.; Karimipour, H.; Dehghantanha, A.; Parizi, R.M. AI and Security of Critical Infrastructure. In *Handbook of Big Data Privacy*; Choo, K.K., Dehghantanha, A., Eds.; Springer: Cham, Switzerland, 2020. [CrossRef]
10. Lampropoulos, G.; Siakas, K.; Anastasiadis, T. Internet of Things in the Context of Industry 4.0: An Overview. *Int. J. Entrep. Knowl.* **2019**, *7*, 4–19. [CrossRef]
11. Du, Y.; Sun, Z.; Hu, H. Security-Aware Collaboration Plan Recommendation for Dynamic Multiple Workflow Processes. *IEEE Trans. Dependable Secur. Comput.* **2023**, *20*, 100–113. [CrossRef]
12. Singh, S.K.; Rathore, S.; Park, J.H. BlockIoTIntelligence: A Blockchain-enabled Intelligent IoT architecture with Artificial Intelligence. *Future Gener. Comput. Syst.* **2019**, *110*, 721–743. [CrossRef]
13. Sgantzos, K.; Grigg, I. Artificial Intelligence Implementations on the Blockchain. Use Cases and Future Applications. *Future Internet* **2019**, *11*, 170. [CrossRef]
14. Liu, W. The data source of this study is Web of Science Core Collection? Not enough. *Scientometrics* **2019**, *121*, 1815–1824. [CrossRef]
15. Gilardoni, S.; Di Mauro, B.; Bonasoni, P. Black carbon, organic carbon, and mineral dust in South American tropical glaciers: A review. *Glob. Planet. Chang.* **2022**, *213*, 103837. [CrossRef]
16. Page, M.J.; McKenzie, J.E.; Bossuyt, P.M.; Boutron, I.; Hoffmann, T.C.; Mulrow, C.D.; Shamseer, L.; Tetzlaff, J.M.; Moher, D. Updating guidance for reporting systematic reviews: Development of the PRISMA 2020 statement. *J. Clin. Epidemiol.* **2021**, *134*, 103–112. [CrossRef]
17. Arksey, H.; O'Malley, L. Scoping studies: Towards a methodological framework. *Int. J. Soc. Res. Methodol.* **2005**, *8*, 19–32. [CrossRef]
18. Ahmed, S.; Alshater, M.M.; Ammari, A.E.; Hammami, H. Artificial intelligence and machine learning in finance: A bibliometric review. *Res. Int. Bus. Financ.* **2022**, *61*, 101646. [CrossRef]
19. Kitchenham, B.A.; Mendes, E.; Travassos, G.H. Cross versus within-company cost estimation studies: A systematic review. *IEEE Trans. Softw. Eng.* **2007**, *5*, 316–329. [CrossRef]
20. van Eck, N.; Waltman, L. Software survey: VOSviewer, a computer program for bibliometric mapping. *Scientometrics* **2010**, *84*, 523–538. [CrossRef]

21. Hosseini, M.R.; Martek, I.; Zavadskas, E.K.; Aibinu, A.A.; Arashpour, M.; Chileshe, N. Critical evaluation of off-site construction research: A Scientometric analysis. *Autom. Constr.* **2018**, *87*, 235–247. [CrossRef]
22. Cobo, M.J.; López-Herrera, A.G.; Herrera-Viedma, E.; Herrera, F. Science mapping software tools: Review, analysis, and cooperative study among tools. *J. Am. Soc. Inf. Sci. Technol.* **2011**, *62*, 1382–1402. [CrossRef]
23. Donthu, N.; Kumar, S.; Mukherjee, D.; Pandey, N.; Lim, W.M. How to conduct a bibliometric analysis: An overview and guidelines. *J. Bus. Res.* **2021**, *133*, 285–296. [CrossRef]
24. Injadat, M.; Moubayed, A.; Nassif, A.B.; Shami, A. Machine learning towards intelligent systems: Applications, challenges, and opportunities. *Artif. Intell. Rev.* **2021**, *54*, 3299–3348. [CrossRef]
25. Dobbe, R.; Gilbert, T.K.; Mintz, Y. Hard choices in artificial intelligence. *Artif. Intell.* **2021**, *300*, 103555. [CrossRef]
26. Kujundžić, M. Analiza Modela Predviđanja Trendova Kretanja Cijena Vrijednosnih Papira. Ph.D. Thesis, Tehnički Fakultet, Sveučilište u Rijeci, Rijeka, Croatia, 2015. Available online: https://urn.nsk.hr/urn:nbn:hr:190:021799 (accessed on 13 April 2024).
27. Siam, A.; Ezzeldin, M.; El-Dakhakhni, W. Machine learning algorithms for structural performance classifications and predictions: Application to reinforced masonry shear walls. *Structures* **2019**, *22*, 252–265. [CrossRef]
28. Shi, B.; Bai, X.; Yao, C. An End-to-End Trainable Neural Network for Image-Based Sequence Recognition and Its Application to Scene Text Recognition. *IEEE Trans. Pattern Anal. Mach. Intell.* **2017**, *39*, 2298–2304. [CrossRef] [PubMed]
29. Taye, M.M. Understanding of Machine Learning with Deep Learning: Architectures, Workflow, Applications and Future Directions. *Computers* **2023**, *12*, 91. [CrossRef]
30. LeCun, Y.; Bengio, Y.; Hinton, G. Deep learning. *Nature* **2015**, *521*, 436–444. [CrossRef] [PubMed]
31. Hou, J.-C.; Wang, S.-S.; Lai, Y.-H.; Tsao, Y.; Chang, H.-W.; Wang, H.-M. Audio-Visual Speech Enhancement Using Multimodal Deep Convolutional Neural Networks. *IEEE Trans. Emerg. Top. Comput. Intell.* **2018**, *2*, 117–128. [CrossRef]
32. Graves, A.; Mohamed, A.R.; Hinton, G. Speech recognition with deep recurrent neural networks. In Proceedings of the 2013 IEEE International Conference on Acoustics, Speech and Signal Processing, Vancouver, BC, Canada, 26–31 May 2013; pp. 6645–6649. [CrossRef]
33. Kochovski, P.; Sakellariou, R.; Bajec, M.; Drobintsev, P.; Stankovski, V. An Architecture and Stochastic Method for Database Container Placement in the Edge-Fog-Cloud Continuum. In Proceedings of the 2019 IEEE International Parallel and Distributed Processing Symposium (IPDPS), Rio de Janeiro, Brazil, 20–24 May 2019; pp. 396–405. [CrossRef]
34. Tyagi, A.K.; Chahal, P. Artificial Intelligence and Machine Learning Algorithms. In *Research Anthology on Machine Learning Techniques, Methods, and Applications*; Information Resources Management Association, Ed.; IGI Global: Hershey, PA, USA, 2022; pp. 421–446. [CrossRef]
35. Dillmann, R.; Friedrich, H. Programming by demonstration: A machine learning approach to support skill acquision for robots. In *Artificial Intelligence and Symbolic Mathematical Computation, Proceedings of the AISMC 1996, Steyr, Austria, 23–25 September 1996*; Lecture Notes in Computer Science; Calmet, J., Campbell, J.A., Pfalzgraf, J., Eds.; Springer: Berlin/Heidelberg, Germany, 1996; Volume 1138, p. 1138. [CrossRef]
36. Gelenbe, E.; Yin, Y. Deep learning with random neural networks. In Proceedings of the 2016 International Joint Conference on Neural Networks (IJCNN), Vancouver, BC, Canada, 24–29 July 2016; pp. 1633–1638. [CrossRef]
37. Antsaklis, P.J. Neural Networks in Control Systems. *IEEE Control Syst. Mag.* **1990**, *10*, 3–5. [CrossRef]
38. Cioffi, R.; Travaglioni, M.; Piscitelli, G.; Petrillo, A.; De Felice, F. Artificial Intelligence and Machine Learning Applications in Smart Production: Progress, Trends, and Directions. *Sustainability* **2020**, *12*, 492. [CrossRef]
39. Nor, R.B.M. AI Applications in Education, Healthcare, and Transportation Trends, Challenges, and Future Directions. *AI IoT Fourth Ind. Revolut. Rev.* **2023**, *13*, 42–51. Available online: https://scicadence.com/index.php/AI-IoT-REVIEW/article/view/50 (accessed on 27 April 2024).
40. Pratici, L.; Singer, P.M. COVID-19 Vaccination: What Do We Expect for the Future? A Systematic Literature Review of Social Science Publications in the First Year of the Pandemic (2020–2021). *Sustainability* **2021**, *13*, 8259. [CrossRef]
41. Xiao, G.; Yang, D.; Xu, L.; Li, J.; Jiang, Z. The Application of Artificial Intelligence Technology in Shipping: A Bibliometric Review. *J. Mar. Sci. Eng.* **2024**, *12*, 624. [CrossRef]
42. Zhu, J.; Liu, W. Comparing like with like: China ranks first in SCI-indexed research articles since 2018. *Scientometrics* **2020**, *124*, 1691–1700. [CrossRef] [PubMed]
43. Tang, L. Halt the ongoing decoupling and reboot US-China scientific collaboration. *J. Informetr.* **2024**, *18*, 101521. [CrossRef]
44. Alzubaidi, L.; Zhang, J.; Humaidi, A.J.; Al-Dujaili, A.; Duan, Y.; Al-Shamma, O.; Santamaría, J.; Fadhel, M.A.; Al-Amidie, M.; Farhan, L. Review of deep learning: Concepts, CNN architectures, challenges, applications, future directions. *J. Big Data* **2021**, *8*, 53. [CrossRef] [PubMed]
45. Karniadakis, G.E.; Kevrekidis, I.G.; Lu, L.; Perdikaris, P.; Wang, S.; Yang, L. Physics-informed machine learning. *Nat. Rev. Phys.* **2021**, *3*, 422–440. [CrossRef]
46. Lei, Y.; Yang, B.; Jiang, X.; Jia, F.; Li, N.; Nandi, A.K. Applications of machine learning to machine fault diagnosis: A review and roadmap. *Mech. Syst. Signal Process.* **2020**, *138*, 106587. [CrossRef]
47. Brunton, S.L.; Noack, B.R.; Koumoutsakos, P. Machine Learning for Fluid Mechanics. *Annu. Rev. Fluid Mech.* **2020**, *52*, 477–508. [CrossRef]

48. Pang, G.; Shen, C.; Cao, L.; Hengel, A.V.D. Deep Learning for Anomaly Detection: A Review. *ACM Comput. Surv.* **2021**, *54*, 38. [CrossRef]
49. Roumeliotis, K.I.; Tselikas, N.D. ChatGPT and Open-AI Models: A Preliminary Review. *Future Internet* **2023**, *15*, 192. [CrossRef]

Disclaimer/Publisher's Note: The statements, opinions and data contained in all publications are solely those of the individual author(s) and contributor(s) and not of MDPI and/or the editor(s). MDPI and/or the editor(s) disclaim responsibility for any injury to people or property resulting from any ideas, methods, instructions or products referred to in the content.

Article

Enhancing IoT Security: Optimizing Anomaly Detection through Machine Learning

Maria Balega [1,2,*], Waleed Farag [1], Xin-Wen Wu [3], Soundararajan Ezekiel [1] and Zaryn Good [1]

1. Department of Mathematical and Computer Sciences, Indiana University of Pennsylvania, Indiana, PA 15705, USA; farag@iup.edu (W.F.); sezekiel@iup.edu (S.E.); zaryngood@outlook.com (Z.G.)
2. Information Networking Institute, Carnegie Mellon University, Pittsburgh, PA 15289, USA
3. Department of Computer Science, University of Mary Washington, Fredericksburg, VA 22401, USA; xwu@umw.edu
* Correspondence: mbalega@andrew.cmu.edu

Abstract: As the Internet of Things (IoT) continues to evolve, securing IoT networks and devices remains a continuing challenge. Anomaly detection is a crucial procedure in protecting the IoT. A promising way to perform anomaly detection in the IoT is through the use of machine learning (ML) algorithms. There is a lack of studies in the literature identifying optimal (with regard to both effectiveness and efficiency) anomaly detection models for the IoT. To fill the gap, this work thoroughly investigated the effectiveness and efficiency of IoT anomaly detection enabled by several representative machine learning models, namely Extreme Gradient Boosting (XGBoost), Support Vector Machines (SVMs), and Deep Convolutional Neural Networks (DCNNs). Identifying optimal anomaly detection models for IoT anomaly detection is challenging due to diverse IoT applications and dynamic IoT networking environments. It is of vital importance to evaluate ML-powered anomaly detection models using multiple datasets collected from different environments. We utilized three reputable datasets to benchmark the aforementioned machine learning methods, namely, IoT-23, NSL-KDD, and TON_IoT. Our results show that XGBoost outperformed both the SVM and DCNN, achieving accuracies of up to 99.98%. Moreover, XGBoost proved to be the most computationally efficient method; the model performed 717.75 times faster than the SVM and significantly faster than the DCNN in terms of training times. The research results have been further confirmed by using our real-world IoT data collected from an IoT testbed consisting of physical devices that we recently built.

Keywords: anomaly detection; DCNN; Internet of Things (IoT); machine learning (ML); SVM; XGBoost

Citation: Balega, M.; Farag, W.; Wu, X.-W.; Ezekiel, S.; Good, Z. Enhancing IoT Security: Optimizing Anomaly Detection through Machine Learning. *Electronics* 2024, 13, 2148. https://doi.org/10.3390/electronics13112148

Academic Editors: Abdussalam Elhanashi and Pierpaolo Dini

Received: 12 March 2024
Revised: 25 May 2024
Accepted: 27 May 2024
Published: 31 May 2024

Copyright: © 2024 by the authors. Licensee MDPI, Basel, Switzerland. This article is an open access article distributed under the terms and conditions of the Creative Commons Attribution (CC BY) license (https://creativecommons.org/licenses/by/4.0/).

1. Introduction

First presented by Kevin Ashton in 1999, the Internet of Things is an extension of the internet to objects or things that are traditionally not interconnected [1,2]. With the technological advances in the decades since, the number of these devices in use has grown exponentially [3]. By 2025, it is projected that there will be around 30.9 billion connected devices which is an increase from the roughly 13.8 billion devices in 2021 [4].

IoT devices are usually connected through wireless communications to various networks for which network nodes can transmit data and interact with one another or with centralized devices. As the Internet of Things continues to evolve, it poses risks and the need for greater security measures. Securing IoT networks and devices is a constant challenge as the development of the IoT is moving faster than the creation of defenses for the devices and users themselves [3,5].

The deployment of IoT applications makes protection more challenging with the increased attack surface as well as the vulnerable and resource-constrained end devices. IoT applications pose a greater attack surface due to the extensive connectivity as well as the deployment of end devices that are not protected or minimally protected by the security protocols commonly applied to traditional computer networks. Additionally, many IoT networks and end devices do not have sufficient computational capabilities to integrate advanced firewalls, antivirus software, and authentication processes. Furthermore, in IoT applications connected with important systems or critical infrastructure, real-time security monitoring and instant risk responses are highly desirable. With these vulnerabilities, constraints, and risk response requirements, securing IoT applications remains a significant challenge [6,7].

Anomaly detection is a crucial procedure in protecting networked systems as it allows for the identification of unusual or abnormal behavior within a network of connected devices. This is particularly important in the context of the IoT, where the sheer volume of data generated by vulnerable devices can make it difficult to identify and respond to potential security threats [8,9].

1.1. Contributions

Machine learning (ML) methods show promise for anomaly detection in IoT systems, as observed in the related works detailed in the next section. Although existing works explored the applications of various classical ML models to IoT anomaly detection, there is a lack of investigations regarding the efficiency of ML-powered anomaly detection. Also, some of the most promising ML models, such as Extreme Gradient Boosting (XGBoost), were not thoroughly investigated with regard to the detection accuracy and efficiency. There is a great gap in this area with regard to identifying optimal (with regard to both effectiveness and efficiency) anomaly detection models for the Internet of Things. To fill the gap, this work thoroughly investigated the effectiveness and efficiency of IoT anomaly detection enabled by several representative machine learning models, namely XGBoost, Support Vector Machines (SVMs), and Deep Convolutional Neural Networks (DCNNs).

Previous studies show that SVMs, DCNNs, and several other ML models are effective in classifying data and detecting anomalies [6,7]. However, these models were not thoroughly studied regarding efficiency. Identifying the optimal anomaly detection models for the IoT is indeed challenging due to the diverse IoT applications and dynamic IoT networking environments. Even if only considering a specific type of IoT application, such as IoT-facilitated smart buildings or intelligent transportations systems, the IoT systems working in dynamic environments produce very different data. Therefore, it is of vital importance to evaluate ML-powered anomaly detection models using multiple datasets collected from different environments. We utilized three reputable datasets to benchmark the machine learning methods, namely, IoT-23, NSL-KDD, and TON_IoT [10–12]. IoT-23 is a dataset that contains 23 different captures of network-based attacks on IoT devices, making it a useful resource for evaluating the performance of anomaly detection algorithms. The NSL-KDD dataset was specifically designed for use in network security research, and it contains a wide range of network-based attacks. TON_IoT is a dataset that is designed to reflect the traffic patterns of IoT devices, making it a valuable resource for understanding the behavior of IoT networks. These datasets have proven to be a great representation of real-world IoT systems and attacks [13].

In this study, each machine learning algorithm was assessed based on accuracy, precision, recall, and F1 score. Our experimental results show that the XGBoost algorithm outperformed both the SVM and DCNN algorithms, achieving accuracies up to 99.98%. We further studied the efficiency of anomaly detection models powered by these learning algorithms. XGBoost proved to be the most efficient method with respect to execution time. While DCNNs and SVMs are more computationally expensive, XGBoost trains data much more quickly. As XGBoost is capable of handling large amounts of data effectively and efficiently, it proves to be the best anomaly detection model when compared against SVMs and DCNNs.

Recently, we have built a testbed with physical IoT devices, including cameras, an Amazon echo, smart plugs, and other IoT sensors and collected data from them. We have tested the machine learning models using our real-world IoT data, and our results confirm and support the research results obtained through using the IoT-23, NSL-KDD, and TON_IoT datasets.

Our evaluation of ML-powered anomaly detection models using real-world data proves that XGBoost can be used to detect anomalies efficiently and accurately in real-world IoT applications. The results of this research are expected to be helpful to individual users and organizations to identify and implement the most effective and efficient anomaly detection systems to secure their IoT applications. Our study offers invaluable insights directly applicable to protecting IoT applications, thereby enhancing the overall security and resilience of these systems and enabling users and organizations to take proactive measures in safeguarding them.

In summary, the contributions of this paper include (1) evaluation of machine learning-powered anomaly detection models for the Internet of Things regarding both effectiveness and efficiency; (2) investigation of XGBoost in IoT anomaly detection applications and proof that it is an effective and efficient model; (3) utilization of distinct datasets in an evaluation of ML-powered IoT anomaly detection models and verification of the models using real-world data collected from an IoT testbed that we recently built; and (4) identification of an optimal anomaly detection model, enabling users and organizations to take proactive measures in safeguarding their IoT applications.

1.2. Paper Layout

The rest of the paper is organized as follows. In Section 2, works related to this study will be presented. In Section 3, the anomaly detection models and datasets used in this study will be described as well as the preprocessing and training of each dataset. In Section 4, the evaluation metrics used for assessing each model will be presented. In Section 5, the results from each of the machine learning models will be compared. Efficiency results, verification of the models using real-world data, and limitations will be discussed in Section 6, and concluding remarks and future work will be presented in Section 7.

2. Related Works

Datasets from real-world IoT applications are not readily available for research purposes, primarily due to privacy concerns. The lack of availability of labeled IoT datasets presents a difficulty in the development of anomaly or intrusion detection systems [14]. Because of this gap, IoT datasets have been produced recently in the research community, including those used in our study.

An attempt to apply machine learning for securing the IoT was reported in [15]. Due to the large amount of network traffic and data collected through various IoT devices, researchers realized that highly effective algorithms are desirable for accurate classification and regression when using machine learning for IoT security [16]. Supervised and unsupervised machine learning methods were further investigated for enhancing IoT security [17].

Several popular machine learning methods and deep learning models, namely, Support Vector Machines (SVMs), k-Nearest Neighbor (kNN), Naïve Bayes (NB), Random Forest (RF), Classification and Regression Trees (CARTs), Logistics Regression (LR), and Linear Discriminant Analysis (LDA) were evaluated regarding the performance of intrusion detection using the TON_IoT dataset [14]. As TON_IoT is made up of smaller datasets, this study compared the performance of the detection models on each dataset as well as a combination of them. The results show that for binary clarification and multiclassification on the combined datasets, CARTs performed best. Specifically, for binary classification, accuracies up to 88% were achieved by this detection model; for multiclassification, accuracies up to 77% were achieved. Random Forest was the second best, achieving accuracies up to 85% for binary classification and 71% for multiclassification. Overall, this study found that CARTs and Random Forest achieved the highest score in all evaluation metrics with separate datasets as well as when the datasets were combined [14].

In 2021, machine learning algorithms such as SVM, Gradient Boosting techniques, Isolation Forest, and Deep Learning Networks were assessed in terms of their capabilities for detecting intrusion using the IoT-23 dataset [18]. In a recent paper [19], the authors provided a review of existing works on developing anomaly detection for the IoT using the KNN, SVM, Bayesian Network, and Neural Network machine learning algorithms.

Recently, interesting results on IoT intrusion detection were published [8,9]. In these papers, the authors examined NetFlow features and assessed their suitability for classifying network traffic. They utilized multiple datasets, containing IoT traffic data represented in a standard set of 43 NetFlow-based features. Reducing the number of features from 43 to 7 can enhance the prediction time and, consequently, the performance in the real world. With the reduced number of features, their network intrusion detection system (based on an ML model boosted by a modified version of the Arithmetic Optimization Algorithm) achieved up to 99% and 98% accuracy for binary and multi-classification, respectively.

However, the efficiency of anomaly detection of different detection models was not thoroughly investigated in these works. In [7], published in 2022, the authors studied the effectiveness as well as efficiency of the following ML models for IoT anomaly detection: Logistic Regression, Decision Tree, Random Forest, Gradient Boosting Machine, and Naïve Bayes. However, like the abovementioned papers, they only evaluated these ML models using a single dataset, that is, a Kaggle dataset. Also, this paper did not evaluate the well-known SVM and DCNN and only studied binary classification anomaly detection.

In our previous work, ML anomaly detection models for IoT were evaluated using the IoT-23 dataset [20], and the models were evaluated using the NSL-KDD dataset in another paper [21]. The present paper will extend these results to include more datasets. A comparison of previous publications and this present paper can be found in Table 1.

Table 1. Comparison of prior publications vs. this paper.

Publications vs. This Paper	Dataset Features Selected (to Increase Detection Accuracy)?	Identified the Optimal Detection Model Regarding Accuracy?	Multiple Datasets Used?	Detection Efficiency Thoroughly Studied?
[7]	N/A *	Compared ML models regarding accuracy and identified the most accurate ML model	No	Compared ML models regarding accuracy and identified the most accurate ML model
[8]	Yes	Studied only one detection model	Yes	N/A
[9]	Yes	Studied only one detection model	Yes	N/A
[15]	No	Yes	No	No
[19]	No	Yes	No	No
[20]	No	Yes	No	No
[21]	No	Yes	No	No
This paper	Yes	Compared ML models regarding accuracy and identified the most accurate ML model	Used multiple datasets to evaluate the ML models' performance in different environments	Compared ML models regarding accuracy and identified the most accurate ML model

* The datasets used in [7] contain a small number of features, therefore feature selection was unnecessary.

With diverse IoT applications and dynamic IoT networking environments, IoT systems may produce very different data. Therefore, it is of vital importance to evaluate ML-powered anomaly detection models using more than one dataset collected from distinct environments. This motivates us to investigate the most accurate anomaly detection models that are effective on different datasets, and we strive to identify the most efficient models that suit IoT applications with resource-constrained devices.

3. Materials and Methods

3.1. Learning Models for Anomaly Detection

Machine learning algorithms can be broken into three primary parts: a decision process that makes a prediction or classification based on some input data, an error function to evaluate the predictions and adjust for accuracy, and a model optimization process that adds or adjusts weights to various factors to reduce discrepancies between the model's estimates and the known examples. Using the input data, whether labeled or unlabeled, a learning algorithm creates an estimate of patterns discovered in the data. The error function then evaluates the prediction of the model, comparing the known examples to assess the model's accuracy. If the model can better fit data in the training set, then the weights are adjusted to optimize the model. This process is repeated until a threshold accuracy is found [22].

In this study, supervised machine learning models were used which apply the use of labeled datasets to classify data. The learning models chosen include XGBoost, SVMs, and DCNNs. These models have proven to be effective as they have been applied in many machine learning challenges. Particularly, XGBoost has been widely used by data scientists and it has shown promising results [23].

3.1.1. Extreme Gradient Boosting

Extreme Gradient Boosting or XGBoost is an implementation of Gradient Boosting which is used for regression and classification problems. Gradient Boosting creates a prediction model through an ensemble of weak prediction models. It generalizes them using the differentiable loss function [24]. XGBoost differs from Gradient Boosting in that it uses a more regularized model to help control overfitting. It considers the complexity of the model, adding terms to limit the growth of the tree which aids in better performance [25].

XGBoost minimizes an objective function (Equation (1)) which combines the loss function, L, and a penalty or regularization term, Ω, for the complexity and overfitting. This is an iterative process where new trees are added to predict the errors of prior trees. These are all combined to make the final prediction. The loss function (Equation (2)) used is the logistic loss, where y_i is the labels.

$$obj(\theta) = L(\theta) + \Omega(\theta) \tag{1}$$

$$L(\theta) = \Sigma_i [y_i \ln(1) + e^{-y_i} + (1 - y_i) \ln(1) + e^{y_i}] \tag{2}$$

XGBoost uses general parameters, booster parameters, and learning task parameters. General parameters relate to the booster being used. This is typically a tree or linear model. Booster parameters are chosen based on the specific type of booster being used. Significant booster parameters are η and max depth. η is the learning rate which helps to prevent overfitting the data. η ranges from zero to one. Max depth refers to the maximum depth of the tree. Increasing this value creates a more complex model. Max depth ranges from zero to infinity. Learning task parameters are based on the learning scenario. With multiclassification problems, the learning task parameter SoftMax is used [26].

3.1.2. Support Vector Machines

Support Vector Machines focus on mapping inputs non-linearly to a high-dimension feature space, Z, where a linear decision surface is constructed. The goal is to find an optimal hyperplane separating the data as generally as possible. In 1965, Vapnik solved an optimal hyperplane for separable classes [27]. The small amount of training data or support vectors determines the margin of the largest separation. The training data are separated into different groups with small training vectors in each proportion while solving the quadratic objective function. This will determine whether either the portion cannot be separated, or an optimal hyperplane has been found.

Typically, there is no exact single line that perfectly divides real data, and a curved boundary is used. For non-linear inputs, the "kernel trick" can be used. The kernel is defined by K(x,x'), where x and x' are the inputs and K is the kernel used [28]. There are a variety of common kernels to choose from. The linear kernel (Equation (3)) is the fastest but least accurate for multiple classes. The polynomial kernel (Equation (4)) is used for non-linear models, where d is the degree of the polynomial. Most often used is the radial basis function (Equation (5)), as it overcomes the space complexity issues of SVMs. Sigmoid or Multi-Layer Perceptron (Equation (6)) uses a single layer, where ρ is the slope and g is the intercept. This kernel is preferred in Neural Networks.

$$K(x,x') = {}^X\langle x,x'\rangle \tag{3}$$

$$K(x,x') = (\langle x,x'\rangle + 1)^d \tag{4}$$

$$K(x,x\prime) = \exp\frac{-||x-x'||^2}{2\sigma^2} \tag{5}$$

$$K(x,x') = \tanh(\rho\langle x,x'\rangle + g) \tag{6}$$

When creating a classifier, tuning parameters C and Gamma are used. C controls the tradeoff between a smooth decision boundary and correctly classifying training points. Gamma defines how far the influence of a single training example reaches. When set to a lower value, this means every point has a far reach. When set to a higher value, this means each point has a close reach. Gamma ranges in value from zero to one [29].

3.1.3. Deep Convolutional Neural Networks

Deep Convolutional Neural Networks or DCNNs use a three-dimensional neural pattern to identify patterns in image and video data. DCNNs are comprised of various types of layers that each perform specific tasks. The first type of layer is the convolutional layer which passes a filter over the image multiple times, performing matrix multiplication with weights and inputs. These values are added together to obtain a value for each filter position.

Another type of layer is the non-linear activation layer which removes negative numbers from the convolutional layer then maps and replaces them with zeros. The pooling layer is used to shrink the image by removing noncritical information. This can be achieved using max pooling or average pooling. With max pooling, only the maximum value in a group of pixels is preserved. With average pooling, only the average value in a group of pixels is preserved.

The fully connected layer is the final type of layer. The fully connected layer takes an input vector of the filtered images and reduced pixels that have been flattened and applies a SoftMax function. This layer is what classifies the image.

The DCNN architectures used in this study were AlexNet, GoogleNet, ResNet-50, VGG16, and VGG19 [30]. The number of layers for each of these networks is presented in Table 2.

Table 2. DCNN network architectures.

Network	Depth/Layers	Parameters	Image Size
AlexNet	8	61	227×227
GoogleNet	22	4	224×224
ResNet-50	50	23	224×224
VGG16	16	138	224×224
VGG19	19	144	224×224

3.2. Datasets

In this study, three datasets containing data from IoT devices and networks were adopted to perform anomaly detection using machine learning algorithms. The first dataset used was the IoT-23 dataset captured by the Stratosphere Laboratory [6]. This dataset was created for researchers to develop and apply machine learning algorithms. The second dataset used was the NSL-KDD dataset, which is an improved version of the KDD '99 dataset. The NSL-KDD dataset is suggested to solve some inherent problems in the KDD '99 dataset. While this new version still suffers from some problems, it is still an effective benchmark dataset that can be used for research [11]. Lastly, the TON_IoT dataset was used. This dataset consists of new generations of IoT and Industrial IoT data for evaluating the fidelity and efficiency of various cybersecurity applications such as artificial intelligence and machine learning [12].

3.2.1. IoT-23 Dataset

The IoT-23 dataset consists of network traffic from IoT devices. It contains 20 malicious captures and three benign captures. The malware captures are from infected IoT devices, and the benign captures are from real IoT devices. For the malicious captures, a specific malware sample was executed in a Raspberry Pi. Data for benign captures were captured through IoT devices, including a Philips HUE smart LED lamp, an Amazon Echo, and a Somfy smart door lock. These three IoT devices are real hardware and were not simulated [10].

Each capture in the dataset has 23 features. Of these 23 features, the use of 13 were focused on in this study. For a complete list of the columns and their descriptions, see [31]. This entire dataset is roughly 90.5% malicious traffic, containing a total of 325,307,990 entries [31]. Multiple types of malicious labels were created by the Stratosphere Lab. The most common labels include PartOfAHorizontalPortScan, Okiru, and DDoS and the least common include C&C-Mirai, PartOfAHorizontalPortScan-Attack, and C&C-HeartBeat-FileDownload [10].

3.2.2. NSL-KDD Dataset

The NSL-KDD dataset was built based on data captured in the DARPA '98 IDS evaluation program and is an improved version of the KDD CUP '99 dataset [32]. DARPA '98 consists of roughly four gigabytes of compressed binary tcpdump data captured from seven weeks of network traffic. The KDD CUP '99 training data consists of around 4,900,000 entries containing 41 features and are labeled as normal or attack [33].

While the KDD CUP '99 dataset has been used for many years, it encounters numerous problems. For example, synthetic data were used for the background and attack data when creating the dataset. The data also do not represent real networks as the workload of data does not seem similar to traffic that occurs in real networks. Further, data are only categorized as normal or attack rather than by the type of specific attack [33]. The KDD CUP '99 dataset also contains a large number of redundant records, as about 78% and 75% are duplicated in the train and test sets, respectively [34]. To address these issues, the NSL-KDD dataset was created.

3.2.3. TON_IoT Dataset

The TON_IoT dataset was created in response to testing the efficiency and preciseness of cybersecurity applications, including the use of machine learning algorithms. This dataset is used for validating and testing applications such as intrusion detection systems, threat intelligence, malware detection, fraud detection, privacy preservation, digital forensics, adversarial machine learning, and threat hunting [12].

The TON_IoT dataset consists of several datasets where data were collected from Telemetry datasets of IoT services, operating systems such as Windows and Linux, and network traffic. The IoT data were collected from more than ten IoT sensors. The Linux data were collected by running a tracing tool on Ubuntu 14 and 18 systems. Finally, the Windows data were captured on Windows 7 and 10 systems. These datasets were preprocessed as csv files with a label indicating normal or attack [35]. The types of attacks present include Backdoor, DDoS, Injection, Normal, Password, Ransomware, Scanning, and XSS.

3.3. Data Preprocessing and Model Training

3.3.1. Preprocessing the IoT-23 Dataset

In this study, we focused on the use of the conn.log.labeled files from the IoT-23 dataset. Each conn.log.labeled file was converted into a csv file and then the twenty-three csv files were combined into one large csv file. The number of benign and malicious items was calculated to create balanced data. As the IoT-23 dataset contains mostly malicious data, a random sample was taken from the malicious data that equates to the amount of benign data. The total number of entries including benign and malicious entries after random sampling was 61,721,287.

Next, the empty values containing a dash were replaced with 'NaN' so that the machine learning algorithms could handle them. Each column was assessed to determine whether it would be useful for predicting anomalies. Columns containing over 70% NaN values were dropped along with columns containing unnecessary and repetitive information. Then, categorical data were label encoded and the detailed label column was encoded to create four total classifications, as shown in Table 3.

Table 3. IoT-23 classifications.

Label Encoding	Type of Capture
0	Benign
1	DDoS
2	Okiru
3	PartOfAHorizontalPortScan

These four specific captures were the most accounted for in our sample. As there were other detailed labels in our sample, these were dropped because there were very few of them. We also included IP addresses where the attack happened and those of the device on which the capture happened. Due to the unique response IPs, we chose to include these features in the data.

As SVMs cannot process NaN values like XGBoost, all rows of data containing any NaN value were dropped, leaving 13,183,092 entries for training. Additionally, float values in the duration column exceeded the values that SVM could process, so it was dropped as well. After preprocessing, 12 features were assessed with 13,183,092 entries of data which were used to train the XGBoost and DCNN models. However, due to technical limitations, samples of 12%, 16%, and 20% were taken for the SVM model. As for the distribution of the dataset, 32.1% of the data were benign, 44.6% were PartOfAHorizontalPortScan, 10.9% were Okiru, and 12.4% were DDoS.

3.3.2. Preprocessing the NSL-KDD Dataset

For the NSL-KDD dataset, the data were already in csv file format. The data were also already split into training and testing files. Empty values that contained a dash were replaced with 'NaN'. Then, each column was assessed to determine whether it would be useful for predicting anomalies. Numerous columns were dropped as they statistically did not contribute to the training data or there was an overwhelming number of NaN values.

Once the unnecessary columns were dropped, categorical data were label encoded, with the label column consisting of 23 classifications of which 22 were malicious and one was normal. There was no inclusion of IP addresses in the preprocessed data to avoid bias. After preprocessing the NSL-KDD dataset, 27 features were assessed with 125,973 entries of data containing no NaN values. As for the distribution of the NSL-KDD dataset, 53% of the data were normal, whereas the other 47% consisted of many types of malicious classifications. A notable classification was malicious—neptune which comprised 33% of the malicious data.

3.3.3. Preprocessing the TON_IoT Dataset

Each of the datasets that make up the entirety of TON_IoT were preprocessed separately, taking into consideration their unique features. A variety of one-hot, label, and cyclical encoding were used as well as scaling techniques. Specifically, min–max scaling was used, which is scaling within a range typically from zero to one. Zero-mean unit-variance scaling was also applied. For each dataset, the time stamp feature was cyclically encoded, and the date and time features were removed as they were not relevant. Each type of attack was label encoded. Each dataset was also checked for duplicate entries. If present, they were removed. With the TON_IoT data, the NaN values were treated differently than the other datasets. The mean value for the column was used to fulfill NaN values rather than drop them. There was no inclusion of IP addresses in the preprocessed data to avoid bias. Once each dataset was preprocessed, they were combined into a single csv file to run through the algorithms. After preprocessing, 24 features were accounted for with 303,855 entries of data. As for the distribution of the TON_IoT dataset, 52% of the data were benign, while the other 48% consisted of malicious data.

3.3.4. Training XGBoost

Using XGBoost, the label of the IoT data establishing whether the network traffic is benign or malicious is set as the control variable so the model can perform classification. The model is then trained using the preprocessed data and a variety of hyperparameters. Since this study focuses on classification, the parameters that were adjusted for the tree booster were η and max depth. η performed best in a range from 0.3 to 0.35. As for max depth, the default value of six was used in our study. SoftMax was also used along with the num class parameter. Additionally, the multiclass classification error rate (Equation (7)) was also calculated.

$$\text{merror} = \frac{\text{wrong cases}}{\text{all cases}} \qquad (7)$$

The IoT-23 data, NSL-KDD data, and TON_IoT data were split into 80% training and 20% testing, 75% training and 25% testing, and 65% training and 35% testing to see how well the XGBoost model could classify anomalies. When training the data, the number of epochs was set to 100.

3.3.5. Training SVMs

Multiple kernels were implemented for the NSL-KDD data while the RBF kernel was mainly focused on using the IoT-23 data and TON_IoT data. C was set to a value of 10, and gamma was set to 0.01.

The IoT-23 data, NSL-KDD data, and TON_IoT data were split into 80% training and 20% testing, 75% training and 25% testing, and 65% training and 35% testing to see how well the SVM model could classify anomalies.

3.3.6. Training DCNN

With XGBoost and SVM, there is a set number of iterations the model conducts to train data, regardless of whether there is an improvement in the performance. DCNNs, however, can stop training before MaxEpochs is reached. Our DCNN is formatted to halt once there is no further improvement in regard to the accuracy. Once the model decides it is optimized, training is complete.

For DCNN parameters, the MiniBatchSize was set to 10 and the MaxEpochs was 5. The InitalLearningRate was set to 0.0001. Our ValidationFrequency was set to 5 and ValidationPatience was set to 10. The IoT-23 data, NSL-KDD data, and TON_IoT data were split into 80% training and 20% testing, 75% training and 25% testing, and 65% training and 35% testing to see how well the XGBoost model could classify anomalies. Additionally, a 70% training and 30% testing set was included for the DCNN.

4. Evaluation Metrics

Various metrics were calculated when running XGBoost, the SVM, and the DCNN. As Python was used for the SVM and XGBoost algorithms, the metrics were calculated using scikit-learn functions. MATLAB was used to run the DCNN, and these metrics were calculated based on confusion matrices.

First, the balanced accuracy (Equation (8)) was calculated as it accounts for imbalanced data. Using balanced accuracy is better in this case since it accounts for both the positive and negative outcome classes. The F1 score (Equation (9)) was another metric taken into consideration. F1 score is the mean of precision (Equation (10)) and recall (Equation (11)). It is a popular metric as it is a combination of two other important metrics. These metrics can be calculated as follows [36]:

$$\text{Balanced Accuracy} = \frac{TPR + TNR}{2} \tag{8}$$

$$\text{F1 Score} = 2\frac{(Precision \times Recall)}{(Precision + Recall)} \tag{9}$$

$$\text{Precision} = \frac{TP}{TP + FP} \tag{10}$$

$$\text{Recall} = \frac{TP}{TP + FN} \tag{11}$$

$$\text{TPR} = \frac{TP}{TN + FP} \tag{12}$$

$$\text{TNR} = \frac{TN}{TN + FP} \tag{13}$$

where TP is True Positive, FP is False Positive, TN is True Negative, FN is False Negative, TPR is the True Positive Rate (Equation (12)), and TNR is the True Negative Rate (Equation (13)).

5. Results

5.1. Detection Performance of XGBoost

Our experimental results proved that XGBoost can predict anomalies very effectively. XGBoost achieved balanced accuracies of up to 99.98% with three different training and testing splits using the IoT-23 data, see Figure 1. With the NSL-KDD dataset, balanced accuracies of up to 80.3% were achieved using three training and testing splits, see Figure 2. As for the TON_IoT data, balanced accuracies of up to 99.90% were achieved using 75% of the data for training and 25% for testing, see Figure 3. In regard to other metrics, precision for IoT-23 was up to 99.99%, up to 89.3% using the NSL-KDD data, and up to 99.78% using the TON_IoT data. Recall was up to 99.98% for IoT-23, up to 80.3% for NSL-KDD, and up to 99.83% for TON_IoT. Further, the F1 score was up to 99.99% for IoT-23, up to 81% for NSL-KDD, and up to 99.84% for TON_IoT.

5.2. Detection Performance of SVM

SVM achieved high balanced accuracies of up to 96.71% using the IoT-23 dataset, as shown in Figure 4. Balanced accuracies of up to 77.07% were achieved with the 75/25 split using the NSL-KDD data, see Figure 5. As different kernels were assessed, the linear kernel achieved the best results using this dataset. As for the TON_IoT data, accuracies of up to 96.72% were achieved using an 80/20 split, see Figure 6. These results were gathered with the application of the RBF kernel.

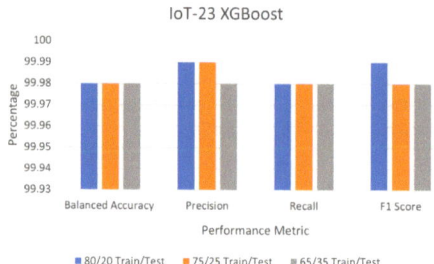

Figure 1. XGBoost performance results using the IoT-23 dataset.

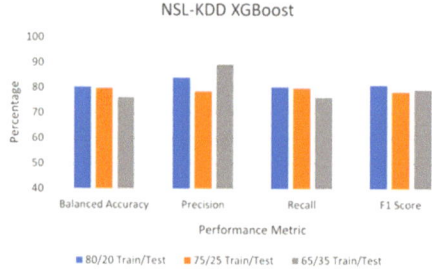

Figure 2. XGBoost performance results using the NSL-KDD dataset.

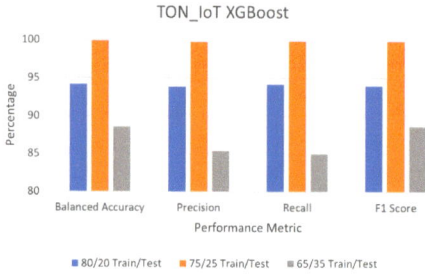

Figure 3. XGBoost performance results using the TON_IoT dataset.

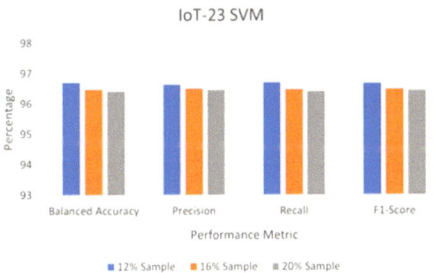

Figure 4. SVM performance results using the IoT-23 dataset.

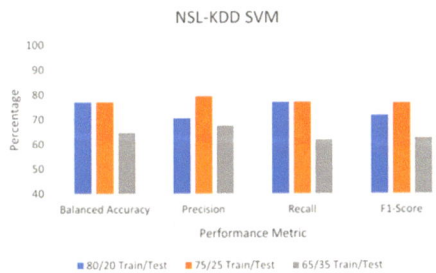

Figure 5. SVM performance results using the NSL-KDD dataset.

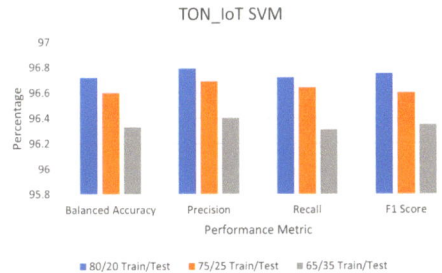

Figure 6. SVM performance results using the TON_IoT dataset.

5.3. Detection Performance of the DCNN

The IoT-23 dataset performed very well across various DCNN architectures. The highest accuracy achieved was 99.90% using the VGG16 architecture. As for the NSL-KDD dataset, it performed the best using the ResNet-50 architecture, achieving accuracies of up to 96.19%. Using the TON_IoT dataset, the ResNet-50 architecture also performed the best, with an accuracy of 92.98%. While the accuracies remained above 85% for each architecture when applying the IoT-23 and NSL-KDD datasets, there was a wider range in accuracy when using the TON_IoT dataset. The best accuracy for each architecture appears in bold in Tables 4–6.

5.4. Comparison of Detection Performance

When comparing the detection of anomalies with the models assessed, XGBoost achieved the best performance, with the highest accuracy of 99.98%. The DCNN performed the second best, achieving the highest accuracy of 99.90%. SVM was the third best algorithm, achieving accuracies up to 96.72%. Although the SVM and DCNN both achieved good accuracies, XGBoost performed well above these two models in our evaluation. In particular, the highest accuracies were found when using the IoT-23 dataset. These results can be seen in Figures 7–9.

Table 4. DCNN performance results using the IoT-23 dataset.

DCNN	Accuracy	Precision	Recall	F1 Score
AlexNet				
80/20	94.59	88.10	72.99	79.84
75/25	95.94	96.64	76.19	85.21
70/30	**97.13 ***	96.43	92.01	88.64
65/35	95.87	97.06	75.99	85.24
GoogleNet				
80/20	98.37	98.61	90.49	94.38
75/25	97.47	98.07	81.77	89.18
70/30	**99.16 ***	96.64	97.41	97.03
65/35	85.96	42.95	49.86	46.15
ResNet-50				
80/20	**99.85 ***	99.28	99.48	99.38
75/25	99.81	99.45	99.11	99.28
70/30	99.79	99.63	98.98	99.30
65/35	99.48	98.60	99.36	98.98
VGG16				
80/20	99.83	99.54	99.35	99.44
75/25	99.64	99.39	98.21	98.80
70/30	99.42	96.65	99.02	97.82
65/35	**99.90 ***	99.86	99.31	99.59
VGG19				
80/20	99.63	99.67	97.42	98.53
75/25	98.22	99.00	91.44	95.07
70/30	**99.69 ***	99.61	97.93	98.76
65/35	99.21	99.45	91.17	96.74

* These were the highest accuracies achieved by each architecture using the IoT-23 dataset.

Table 5. DCNN performance results using the NSL-KDD dataset.

DCNN	Accuracy	Precision	Recall	F1 Score
AlexNet				
80/20	90.60	90.53	90.63	90.58
75/25	90.53	90.50	90.70	90.60
70/30	90.79 *	91.23	90.45	90.84
65/35	90.59	90.82	90.34	90.58
GoogleNet				
80/20	93.49	93.42	93.57	93.49
75/25	92.81	92.84	92.71	92.77
70/30	92.15	92.16	92.36	92.26
65/35	93.75 *	93.69	93.80	93.74
ResNet-50				
80/20	96.19 *	96.15	96.21	96.18
75/25	94.86	94.84	94.83	94.84
70/30	94.71	94.86	94.55	94.70
65/35	94.66	94.60	94.77	94.69
VGG16				
80/20	95.20 *	95.41	95.01	95.21
75/25	94.33	94.38	94.23	94.30
70/30	93.20	93.23	93.44	93.33
65/35	94.78	94.96	94.60	94.78
VGG19				
80/20	92.97	93.02	93.22	93.11
75/25	93.95	94.03	93.83	93.93
70/30	95.02	94.96	95.07	95.02
65/35	95.23 *	95.21	95.19	95.20

* These were the highest accuracies achieved by each architecture using the NSL-KDD dataset.

Table 6. DCNN performance results using the TON_IoT dataset.

DCNN	Accuracy	Precision	Recall	F1 Score
AlexNet				
80/20	64.08	69.66	65.07	67.29
75/25	67.02	75.48	68.12	71.61
70/30	92.29 *	92.71	92.11	92.41
65/35	89.14	91.30	88.70	89.98
GoogleNet				
80/20	54.80	55.98	55.42	55.70
75/25	61.03	63.15	61.71	62.42
70/30	87.83 *	89.26	87.46	88.35
65/35	81.04	81.01	81.01	81.01
ResNet-50				
80/20	92.98 *	92.96	93.01	92.98
75/25	90.82	91.38	90.6	90.99
70/30	88.93	91.13	88.48	89.78
65/35	89.03	91.22	88.57	89.88
VGG16				
80/20	84.98 *	86.24	84.60	85.41
75/25	63.39	68.40	64.35	66.32
70/30	70.98	80.84	72.08	76.21
65/35	74.81	78.53	72.08	76.21
VGG19				
80/20	69.88	73.52	70.62	71.04
75/25	71.79	72.71	72.14	72.43
70/30	86.76	88.41	86.34	87.37
65/35	88.85 *	91.12	88.39	89.74

* These were the highest accuracies achieved by each architecture using the TON_IoT dataset.

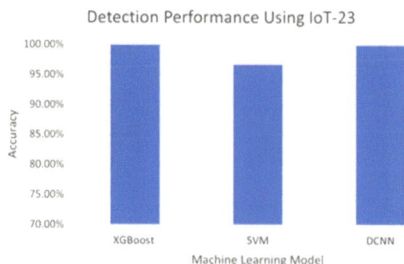

Figure 7. Machine learning model comparison using the IoT-23 dataset.

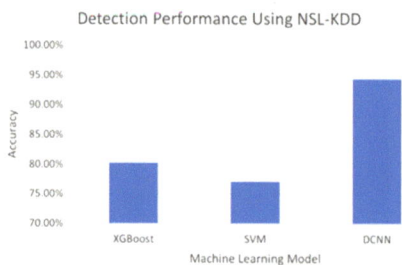

Figure 8. Machine learning model comparison using the NSL-KDD dataset.

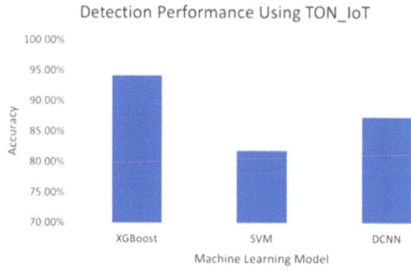

Figure 9. Machine learning model comparison using the TON_IoT dataset.

6. Efficiency, Real-World Data, and Limitation Discussion

6.1. Efficiency

In terms of CPU time, XGBoost proved to be the most efficient algorithm. By changing the parameters of the machine learning models, the models might obtain slightly better results; however, training takes an extremely long amount of time, particularly for the DCNN and SVM.

Training the DCNN is much more time-intensive than training both the SVM and XGBoost. One reason for this is because the training inputs are image data rather than raw numerical data. The larger size of the images means that it is more computationally expensive to use than relatively small-sized numerical data. Additionally, the DCNN uses a multilayer approach, which requires more time for each input. Some images will also be used in the training process multiple times. However, once the training is complete, testing the trained network is much quicker and more efficient, taking a small fraction of the time it takes to originally train it.

Since the TON_IoT and NSL-KDD preprocessed data were consistently used across all three models, the averages of the 80/20 training and testing split times from these datasets were calculated to compare each model's efficiency, as shown in Table 7. Each of these models were run on the same machine under equivalent conditions. XGBoost took an average of 0.7748 s for training and only 0.0032 s for testing. The SVM took an average of 556.11 s for training and 79.06 s for testing. As for the average across all DCNN architectures, the training time was about 9916.9 s and the testing time was 342.49 s. Therefore, XGBoost proves to be 717.75 times faster than the SVM and 12,799.3 times faster than the DCNN in regard to training times.

Table 7. ML model efficiency results.

	XGBoost	SVM	DCNN
Training (s)	0.7748	556.11	9916.9
Testing (s)	0.0032	79.06	342.49

6.2. Verification with Real-World Data

As part of our ongoing project, we recently built an IoT testbed as previously mentioned. Using our real data collected from the IoT testbed, we applied the machine learning models used in this study. Balanced data was used, where there was a total of 320,000 captures and 12 features. Various training and testing splits were assessed including 80/20, 75/25, 70/30, and 65/35. XGBoost proved to perform the best overall, achieving accuracies up to 99.97%, whereas the SVM achieved accuracies of up to 95.55%. Using the ResNet-50 architecture, the DCNN achieved accuracies of up to 98.46%.

As far as execution times are concerned, training the data using XGBoost took 1.71 s on average, while testing took about 0.0058 s. The SVM took about 981.02 s for training and 220.74 s for testing. Lastly, the DCNN took an average of 23,598.96 s for training and 342.21 s for testing. The results using our testbed data strongly support our results previously discussed in Section 5.

6.3. Limitation Discussion

In this paper, we studied IoT anomaly detection powered by ML models using multiple publicly available datasets (as well as real-world data collected from the smart building testbed that we recently built) from distinct IoT systems to evaluate the effectiveness of the models regarding dynamic IoT environments. Not only did we examine the accuracy of anomaly detection, but we also thoroughly assessed the efficiency that is highly desired for resource constrained IoT applications (but not systematically addressed by any prior works). We focused on three representative ML models, namely, Extreme Gradient Boosting, Support Vector Machines, and Deep Convolutional Neural Networks, to identify the optimal model with regard to both the detection accuracy and efficiency based on our selected datasets. However, as discussed in previous sections, IoT systems are applied to various dynamic environments that produce very different data. We do not expect any single detection model (such as XGBoost that we identified in this research as the optimal model for the datasets that we used) will work most effectively for all IoT systems with different devices and for different purposes. Therefore, it is important to study as many ML models as possible, evaluating their effectiveness and efficiency regarding various real-world datasets so that users can identify the most appropriate models based on their unique business nature and security requirements. Due to the unavailability of appropriate datasets and the scope of our current project, we did not consider other machine learning models, such as Elastic Regression, Lasso Regression, Ridge Regression, Random Forest, and Naive Bayes which have been studied in some recent publications [7].

7. Conclusions and Future Work

The vulnerable and resource-constrained end devices that are used in emerging IoT applications make the security of these applications a great challenge. Anomaly detection is a crucial security procedure that protects IoT applications through the identification of unusual or abnormal behavior. In this paper, we investigated anomaly detection powered by machine learning algorithms. In view of the lack of studies identifying optimal anomaly detection models for the IoT, our work investigates the effectiveness and efficiency of various models. Three representative machine learning models were studied in this paper, namely XGBoost, SVMs and DCNNs. Evaluating ML-powered anomaly detection models using distinct datasets collected from different environments is vital. Therefore, multiple datasets containing data from diverse environments were utilized. These datasets included the IoT-23 dataset, NSL-KDD dataset, and TON_IoT dataset.

The experimental results proved that XGBoost is the most effective method, achieving accuracies of up to 99.98%. The DCNN performed second best, achieving the highest accuracy of 99.90%. The SVM was the third best algorithm, achieving accuracies of up to 96.71%. Although the SVM and DCNN both achieved good accuracies, XGBoost performed well above these two models in our evaluation. In particular, the highest accuracies were found when executing these methods on the IoT-23 dataset. XGBoost also proved to be the most efficient regarding the training time. XGBoost took an average of 0.7748 s for training and only 0.0032 s for testing. The SVM took an average of 556.11 s for training and 79.06 s for testing. As for the average across all DCNN architectures, the training time was about 9916.9 s and the testing time was 342.49 s.

These models have further been assessed using our real-world IoT data collected from an IoT testbed consisting of various physical devices. Our evaluation of the anomaly detection models using real-world data proves that XGBoost can efficiently and accurately detect anomalies in real-world IoT applications. However, with the many ML models and unique IoT environments, it is difficult to conclude which model works most effectively in all IoT systems with different devices. We may extend our research to include more well-known ML models in our future projects. We believe more research findings will be achieved which will be beneficial for industrial and organizational IoT users.

Author Contributions: Conceptualization, M.B., W.F., X.-W.W., S.E. and Z.G.; methodology, M.B. and Z.G.; software, M.B. and Z.G.; validation, M.B. and Z.G.; formal analysis, M.B. and Z.G.; investigation, M.B. and Z.G.; resources, W.F., X.-W.W. and S.E.; data curation, M.B.; writing—original draft preparation, M.B.; writing—review and editing, M.B.; visualization, M.B.; supervision, W.F., X.-W.W. and S.E.; project administration, W.F., X.-W.W. and S.E.; funding acquisition, W.F. All authors have read and agreed to the published version of the manuscript.

Funding: This work was supported by the NSA-NCAE-C grants H98230-20-1-0296 and H98230-22-1-0315.

Data Availability Statement: Publicly available datasets were analyzed in this study. These data can be found here: https://www.stratosphereips.org/datasets-iot23 (accessed on 18 February 2021) [6], https://www.unb.ca/cic/datasets/nsl.html (accessed on 18 February 2021) [7], and https://research.unsw.edu.au/projects/toniot-datasets (accessed on 18 February 2021) [8].

Acknowledgments: The authors extend their thanks to Larry Pearlstein, TCNJ, for the use of Zelda. In addition, the authors would like to thank members of the IUP IoT Research Team.

Conflicts of Interest: The authors declare no conflicts of interest.

References

1. Hossain, M.; Kayas, G.; Hasan, R.; Skjellum, A.; Noor, S.; Islam, S.M.R. A Holistic Analysis of Internet of Things (IoT) Security: Principles, Practices, and New Perspectives. *Future Internet* **2024**, *16*, 40. [CrossRef]
2. Cole, T. Interview with Kevin Ashton—Inventor of IoT: Is Driven by the Users. Available online: https://www.avnet.com/wps/portal/silica/resources/article/interview-with-iot-inventor-kevin-ashton-iot-is-driven-by-the-users/ (accessed on 1 April 2022).
3. Al-Hejri, I.; Azzedin, F.; Almuhammadi, S.; Eltoweissy, M. Lightweight Secure and Scalable Scheme for Data Transmission in the Internet of Things. *Arab. J. Sci. Eng.* **2024**. [CrossRef]

4. Vailshery, L.S. Global IoT and Non-IoT Connections 2010–2025. Available online: https://www.statista.com/statistics/1101442/iot-number-of-connected-devices-worldwide/ (accessed on 1 April 2022).
5. Posey, B.; Shea, S. What Are IoT Devices?—Definition from Techtarget.com. Available online: https://internetofthingsagenda.techtarget.com/definition/IoT-device (accessed on 1 April 2022).
6. Shea, S.; Wigmore, I. IoT Security (Internet of Things Security). Available online: https://www.techtarget.com/iotagenda/definition/IoT-security-Internet-of-Things-security (accessed on 1 April 2022).
7. Wu, X.W.; Cao, Y.; Dankwa, R. Accuracy vs Efficiency: Machine Learning Enabled Anomaly Detection on the Internet of Things. In Proceedings of the IEEE International Conference on Internet of Things and Intelligence Systems, Bali, Indonesia, 24–26 November 2022; pp. 245–251.
8. Fraihat, S.; Makhadmeh, S.; Awad, M.; Al-Betar, M.A.; Al-Redhaei, A. Intrusion detection system for large-scale IoT NetFlow networks using machine learning with modified Arithmetic Optimization Algorithm. *Internet Things* **2023**, *22*, 100819. [CrossRef]
9. Awad, M.; Fraihat, S.; Salameh, K.; Al Redhaei, A. Examining the Suitability of NetFlow Features in Detecting IoT Network Intrusions. *Sensors* **2022**, *22*, 6164. [CrossRef] [PubMed]
10. Garcia, S.; Parmisano, A.; Erquiaga, M.J. IoT-23: A Labeled Dataset with Malicious and Benign IoT Network Traffic (Version 1.0.0). 2020. Available online: https://www.stratosphereips.org/datasets-iot23 (accessed on 18 February 2021).
11. NSL-KDD Dataset. Available online: https://www.unb.ca/cic/datasets/nsl.html (accessed on 18 February 2021).
12. TON_IoT Datasets. Available online: https://research.unsw.edu.au/projects/toniot-datasets (accessed on 18 February 2021).
13. Hossain, M.T.; Imran, M.A. ToN-IoT: A dataset for traffic analysis of IoT devices. In Proceedings of the IEEE International Conference on Communications, Kansas City, MO, USA, 20–24 May 2018; pp. 1–6.
14. Alsaedi, A.; Moustafa, N.; Tari, Z.; Mahmood, A.; Anwar, A. TON_IoT Telemetry Dataset: A New Generation Dataset of IoT and IIoT for Data-Driven Intrusion Detection Systems. *IEEE Access* **2022**, *8*, 165130–165150. [CrossRef]
15. Cañedo, J.; Skjellum, A. Using Machine Learning to secure IoT systems. In Proceedings of the 2016 14th Annual Conference on Privacy, Security and Trust (PST), Auckland, New Zealand, 12–14 December 2016; pp. 219–222.
16. Hussain, F.; Hussain, R.; Hassan, S.A.; Hossain, E. Machine Learning in IoT security: Current solutions and future challenges. *IEEE Commun. Surv. Tutor.* **2020**, *22*, 1686–1721. [CrossRef]
17. Dalal, K.R. Analyzing the role of supervised and unsupervised Machine Learning in IoT. In Proceedings of the 2020 International Conference on Electronics and Sustainable Communication Systems (ICESC), Coimbatore, India, 2–4 July 2020; pp. 75–79.
18. Vitorino, J.; Andrade, R.; Praca, I.; Sousa, O.; Maia, E. A Comparative Analysis of Machine Learning Techniques for IoT Intrusion Detection. In Proceedings of the 14th International Symposium on Foundations and Practice of Security (FPS 2021), Paris, France, 7–10 December 2021; pp. 191–207.
19. Diro, A.; Chilamkurti, N.; Nguyen, V.D.; Heyne, W. A Comprehensive Study of Anomaly Detection Schemes in IoT Networks Using Machine Learning Algorithms. *Sensors* **2021**, *21*, 8320. [CrossRef] [PubMed]
20. Balega, M.; Farag, W.; Ezekiel, S.; Wu, X.-W.; Deak, A.; Good, Z. IoT Anomaly Detection Using a Multitude of Machine Learning Algorithms. In Proceedings of the 2022 IEEE Applied Imagery Pattern Recognition Workshop, Washington, DC, USA, 11–13 October 2022; pp. 1–6.
21. Good, Z.; Farag, W.; Wu, X.-W.; Ezekiel, S.; Balega, M.; May, F.; Deak, A. Comparative Analysis of Machine Learning Techniques for IoT Anomaly Detection Using the NSL-KDD Dataset. *Int. J. Comput. Sci. Netw. Secur.* **2023**, *23*, 46–52.
22. What Is Machine Learning? Available online: https://www.ibm.com/cloud/learn/machine-learning (accessed on 1 April 2022).
23. Chen, T.; Guestrin, C. XGBoost: A scalable tree boosting system. In Proceedings of the 22nd ACM SIGKDD International Conference on Knowledge Discovery and Data Mining, San Francisco, CA, USA, 13–17 August 2016; pp. 785–794.
24. Hastie, T.; Tibshirani, R.; Friedman, J.H.; Friedman, J.H. *The Elements of Statistical Learning: Data Mining, Inference, and Prediction*; Springer: New York, NY, USA, 2009; Volume 2.
25. Chang, W.; Liu, Y.; Xiao, Y.; Yuan, X.; Xu, X.; Zhang, S.; Zhou, S. A Machine Learning based prediction method for hypertension outcomes based on medical data. *Diagnostics* **2019**, *9*, 178. [CrossRef] [PubMed]
26. XGBoost Documentation. Available online: https://xgboost.readthedocs.io/en/stable/index.html (accessed on 1 June 2021).
27. Vapnik, V. *Estimation of Dependences Based on Empirical Data*; Springer: New York, NY, USA, 2006. [CrossRef]
28. Jakkula, V. *Tutorial on Support Vector Machine (SVM)*; School of EECS, Washington State University: Pullman, WA, USA, 2006; Volume 37, p. 3.
29. Pupale, R. Support Vector Machines (SVM)—An Overview. Available online: https://towardsdatascience.com/https-medium-com-pupalerushikesh-svm-f4b42800e989 (accessed on 1 April 2022).
30. Deep Convolutional Neural Networks. Available online: https://www.run.ai/guides/deep-learning-for-computer-vision/deep-convolutional-neural-networks (accessed on 1 April 2022).
31. Stoian, N. Machine Learning for Anomaly Detection in IoT Networks: Malware Analysis on the IoT-23 Dataset. Bachelor's Thesis, University of Twente, Enschede, The Netherlands, 2020.
32. Lippmann, R.; Fried, D.; Graf, I.; Haines, J.; Kendall, K.; McClung, D.; Weber, D.; Webster, S.; Wyschogrod, D.; Cunningham, R.; et al. Evaluating intrusion detection systems: The 1998 darpa offline intrusion detection evaluation. In Proceedings of the DARPA Information Survivability Conference and Exposition, DISCEX'00, Hilton Head, SC, USA, 25–27 January 2000; Volume 2, pp. 12–26.

33. Tavallaee, M.; Bagheri, E.; Lu, W.; Ghorbani, A.A. A detailed analysis of the KDD Cup 99 dataset. In Proceedings of the IEEE Symposium on Computational Intelligence for Security and Defense Applications, Ottawa, ON, Canada, 8–10 July 2009; pp. 1–6.
34. Revathi, S.; Malathi, A. A detailed analysis on the NSL-KDD dataset using various machine learning techniques for intrusion detection. *Int. J. Eng. Res. Technol.* **2013**, *2*, 1848–1853.
35. Moustafa, N.; Keshky, M.; Debiez, E.; Janicke, H. Federated TON_IoT Windows Datasets for Evaluating AI-Based Security Applications. In Proceedings of the IEEE 19th International Conference on Trust, Security and Privacy in Computing and Communications (TrustCom), Guangzhou, China, 29 December 2020; pp. 848–855.
36. Hale, J. The 3 Most Important Composite Classification Metrics. Available online: https://towardsdatascience.com/the-3-most-important-composite-classification-metrics-b1f2d886dc7b (accessed on 1 April 2022).

Disclaimer/Publisher's Note: The statements, opinions and data contained in all publications are solely those of the individual author(s) and contributor(s) and not of MDPI and/or the editor(s). MDPI and/or the editor(s) disclaim responsibility for any injury to people or property resulting from any ideas, methods, instructions or products referred to in the content.

Article

Effects of RF Signal Eventization Encoding on Device Classification Performance

Michael J. Smith, Michael A. Temple * and James W. Dean

Department of Electrical and Computer Engineering, US Air Force Institute of Technology, Wright-Patterson AFB, Dayton, OH 45433, USA; michael.smith2@afit.edu (M.J.S.); james.dean@afit.edu (J.W.D.)
* Correspondence: michael.temple@afit.edu

Abstract: The results of first-step research activity are presented for realizing an envisioned "event radio" capability that mimics neuromorphic event-based camera processing. The energy efficiency of neuromorphic processing is orders of magnitude higher than traditional von Neumann-based processing and is realized through synergistic design of brain-inspired software and hardware computing elements. Relative to event-based cameras, the development of event-based hardware devices supporting Radio Frequency (RF) applications is severely lagging and considerable interest remains in obtaining neuromorphic efficiency through event-based RF signal processing. In the Operational Technology (OT) protection arena, this includes efficient software computing capability to provide reliable device classification. A Random Forest (RndF) classifier is considered here as a reliable precursor to obtaining Spiking Neural Network (SNN) benefits. Both 1D and 2D eventized RF fingerprints are generated for bursts from N_{Dev} = 8 WirelessHART devices. Average correct classification (%C) results show that 2D fingerprinting is best overall using detected events in burst Gabor transform responses. This includes %C $\geq$ 90% under multiple access interference conditions using an average of $N_{EPB} \geq$ 400 detected events per burst. This is sufficiently promising to motivate next-step activity aimed at (1) reducing fingerprint dimensionality and minimizing the required computational resources, and (2) transitioning to a neuromorphic-friendly SNN classifier—two significant steps toward developing the necessary computing elements to achieve the full benefits of neuromorphic processing in the envisioned RF event radio.

Keywords: device fingerprinting; neuromorphic; spiking neural network; eventization; encoding; SNN; operational technology; OT; random forest; WirelessHART

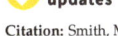

Citation: Smith, M.J.; Temple, M.A.; Dean, J.W. Effects of RF Signal Eventization Encoding on Device Classification Performance. *Electronics* **2024**, *13*, 2020. https://doi.org/10.3390/electronics13112020

Academic Editors: Abdussalam Elhanashi and Pierpaolo Dini

Received: 8 April 2024
Revised: 16 May 2024
Accepted: 17 May 2024
Published: 22 May 2024

Copyright: © 2024 by the authors. Licensee MDPI, Basel, Switzerland. This article is an open access article distributed under the terms and conditions of the Creative Commons Attribution (CC BY) license (https:// creativecommons.org/licenses/by/ 4.0/).

1. Introduction

As the use of wireless Operational Technology (OT) expands to support nearly all Critical Infrastructure (CI) sectors, so does the need for increased communication network reliability, security, and resiliency [1–3]. A majority of today's OT technology evolved from information technology applications, with cost reduction and performance benefits of information technology adaptation leading to the development and deployment of various smart technologies throughout numerous CI sectors. This includes the complementary use of Internet of Things (IoT) and Industrial Internet of Things (IIoT) devices to produce an integrated mesh of physical systems supporting (1) healthcare, transportation, energy, and industrial control system sectors [4], and (2) other OT-dominated sectors that include critical manufacturing, defense industrial base, emergency service, and water/wastewater treatment sectors [5]. Enabling OT is vital to cross-sector critical infrastructure operations and the communications interconnectivity has driven the need for ensuring resilience, safety, and security.

Various Radio Frequency (RF) fingerprinting methods have been considered over the past decade to improve the security of communication systems supporting OT networks. Among these methods are those identified in [6] where the authors have aptly summarized different RF fingerprint generation, extraction, and discrimination methods that can be considered to boost physical layer protection. A majority of the methods noted in [6] can be categorized as passive methods where the fingerprint features are extracted from collected RF emissions of operationally installed devices performing their by-design function. This differs from active fingerprinting methods, whereby the features are extracted from externally stimulated RF emissions collected from uninstalled devices that may or may not be functionally operating.

Fingerprinting methods based on Distinct Native Attribute (DNA) features are duly noted in [6] and have demonstrated potential for enhancing physical layer security in wireless OT networks [7–9]. This includes demonstrations using WirelessHART communication devices, which (1) have rapidly proliferated throughout OT application spaces, (2) are among the two most widely used industrial international standards, and (3) have fielded numbers approaching tens of millions [10]. The statistical DNA-based fingerprinting methods have employed both passive [11–13] and active [8,14] approaches for exploiting WirelessHART DNA features. These works have predominantly employed Multiple Discriminant Analysis (MDA) and ensemble-based Random Forest (RndF) classifiers to perform device discrimination.

1.1. Research Motivation

Advances in neuromorphic computing [15] and the use of Spiking Neural Network (SNN) classification architectures [16] have enabled near-equivalent classification performance as obtained with traditional von Neumann-based artificial neural network architectures at a fraction of the energy consumption. While neuromorphic computing has shown energy reduction approaching $1000\times$ in selected applications [17–19], its impact on radio-frequency (RF) applications remains largely uninvestigated.

RF sensing systems stand to gain considerable benefit from new energy-efficient neuromorphic processing. This includes upgrades to RF receiver processing that historically requires analog-to-digital conversion of continuous signals and sample decimation to reduce the amount of data presented to downstream processing elements. RF signal "eventization" holds promise for decreasing the overall required power by reducing the amount of data required by downstream processes. For RF fingerprinting applications [14], this includes reducing the amount of data required for fingerprint generation and device discrimination by using features extracted from information-bearing event data.

Neuromorphic event-based camera processing is orders of magnitude more energy efficient than traditional frame-based camera processing [20]. The envisioned "event radio" concept seeks to bring these advantages into the RF processing arena. The belief that similar benefits may be realized for an RF event radio is motivated by (1) the existence of hardware such as the Intel Loihi neuromorphic chip [21] that provides an energy-efficient speed improvement when processing RF signals [22], and (2) the use of computationally efficient deep machine learning algorithms in RF electronic warfare applications [23].

Potential benefits of neuromorphic computing are considered here in light of previously demonstrated DNA fingerprinting methods. This includes comparing non-eventized and eventized RF fingerprinting performance using features from experimentally collected signals for eight WirelessHART devices—some of the same devices that are proliferating throughout various OT application spaces. The WirelessHART signals are adopted from [14] and are representative of communications using Frequency-Hopped, Time Division Multiple Access (FH-TDMA) to provide multi-user, multi-channel operation. The WirelessHART FH-TDMA demonstration signals include the effects of multi-device, multi-channel interference, which significantly increases the fingerprint discrimination challenge when compared with single-device, single-channel operation.

The RF eventization work here represents the first step toward transitioning to a neuromorphic-friendly, low power-consuming, SNN classification capability of enhancing edge device operational protection. This includes consideration for implementing SNN-based neuromorphic processing in Field Programmable Gate Array (FPGA) hardware being hosted on Software-Defined Radio (SDR) platforms. Several RF eventization encoding methods are considered here using WirelessHART one-dimensional Time Domain (1D-TD) signals and their corresponding two-dimensional Gabor transform (2D-GTX) responses. Classification results are generated for N_{Dev} = 8 devices using fingerprints from selected eventization methods to highlight the potential for realizing event radio capability—this is envisioned as being an SDR-based solution hosting FPGA hardware to achieve low-power SNN-based neuromorphic processing objectives.

1.2. Relationship to Prior Research

The concept of an "event" occurring within time-series data has been studied in multiple applications and includes events at the signal waveform level as well as the machine-processing level [24–26]. A key issue with time-series processing is dealing with vast amounts of data generated by continuously sampled univariate or multivariate processes. For the purpose of RF eventization, as defined in this work, a large sequence of real-valued numbers (raw signal samples) is transformed into a shorter length sequence of binary-valued "spikes", which are either positive or negative, for use in neuromorphic processing. Various methods have been devised to implement this type of eventization encoding and are broadly categorized as rate coding and temporal coding methods [27]. For a given application, a preferred encoding method may be empirically selected based on information loss of the original signal [28], classifier accuracy [29], or some other metric. As achieved in the event camera arena [20], key eventization advantages include very high temporal resolution, low latency, very high dynamic range, and low power consumption [20]. It is desired to achieve similar advantages in neuromorphic RF systems.

For the purpose of this work, we used a threshold-based temporal coding method given its successful application in the event-based classification of Free-Spoken Digit (FSD) audio datasets in [30]. Audio is perhaps the most similar signal type to the WirelessHART data used here. This earlier work addresses pre-processing encoding techniques for lower-frequency, narrowband (100 < f < 1000 Hz), and time-varying audio signals. The pre-processing of raw input signal samples includes multi-channel Butterworth filtering followed by temporal spike encoding prior to feature extraction and classification. Of the various temporal encoding methods considered in [30], it was concluded that Temporal Contrast encoding was best overall and yielded the highest average classification accuracy with minimum accuracy variance. Temporal Contrast encoding generates spikes by comparing absolute signal variation with a fixed detection threshold and assigning resultant positive (+1) and negative (−1) spikes to the encoded sequence.

Given its success in [30] with audio signals, threshold-based contrast encoding was adopted here for initial RF eventization demonstration using the higher-frequency/higher-bandwidth WirelessHART signals detailed in Section 2.1. Contributions of the work here include (1) the first application of Temporal Contrast encoding to one-dimensional Time Domain (1D-TD) WirelessHART signals as detailed in Section 2.2.1, and (2) the introduction and demonstration of a Temporal–Spectral Contrast encoding approach using two-dimensional Gabor Transform (2D-GTX) responses of WirelessHART signals in Section 2.2.2.

1.3. Paper Contribution

There is a wide range of RF fingerprinting methods surveyed in [6] that can be used for device classification. The effectiveness of these methods is directly related to fingerprint "profile" features, which yield best-case performance when the features are ideally unique across the pool of devices to be classified. While the numerous fingerprint features noted in [6] are influenced by waveform level "events" (turn-on/turn-off transients, intentional/unintentional amplitude, phase, frequency modulation, etc.), none of the noted methods in [6] use RF fingerprints comprised of event-based features such as used here.

As completed here, the use of RF signal eventization coding and characterizing its effect on device classification performance is believed to be a first-of-a-kind activity for higher RF frequency (10 s of MHz) event processing. The work in [30] is perhaps the most related but considered much lower frequency signals, including very low-frequency signals (10 Hz to 1 KHz) as occurs in smart device biometric applications and mid-audio frequency signals (500 Hz to 2 KHz) as occurs in human voice recognition applications. As noted in Section 1.2, the success in [30] with these lower frequency signals is achieved using a threshold-based contrast encoding method, which was adopted here for higher frequency RF signal consideration.

Of equal importance, the RndF accuracy benchmark (%C $\geq$ 90%) established here and the identification of the average number of RF Events-Per-Burst ($N_{EPB} \geq 400$) required to achieve this accuracy sustain motivation for continuing the development and demonstration of an RF event radio capability. The RndF classifier is a reliable precursor for subsequent RndF-to-Convolutional Neural Network (CNN) classifier transition activity that is currently underway. The RndF-to-CNN transition activity will provide the baseline for the final CNN-to-SNN transition activity with the resultant neuromorphic-friendly SNN classifier holding the most promise for meeting high accuracy, low latency, and high energy efficiency objectives when implemented in hardware.

1.4. Paper Organization

The remainder of this paper is organized as follows. The Demonstration Methodology is presented in Section 2 and includes selected details on WirelessHART signals, RF signal eventization encoding, and fingerprint formation using non-eventized and eventized features. Summary details for the implemented MDA/ML and RndF classifiers are presented in Section 3. The device classification results are presented in Section 4 and predominantly focus on RndF classifier performance, which provides the basis for drawing conclusions and declaring demonstration success. The paper ends with a research summary and conclusions in Section 5.

2. Demonstration Methodology

This section provides selected details for key processes and activities used for conducting the experimental demonstrations and generating the classification results presented in Section 4. Selected details are provided for the following:

1. WirelessHART Signals in Section 2.1—provides details for the WirelessHART adapters being discriminated, the Frequency-Hopped Time Division Multiple Access (FH-TDMA) process used to add multi-channel Cross-Channel Interference (CCI) effects, and the Signal-to-Interference-plus-Noise (SINR) ratio scaling process used to induce varying channel effects. The use of CCI-laden bursts here for the initial demonstration was (i) a matter of experimental convenience given that the single-channel WirelessHART bursts and the FH-TDMA process for coherently combining them were available [14]; and (ii) it was believed that discrimination using CCI-laden bursts presented a greater challenge when compared with single-channel discrimination;
2. RF Signal Eventization Encoding in Section 2.2—provides details for the threshold-based contrast encoding used for one-dimensional Time Domain (1D-TD) and two-dimensional Gabor Transform (2D-GTX) fingerprinting demonstrations. This includes encoding the WirelessHART FH-TDMA signal responses into event sequences used for fingerprint generation;
3. Fingerprint Formation in Section 2.3—provides details for forming the non-eventized statistical DNA and eventized fingerprint vectors that are input to the classifiers to assess device discriminability. These vectors are often referred to as input "samples" within the machine learning community. However, the term fingerprints is used exclusively herein to help minimize potential confusion between classifier input samples and experimentally collected time domain samples in the FH-TDMA signals.

2.1. WirelessHART Signals

The experimental signals used here for eventization demonstrations were originally collected in support of work in [13] and subsequently processed to support work in [14] that added multi-channel Cross-Channel Interference (CCI) effects. Selected experimental collection details from these previous works are included here for completeness. The signal collections were performed in a typical laboratory environment using the four Siemens AW210 [31] and four Pepperl+Fuchs Bullet [32] devices listed in Table 1. Note that while two different manufacturers are listed, which is based on labels affixed to the devices, all devices were actually manufactured in a single plant located in Twinsburg Ohio, USA. The change in labeling and serial number sequencing is a result of Pepperl+Fuchs acquiring the manufacturing facility from Siemens.

Table 1. Details for the N_{Dev} = 8 WirelessHART adapters (D1, D2, ..., D8) used for demonstration [13,14] showing two different manufacturers and models with the indicated serial number (S/N).

Device ID	Manufacturer	Model	S/N
D1	Siemens	AW210	003095
D2	Siemens	AW210	003159
D3	Siemens	AW210	003097
D4	Siemens	AW210	003150
D5	Pepperl+Fuchs	Bullet	1A32DA
D6	Pepperl+Fuchs	Bullet	1A32B3
D7	Pepperl+Fuchs	Bullet	1A3226
D8	Pepperl+Fuchs	Bullet	1A32A4

As illustrated in Figure 1, the devices communicate directly with an Emerson 1410 gateway [33] one at a time while a USRP X310 Software Defined Radio (SDR) [34] collects the device emissions. The devices were placed 8 ft. from the gateway, and the SDR collection antenna was positioned 18 in. from the operating device. Each device was operated in non-overlapping 5.0 MHz WirelessHART channels (Chn#1–Chn#15) in the 2.4 GHz industrial, scientific, and medical frequency band [35]. The channel center frequencies include $f_c(m) = 2405 + 5m$ MHz for $m \in \{1, 2, \ldots, 15\}$ and span $[f_c(m) - 2.5, f_c(m) + 2.5]$ MHz. The devices use preamble-based signaling with Offset Quadrature Phase-Shift Keyed (O-QPSK) data modulation.

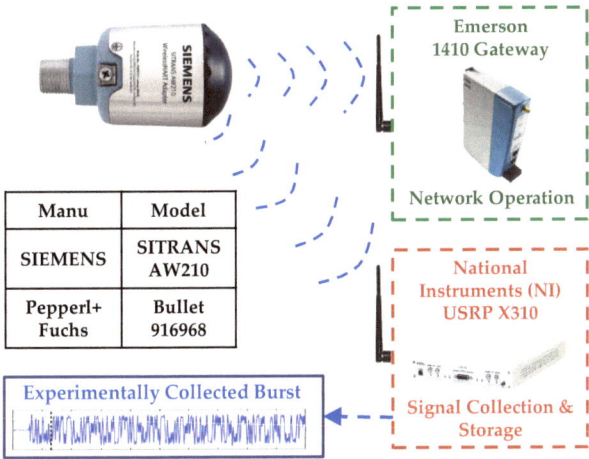

Figure 1. Experimental setup used for collecting the WirelessHART bursts that were input to the FH-TDMA process to induce adverse Cross-Channel Interference (CCI) effects.

The X310 collection receiver was manufactured by Ettus Research in Austin, TX, USA. It uses an RF bandwidth of BW_{Col} = 100 MHz and was set to a center frequency of f_{Ctr} = 2.440 GHz, which was sufficient for collecting signals in all WirelessHART frequency channels. The collection receiver employed an f_S = 100 Mega-Samples per second (MSps) sample rate and a total of 2500 experimental bursts were collected per device in each of the 15 WirelessHART channels. Post-collection Signal-to-Noise Ratio (SNR) analysis in [13] showed that an average collected $SNR_{Col} \approx 41.4$ dB was realized across all devices, all channels, and all collected bursts. The effects of real-world CCI were added post-collection through coherent re-combination using a Frequency-Hopped, Time Division Multiple Access (FH-TDMA) process [14]. This process was used to (1) mimic WirelessHART network operation that includes multiple adapters simultaneously transmitting in multiple frequency channels [36], and (2) enable classification performance assessment under conditions that are expected to increase the device discrimination challenge. As detailed below, the induction of CCI necessitated the need for an alternate Signal-to-Interference-plus-Noise Ratio (SINR) metric.

The FH-TDMA processing in Figure 2 was implemented such that N_{Frq} = 8 adjacent Frequency Slots (FS) were occupied within each Time Slot (TS). The frequency assignment of device D4 across time slots is highlighted in blue to illustrate the frequency hopping nature of the process when considering the FH-TDMA assignment for a given device. The FH sequential TS-by-TS processing included (1) random assignment of collected signals from all N_{Dev} = 8 devices to 1-of-8 FH frequency slots corresponding to WirelessHART Chn#8–Chn#15 operation; (2) coherent summation of the eight assigned bursts, one from each of the users operating in their assigned frequency channel—this effectively induces the cross-device CCI effects; (3) FS-by-FS slot (channel-by-channel) filtering using a 5.0 MHz bandpass channel filter to extract and segregate individual user signals; (4) device-by-device down-conversion and filtering using a 2.5 MHz baseband filter; and (5) factor-of-20 sample decimation for a final sample rate of f_S = (100 MSps)/20 = 5 MSps.

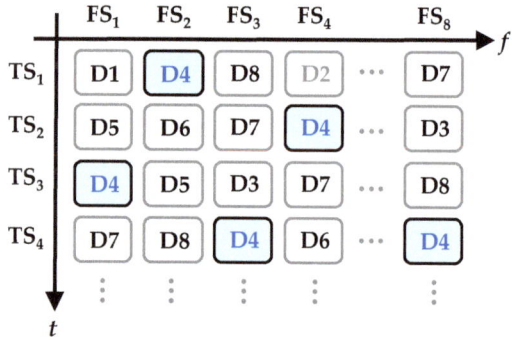

Figure 2. Frequency-Hopped, Time Division Multiple Access (FH-TDMA) process used with collected bursts from [13] to induce adverse Cross-Channel Interference (CCI) effects.

The FH-TDMA processing was repeated until a total of N_{Brst} = 8576 CCI-bearing responses, denoted as $s_{FhTd}(t)$, were generated for each of the eight devices. The SINR of the channel-filtered $s_{FhTd}(t)$ was scaled prior to eventization and DNA fingerprint generation. The input signal to the eventization and fingerprinting processes is given by

$$s_{FhTd}(t) = filter[\, s_{WH}(t) + s_{CCI}(t) + n_{Bck}(t)\,] \,, \qquad (1)$$

where $filter[\,]$ denotes the WirelessHART channel filtering used to segregate individual user signals, $s_{WH}(t)$ is the experimentally collected WirelessHART signal, $s_{CCI}(t)$ accounts for the induced CCI contribution, and $n_{Bck}(t)$ represents background noise that was present during the collection. The SINR of the $s_{FhTd}(t)$ responses in Equation (1) were scaled on a channel-by-channel basis using channel-filtered Additive White Gaussian Noise (AWGN) to induce channel variation effects according to

$$\text{SINR} \equiv \frac{S_{WH}}{(I_{CCI} + N_{Bck} + N_{AWGN})}, \qquad (2)$$

$$\text{SINR(dB)} = S_{WH}(\text{dB}) - 10 \times \text{Log}_{10}(I_{CCI} + N_{Bck} + N_{AWGN}), \qquad (3)$$

where S_{WH} is the average collected WirelessHART signal power, I_{CCI} is the average induced CCI power, N_{Bck} is the average collected background power, and N_{AWGN} is the average channel-filtered noise power that is added to set the desired analysis SINR. It was empirically determined that the average (cross-device, cross-burst, cross-channel) power levels before FH-TDMA processing (N_{Bck} only contributes) and after FH-TDMA processing (both $I_{CCI} + N_{Bck}$ contribute) resulted in an approximate +8.5 dB increase in the noise floor of $s_{FhTd}(t)$. The noise floor was further increased by varying N_{AWGN} to set the simulated range of +6.5 > SINR(dB) > +26.5 dB used for results presented in Section 4.

2.2. RF Signal Eventization Encoding

Threshold-based contrast encoding [30] was adopted here for initial RF eventization demonstration using the WirelessHART signals detailed in Section 2.1. The methodology here includes (1) the first application of Temporal Contrast encoding to one-dimensional Time Domain (1D-TD) WirelessHART signals in Section 2.2.1, and (2) the introduction of a Temporal–Spectral Contrast encoding approach using two-dimensional Gabor Transform (2D-GTX) responses in Section 2.2.2.

Effects of RF eventization on device classification performance were investigated using the experimentally collected WirelessHART burst responses. Figure 3a,b show the 1D-TD amplitude and 2D-GTX magnitude responses generated from the sequence of samples $\{b_{Col}(n)\}$ for a representative WirelessHART burst. Elements of the $b_{Col}(n)$ sequence are sample values of the real-valued continuous burst response $b_{Col}(t)$ taken at the t_n-th time instant. There are various eventization processes that could be considered and a select few were considered here for initial assessments. As detailed in the next two sub-sections, the threshold-based contrast encoding processes used here include (1) eventization of 1D-TD non-transformed $\{b_{Col}(n)\}$ sample sequences, and (2) eventization of transformed 2D-GTX sequences of the $\{b_{Col}(n)\}$ sequence.

For both the 1D-TD and 2D-GTX demonstrations, a threshold-based eventization process is used. For 1D-TD, a given threshold value (Thr) is applied to centered (mean removed) and normalized (values scaled to span [−1 +1]) responses. For all of the 1D-TD and 2D-GTX responses to be eventized, with CtrNrm(m) representing a given element of the centered-normalized response, the detection of an event using CtrNrm(m) includes comparing abs[CtrNrm(m)] with the Thr value and declaring one of two conditions: (1) for abs[CtrNrm(m)] ≥ Thr, a detection is declared and an eventization value of sign[CtrNrm(m)] = ±1 is assigned to the mth element where sign[x] is the sign of x—this is the detection of what are called positive (+1) and negative (−1) events; and (2) for abs[CtrNrm(m)] < Thr, a value of zero is assigned for the mth element to indicate no event was detected.

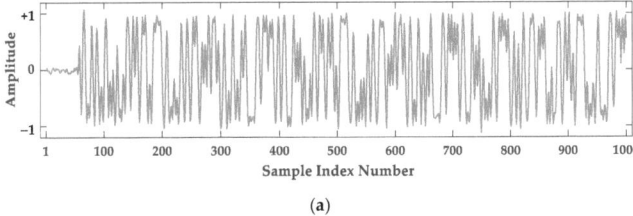

(a)

Figure 3. *Cont.*

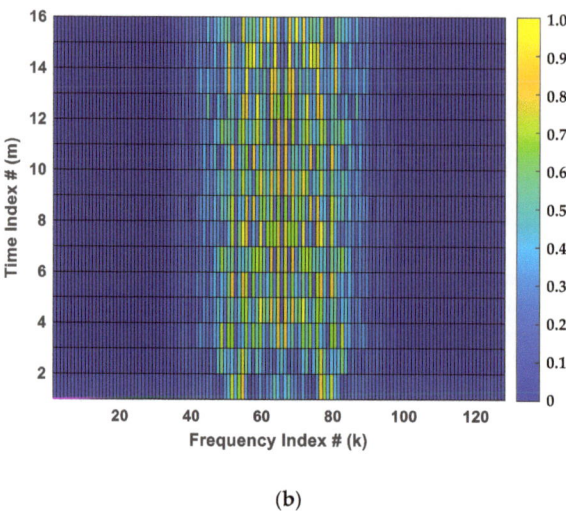

(b)

Figure 3. Illustration showing (**a**) one-dimensional Time Domain (1D-TD) amplitude and (**b**) two-dimensional Gabor transform (2D-GTX) magnitude responses for a representative WirelessHART burst. The normalized 2D-GTX magnitude response was generated using the processing in Section 2.2.2 with MTim = 16, KFrq = 128, and N_Δ = 64 time sample shifts between transforms.

2.2.1. One-Dimensional Time Domain (1D-TD) Eventization

The 1D-TD eventization processing included (1) Direct Eventization using the centered-normalized $\left\{b_{\text{Ctr-Nrm}}^{\text{1D-Dir}}(n)\right\}$ sequence samples for threshold-based event detection; (2) Integral Eventization using the centered-normalized integrated $\left\{b_{\text{Ctr-Nrm}}^{\text{1D-Int}}(n)\right\}$ sequence samples for threshold-based event detection; and (3) Derivative Eventization using the centered-normalized first derivative $\left\{b_{\text{Ctr-Nrm}}^{\text{1D-Der}}(n)\right\}$ sequence samples for threshold-based event detection. The 1D-TD centering (mean removal), normalization, thresholding, and eventization process are illustrated in Figure 4 using the first derivative $\left\{b_{\text{Ctr-Nrm}}^{\text{1D-Der}}(n)\right\}$ sequence of samples for the representative WirelessHART response in Figure 3.

The threshold-based eventization process illustrated in Figure 4 was applied to the 1D-TD direct $\left\{b_{\text{Ctr-Nrm}}^{\text{1D-Dir}}(n)\right\}$, integrated $\left\{b_{\text{Ctr-Nrm}}^{\text{1D-Int}}(n)\right\}$, and first derivative $\left\{b_{\text{Ctr-Nrm}}^{\text{1D-Der}}(n)\right\}$ sequences using selected threshold values (+/− Thr). The event declaration assignments of $e(n) \in [-1\ 0\ +1]$ were made and used to form eventized fingerprints per Equation (7) in Section 2.3.2 and used for RndF classifier training and testing. Apart from the actual eventization detections, i.e., assignment of ±1 detection values at specific n indices, results for the other 1D-TD $\left\{b_{\text{Ctr-Nrm}}^{\text{1D-Dir}}(n)\right\}$ direct and $\left\{b_{\text{Ctr-Nrm}}^{\text{1D-Int}}(n)\right\}$ integral sequences would appear similar.

2.2.2. Two-Dimensional Gabor Transform (2D-GTX) Eventization

A detailed discussion of GTX processing is omitted here for brevity. The reader is referred to the many works that have addressed GTX development and its application for improving signal joint time–frequency resolution. The GTX application trail is diverse and spans from some of the earliest signal processing development activity [37–39] to some of the most recent work [40,41]. Of most relevance here is the Gabor-based DNA fingerprinting work, with early demonstrations occurring in [42,43] and the most recent demonstrations being performed in [8]—the GTX processing implemented here was based on these works. Elements of the complex GTX matrix are denoted by GTX(k,m) for k = 1, 2, ..., KTim and m = 1, 2, ..., MFrq. The full GTX matrix generation parameters included KTim = 16,

MFrq = 128, with N_Δ = 64 time sample shifts between transformations. The analysis Gaussian window width was set to W_{GW} = 0.01. Following the generation of GTX(m,k) for a given input burst, there were two 2D-GTX eventization encoding processes considered:

1. Direct Eventization: Elements of the (MTim × KFrq)-dimensional GTX matrix such as shown in Figure 5a were generated from collected $\{b_{Col}(n)\}$ sequence samples. Centering and normalization were applied to the resultant GTX matrix such that $-1 \leq \text{abs}\left[\text{GTX}^{Dir}_{Ctr-Nrm}(m,k)\right] \leq +1$ for all m, k.

2. Derivative Eventization: The (MTim × KFrq)-dimensional GTX matrix was used with row-wise (row-by-row) differencing followed by centering and normalization such that $-1 \leq \text{abs}\left[\text{GTX}^{Der}_{Ctr-Nrm}(m,k)\right] \leq +1$ for all m, k.

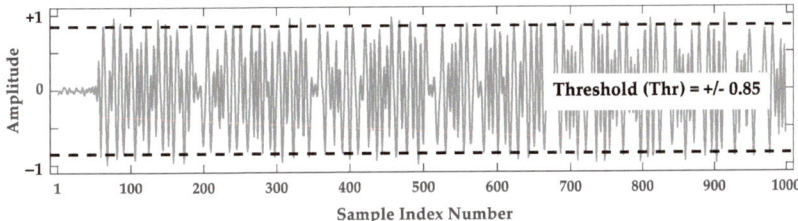

(a)

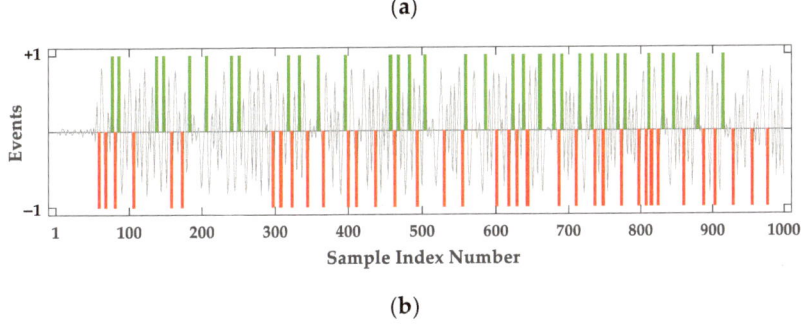

(b)

Figure 4. Illustration of 1D-TD Derivative Eventization of the WirelessHART burst in Figure 3 showing (**a**) the centered-normalized First Derivative response with a threshold value of Thr = ±0.85 (dashed lines) overlaid, and (**b**) the corresponding eventized response showing the detection of 35 Positive (green +1 indices) and 41 Negative (red −1 indices) events at the indicated time indices.

The previously detailed threshold-based eventization detection process was applied to elements of the centered-normalized matrices, i.e., CtrNrm(m,k) = $\text{abs}\left[\text{GTX}^{Dir}_{Ctr-Nrm}(m,k)\right]$ and CtrNrm(m,k) = $\text{abs}\left[\text{GTX}^{Der}_{Ctr-Nrm}(m,k)\right]$, respectively. These CtrNrm(m,k) were compared with specified threshold values (0 < Thr < 1) and event declaration assignments of $e(n) \in [-1\ 0 +1]$ made. The e(n) assignments were used to form eventized fingerprints per Equation (7) in Section 2.3.2 and used for RndF classifier training and testing.

Sequential stages of the GTX eventization process are illustrated in Figure 5 for derivative eventization of the GTX matrix in Figure 3b—the graphic illustration would be similar for direct eventization. The illustration includes (1) Figure 5a, which is the derivative GTX matrix response with the event detection ROI highlighted; (2) Figure 5b, which is an expanded view of the event detection ROI in Figure 5a; (3) Figure 5c, which is the eventized GTX derivative matrix for the ROI using a threshold value of Thr = 0.6; and (4) Figure 5d, which is the vectorized form of the GTX event matrix in Figure 5c.

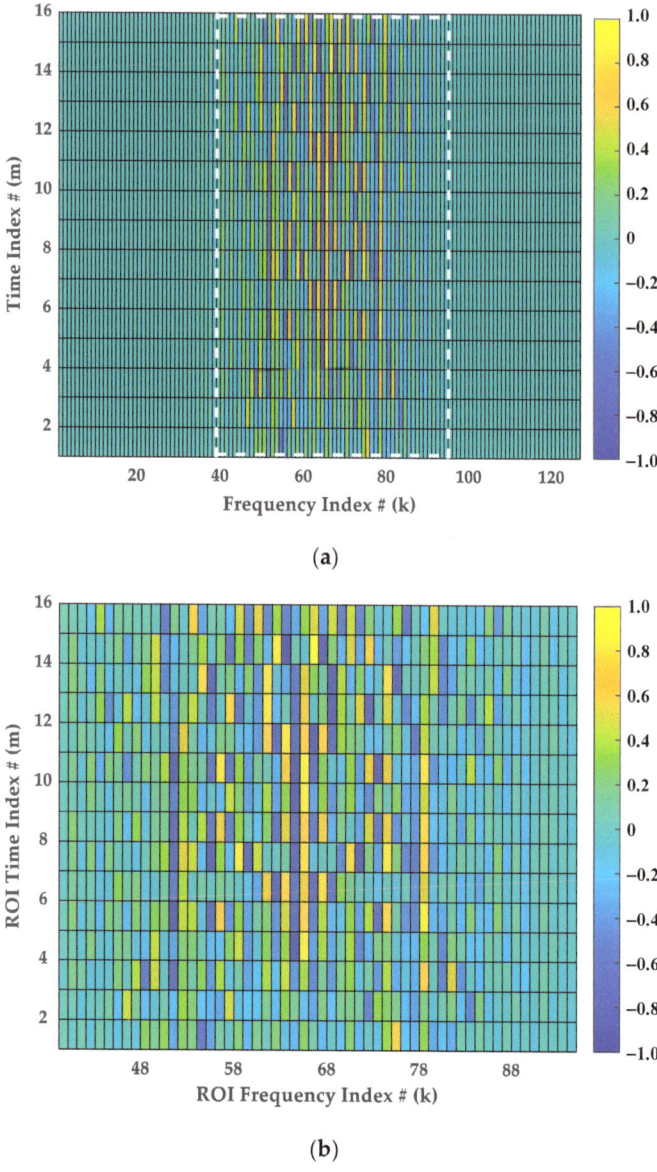

(a)

(b)

Figure 5. *Cont.*

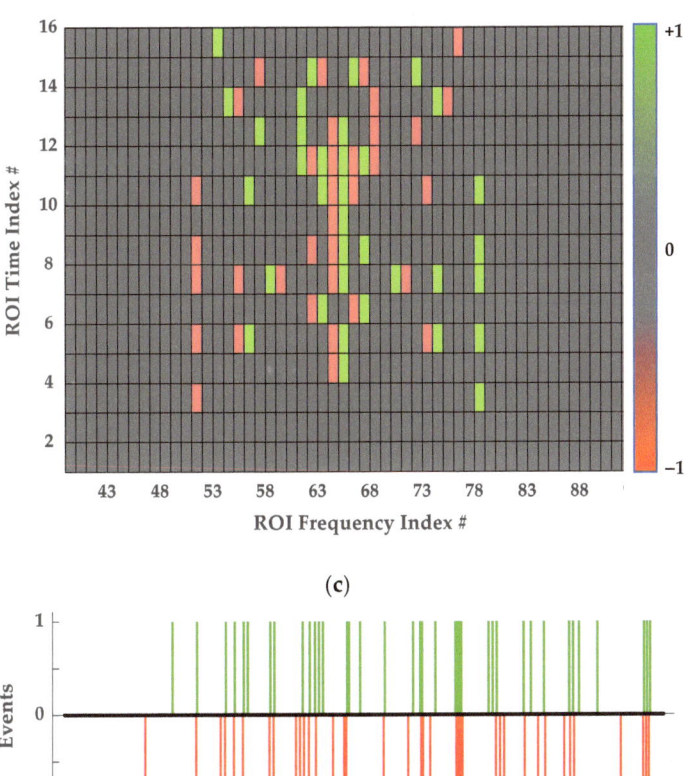

Figure 5. Illustration of 2D-GTX Derivative Eventization of the WirelessHART GTX burst response in Figure 3b. Results for (KTim × MFrq) = (16 × 128)-dimensional transform using a threshold value of Thr = 0.60. A total of 76 events were detected, including 38 Positive (+1) and 38 Negative (−1). Graphic shows (**a**) centered-normalized GTX first derivative matrix with the event detection ROI highlighted; (**b**) expanded view of the detection ROI; (**c**) three-valued (−1, 0, +1) eventized GTX matrix with locations of Positive (green elements), Negative (red elements) and no event detections (gray elements) highlighted; and (**d**) vectorized form of the GTX event matrix in (**c**) showing Positive (green +1 indices), Negative (red −1 indices) and no event (gray 0 indices). The vector elements in (**d**) are taken sequentially from (**c**) in a row-wise bottom-to-top, left-to-right, order.

2.3. Fingerprint Formation

For consistency with a majority of AFIT's historically related fingerprinting work [8,9,11,13,14,42,43], the classifier input "samples" are referred to herein as fingerprints—this helps minimize the potential confusion between classifier input "samples" and experimentally collected time domain samples in the $\{b_{Col}(n)\}$ sequence. Three different types of fingerprints were formed using N_{Col} samples of the collected burst sequence $\{b_{Col}(n)\}$. These include (1) the non-eventized as-collected $\left(\vec{\mathbf{F}}^{Col}_{NonEv}\right)$ and non-eventized statistical $\left(\vec{F}_{NonEv}^{Stat}\right)$ fingerprints detailed in Section 2.3.1, and (2) the eventized $\left(\vec{F}_{Ev}\right)$ fingerprints detailed in Section 2.3.2.

2.3.1. Non-Eventized Fingerprint Formation

Classification assessments were performed using two different Non-Eventized (NonEv) fingerprint types. The non-eventized as-collected fingerprints $\left(\vec{F}_{NonEv}^{Col}\right)$ were formed as vectors having elements equaling the N_{Col} sample values of the collected burst sequence $\{b_{Col}(n)\}$ according to

$$\vec{F}_{NonEv}^{Col} = \left[\, b_{Col}(1)\ \ b_{Col}(2)\ \ \ldots\ \ b_{Col}(n)\ \ \ldots\ \ b_{Col}(n_{N_{Col}-1})\ \ b_{Col}(n_{N_{Col}})\, \right]_{1 \times N_{Feat}} \quad (4)$$

where $b_{Col}(n)$ is the nth collected sample for $n = 1, 2, \ldots, N_{Col}$, and $N_{Feat} = N_{Col}$ is the total number of fingerprint features. This produces the highest-dimensional non-eventized fingerprints and generally requires the highest computational intensity (energy consumption) to perform classification.

The non-eventized statistical DNA fingerprints $\left(\vec{F}_{NonEv}^{DNA}\right)$ were formed from statistical features of non-transformed 1D-TD and 2D-GTX responses as illustrated in Figures 3a and 3b, respectively. This process is consistent with previous DNA fingerprint generation where the fingerprint features include statistical metrics of variance (σ^2), skewness (γ), and kurtosis (κ). These statistics are calculated using samples within selected sub-regions of (1) 1D-TD amplitude (Amp), phase (Phz), and/or frequency (Frq) responses as detailed in [14], and (2) 2D-GTX transform matrices as detailed in [8,42,43]. For each of the selected 1D-TD responses (centered-normalized amplitude, phase, and/or frequency) and 2D-GTX response (centered-normalized magnitude), the response is divided into N_{SRgn} sub-regions, and the non-eventized statistical DNA fingerprint is formed according to

$$\vec{F}_{NonEv}^{DNA} = \left[\, f_{SR_1} \vdots f_{SR_2} \vdots f_{SR_3} \vdots \ldots\ f_{SR_i}\ \ldots \vdots f_{SR_{N_{SRgn}-1}} \vdots f_{SR_{N_{SRgn}}} \, \right]_{1 \times N_{Feat}}, \quad (5)$$

where the $\vdots$ symbol denotes vector concatenation, index $i = 1, 2, \ldots, N_{SRgn}$, and N_{Feat} is the total number of fingerprint features. Considering the noted $N_{Stat} = 3$ statistical features (σ^2, γ, κ), the f_{SR_i} in Equation (5) are calculated using ith sub-region samples according to

$$f_{SR_i} = \left[\, \sigma^2_{SR_i}\ \gamma_{SR_i}\ \kappa_{SR_i}\, \right]_{1 \times 3}. \quad (6)$$

This yields a total of $N_{Feat} = N_{SRgn} \times N_{Stat} \times N_{Resp}$ features when substituting Equation (6) back into Equation (5) where N_{Resp} is the total number of instantaneous responses used (amplitude, phase, and/or frequency). Statistical DNA fingerprints generated per Equation (6) are the lowest-dimensional non-eventized fingerprints and require lower computational intensity (energy consumption) to perform classification.

2.3.2. Eventized Fingerprint Formation

The Eventized (Ev) classification assessments were performed using fingerprints formed directly from $e(n) \in [-1\ 0\ +1]$ assignments such as illustrated in the eventization vectors shown in Figures 2 and 5. The eventized fingerprint is formed according to

$$\vec{F}_{Ev} = \left[\, e(1)\ e(2)\ e(3) \ldots\ e(n)\ \ldots\ e(N_{Ev})\, \right]_{1 \times N_{Feat}}, \quad (7)$$

where $n = 1, 2, \ldots N_{Ev} = N_{Feat}$. Given a total number of N_{Ev} detected events, the nth fingerprint feature in Equation (7) is assigned as either (1) $e(n) = +/-1$ for a Positive/Negative event or (2) $e(n) = 0$ when no event is detected. The N_{Ev} events used to form the $\vec{F}_{Ev}$ fingerprint in Equation (7) equals either (1) the total number of 1D-TD eventized response samples, e.g., the total number of samples in Figure 4b, or (2) the total number of elements in the eventized 2D-GTX response, e.g., the total number of 2D-GTX elements (KTim × MFrq) in Figure 5d.

3. Device Classification

Both 1D-TD and 2D-GTX fingerprints were generated per Section 2.3 for the eventization methods in Section 2.2 and used to discriminate the WirelessHART devices listed in Table 1. For consistency with prior non-eventized DNA fingerprinting works [8,9,11,13,14,42,43], classification results are first generated for a Multiple Discriminant Analysis, Maximum Likelihood (MDA/ML) classifier adopted from these earlier works. Some fundamental details for MDA/ML processing are provided next in Section 3.1—the reader is referred to cited works therein for additional details. A Random Forest (RndF) classifier is then introduced and used to generate the primary results used for drawing conclusions related to similarity and/or differences between Non-Eventized (NonEv) and Eventized (Ev) fingerprinting performances. Fundamental details for the RndF classifier used here are provided below in Section 3.2—the reader is referred to cited works therein for additional details.

Regardless of the classification method used, the classification results reported herein are based on a classification confusion matrix, as illustrated in Table 2. This shows classification testing results using fingerprints for the 2D-GTX derivative eventization process illustrated in Figure 5. The entries are used to calculate the average cross-class percent correct %C, which is a less rigorous metric than that used herein to enhance appreciation for the work across a broader, cross-discipline readership [11]. A more detailed analysis of confusion matrix results for a multi-class classifier may be performed using the methods in [44]. The Table 2 classification results are for an N_{Cls} = 8 class (device) model using a total of N_{TST} = 1716 held-out testing fingerprints for each device. The sum of the bold diagonal entries is divided by the total number of trials represented in the confusion matrix to provide an estimated $\%C \approx [13{,}094/(8 \times 1716)] \times 100 \approx 95.38\%$. The last row in Table 2 also provides the per-class average percent correct $\%C_{Cls}$, which is calculated using the number of correct estimates for the class divided by N_{TST} = 1716.

Table 2. Classification confusing matrix for 2D-GTX derivative eventized fingerprints for an N_{Cls} = 8 class (device) classifier using N_{TST} = 1716 testing fingerprints per device. The bold diagonal entries are correct device estimates and yield an overall average cross-class percent correct of $\%C \approx 95.38$. The bottom row shows the individual per-class $\%C_{Cls}$ performance.

Actual Class	Estimated Class (%)							
	D1	D2	D3	D4	D5	D6	D7	D8
D1	**1708**	2	5	0	0	0	0	1
D2	3	**1695**	13	0	0	3	0	2
D3	0	31	**1677**	0	0	7	1	0
D4	0	0	0	**1695**	2	0	19	0
D5	0	0	1	4	**1630**	64	14	3
D6	6	21	30	0	109	**1494**	9	47
D7	0	0	63	30	65	19	**1539**	0
D8	12	10	0	0	1	37	0	**1656**
$\%C_{Cls}$	99.53%	98.78%	97.73%	98.78%	94.99%	87.06%	89.69%	96.50%

3.1. MDA/ML-Based Classification

The MDA/ML designation is used to reinforce that there are two fundamental processes involved, including MDA model development and ML class estimation. A few summary details of MDA and ML processing are provided here for completeness and are primarily taken from [11]. MDA is a multi-class ($N_{Cls} > 2$) form of Fisher's linear discriminant processing. The process takes in N_{TNG} ($1 \times N_{Feat}$)-dimensional training fingerprints ($\mathbf{F_{TNG}}$) and outputs an [$N_{Feat} \times (N_{Cls} - 1)$]-dimensional projection matrix ($\mathbf{W}$) and training fingerprint statistics—the collection of ($\mathbf{W}$, $\boldsymbol{\mu_F}$, $\boldsymbol{\sigma_F}$, $\boldsymbol{\mu}_k$, $\boldsymbol{\Sigma}_k$) for $k = 1, 2, \ldots, N_{Cls}$ that is returned from MDA processing is known as the model.

The generation of $\mathbf{W}$ is analytically driven by optimization conditions that include the pool of input N_{TNG} fingerprint features being normally distributed. The resultant matrix $\mathbf{W}$ effectively projects the N_{TNG} fingerprints into an ($N_{Cls} - 1$) decision space where the inter-class mean separation distance is maximized and the intra-class spread is minimized [45]. Given the trained MDA model, the MDA projection vector for a given testing fingerprint $\mathbf{F_U}$, i.e., an "unknown" fingerprint for one of the classes that was not used for model development, the ($N_{Cls} - 1$)-dimensional projection vector is calculated using $\mathbf{p_U} = \left[(\mathbf{F_U} - \boldsymbol{\mu_F}) \odot \boldsymbol{\sigma}_F^{-1}\right]\mathbf{W}$ where $\odot$ denotes a Hadamard product.

The resultant projection vector $\mathbf{p_U}$ is used for class estimation, a one-versus-all best match assessment where $\mathbf{p_U}$ is declared (rightly or wrongly) as coming from/belonging to the kth-class ($k = 1, 2, \ldots,$ NCls). The declaration of a class estimate can be made using many measures of similarity, including (1) a distance-based Euclidean distance measure where the class estimate is based on the minimum geometric distance between fingerprint projection $\mathbf{p_U}$ and the model training class mean $\boldsymbol{\mu}_k$ across all classes; and (2) probability-based measure where $\mathbf{p_U}$ is mapped to the a-posterior probabilities (likelihoods) of all classes and the class yielding the maximum likelihood is the estimated class. When performed under Bayesian conditions of equal a priori probabilities for all classes and equal costs for making an estimation error, the resultant probability-based estimate is referred to as a Maximum Likelihood (ML) estimate.

3.2. RndF-Based Classification

For RndF classification decisions each "unknown" $\mathbf{F_U}$ testing fingerprint is classified by each tree in the ensemble. The final output class estimate is made through a plurality vote, whereby the class receiving the most votes across the ensemble is declared as the estimate [46]—a majority of the ensemble classifiers do not have to agree with the estimate. This requires that a large ensemble of classifiers be used to ensure an unbiased and accurate classification estimate is made. In general, the use of more trees can increase classification performance but at the expense of increased estimation time. To minimize both bias and between-tree correlation across the ensemble, the trees can be grown to be full-length (maximum depth). For all RndF results presented in Section 4, the RndF classifier was implemented in a Python-based Scikit-Learn environment using the following:

- Number of Trees: 100;
- Epoch Termination Criterion: Gini Impurity;
- Maximum Tree Depth: None; resulting in full-length trees with pure leaves;
- Minimum Number of Fingerprints to Split: 2;
- Minimum Number of Fingerprints in a Leaf: 1;
- Number of features for random subspace selection: $\sqrt{N_{Feat}} = \sqrt{1000} \approx 32$.

One unique benefit is that RndF processing enables the calculation and output of Gini index relevance values for all features. The Gini index value reflects the relative importance of a given feature on the final classification estimate—a higher Gini Index value indicates a greater influence on the final class estimate. Gini index ranking was successfully exploited in previous DNA-based fingerprinting work [12] using the same WirelessHART devices being used here. This work demonstrated fingerprint dimensional reduction that included Gini-based identification and removal of the lowest-ranked, most irrelevant 80% of the fingerprint features while sacrificing a marginal $1\% < \%C_\Delta < 2\%$ degradation in overall classification performance.

Representative Gini index values are presented in Figure 6 for an $N_{Cls} = 8$ RndF classifier using 1D-TD eventized fingerprints. The eventized fingerprints from Equation (7) included a total of $N_{Feat} = 1000$ features and an average number of detected Events-Per-Burst equaling $N_{EPB} = 206$ events—as introduced here the N_{EPB} average is calculated across eventized fingerprints for all $N_{Brst} = 8576$ bursts from each of the $N_{Dev} = 8$ devices. Figure 6a shows the sorted (highest-to-lowest) Gini values and Figure 6b shows the non-sorted Gini values for each feature index number. The feature-by-feature retention/removal of higher/lower significant features is clearly supported and suggests that the dimensional reduction benefit demonstrated in [12] may be realizable for the eventized fingerprints being considered here—this investigation remains a topic for future research.

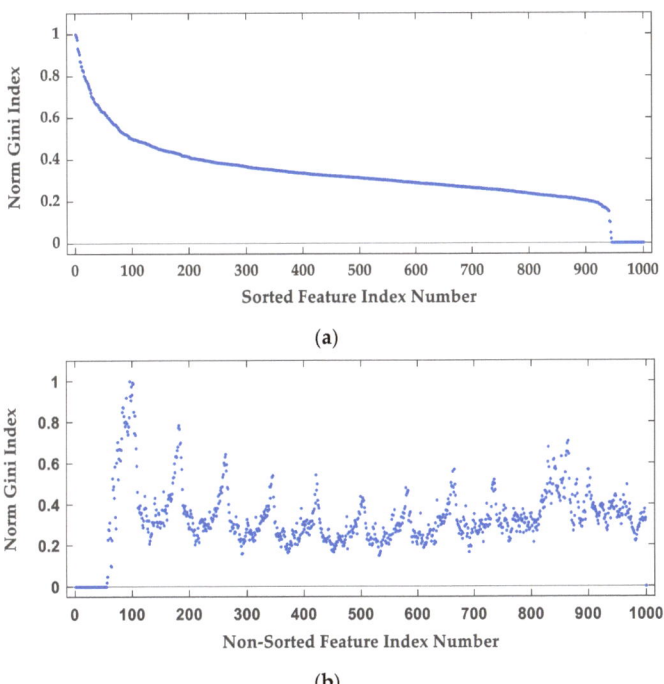

Figure 6. RndF Gini index relevance using 1D-TD derivative eventized fingerprints having $N_{Feat} = 1000$ features: (**a**) Sorted rank-ordered relevance, and (**b**) Non-sorted per-feature relevance.

4. Device Classification Results

Both 1D-TD and 2D-GTX fingerprints were generated per Section 5 and used to discriminate WirelessHART signals. For consistency with prior non-eventized DNA fingerprinting works [8,9,11,13,14,42,43], classification was first performed with a Multiple Discriminant Analysis, Maximum Likelihood (MDA/ML) classifier—the reader is referred to the noted references for MDA/ML-based device discrimination details. The RndF classifier detailed in Section 3.2 was then introduced and an MDA/ML vs. RndF classification comparison was performed.

Figure 7 shows classification results for both classifiers and $N_{Cls} = N_{Dev} = 8$ classes using both non-eventized (NonEv) 1D-TD and 2D-GTX DNA fingerprint features. The sets of $N_{FP} = N_{Brst} = 8576$ 1D-TD and 2D-GTX DNA fingerprints per device were generated using Equations (5) and (6). The total number of available fingerprints was then divided into an 80% training subset and a 20% testing subset. Each subset was stratified and balanced to ensure an equal number of class representations. Ultimately, 54,880 fingerprints were used for training, and 13,728 fingerprints were used for testing.

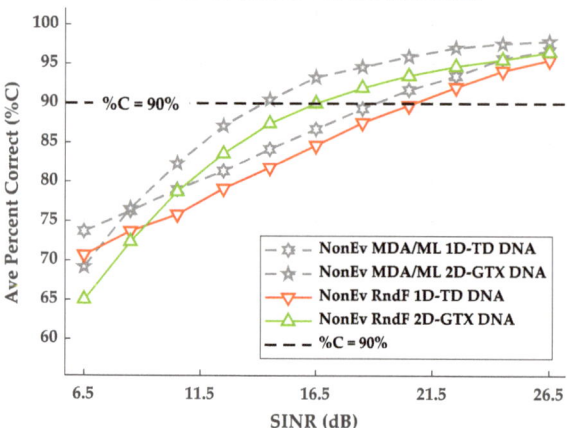

Figure 7. MDA/ML and RndF classification performance using Non-Eventized (NonEv) statistical 1D-TD DNA and 2D-GTX DNA fingerprints generated from Equation (6).

For 1D-TD DNA fingerprinting results in Figure 7, the statistical $\vec{F}_{NonEv}^{Stat}$ fingerprints in Equation (6) were formed using (1) N_{Resp} = 3 (Amp, Phz, Frq) non-eventized burst responses; (2) N_{SRgn} = 16 fingerprinting regions per response; and (3) N_{Stat} = 3 statistics (σ^2, γ, κ) to form vectors per Equation (5). The statistical vectors from Equation (6) were used in Equation (5) to form the final composite $\vec{F}_{NonEv}^{1D-TD} = \vec{F}_{NonEv}^{Stat}$ fingerprints. The resultant $\vec{F}_{NonEv}^{1D-TD}$ fingerprints used for Figure 7 results had a total of $N_{Feat}^{1D-TD} = 3 \times 16 \times 3 = 144$ features.

For 2D-GTX DNA fingerprinting results in Figure 7, the statistical $\vec{F}_{NonEv}^{Stat}$ fingerprints in Equation (5) were formed by (1) dividing non-eventized 2D-GTX matrices into $\left(N_{TimBlk} \times N_{FrqBlk}\right)$-dimensional sub-matrices where N_{TimBlk} and N_{FrqBlk} denote the number of time and frequency blocks, respectively; (2) calculating N_{Stat} = 3 statistics (σ^2, γ, κ) for each of the $(N_{TimBlk} = 9) \times \left(N_{FrqBlk} = 6\right) = 54$ sub-matrices; and (3) forming the statistical feature vectors per Equation (6). The statistical vectors from Equation (6) were used in Equation (5) to form the final composite $\vec{F}_{NonEv}^{2D-GTX} = \vec{F}_{NonEv}^{Stat}$ fingerprints. The resultant $\vec{F}_{NonEv}^{2D-GTX}$ fingerprints used for Figure 7 results had a total of $N_{Feat}^{2D-GTX} = 9 \times 6 \times 3 = 162$ features.

In light of next-step performance assessments in the following sub-sections using eventized fingerprints, there are two conclusions to be considered from Figure 7:

- 2D-GTX fingerprinting outperforms 1D-TD fingerprinting for both classifiers across a majority of the SINR ≥ 10.5 dB considered—the 2D-GTX vs. 1D-TD benefit here is consistent with prior 2D-GTX DNA fingerprinting findings [8,42,43];
- For best-case 2D-GTX fingerprints, MDA/ML (unfilled ★) outperforms RndF (unfilled ▽) by an average of $\%C_\Delta^{GTX-TD} \approx +2.97\%$ across the range of SINR(dB) considered. This includes operating under lower $SINR_\Delta^{GTX-TD} \approx -4$ dB channel conditions to achieve an arbitrary performance benchmark of $\%C = 90\%$.

In transitioning to eventized classification performance results in the following sub-sections, it is noted that the eventized $\vec{F}_{Ev}$ fingerprints are formed using Equation (7) and contain zero-valued elements. Thus, the statistical distribution of eventized fingerprint features is binary and not well-suited for MDA/ML classification—the development of MDA models and classification performance is dependent upon the input features being normally distributed. This is accounted for in MDA projection matrix formation, which requires the inversion of a composite input fingerprint matrix. As the number of zero-valued fingerprint features increases, the matrix becomes non-invertible (determinant approaches zero) and projection matrix formation is not possible. Thus, all results in the following sub-sections were obtained using an RndF classier with the parameters noted in Section 3.2.

4.1. Eventized Classification

Classification performance using 1D-TD and 2D-GTX eventized fingerprints was next considered for the eventization methods detailed in Section 2.2. The results were generated using the same RndF classifier detailed in Section 3.2 and used for the non-eventized results in Figure 7. The 1D-TD vs. 2D-GTX RndF classification results are presented in Figure 8. The eventized $\vec{\mathbf{F}}_{Ev}^{1D-TD}$ and $\vec{\mathbf{F}}_{Ev}^{2D-GTX}$ fingerprints were formed using Equation (7) for each of the Direct, Derivative, and Integral eventization methods considered. The fingerprint features included the detected eventization values (-1, $+1$) such as those illustrated in Figure 4b (TD-DNA) and Figure 5d (GTX-DNA), and zero values at remaining indices where no event is detected.

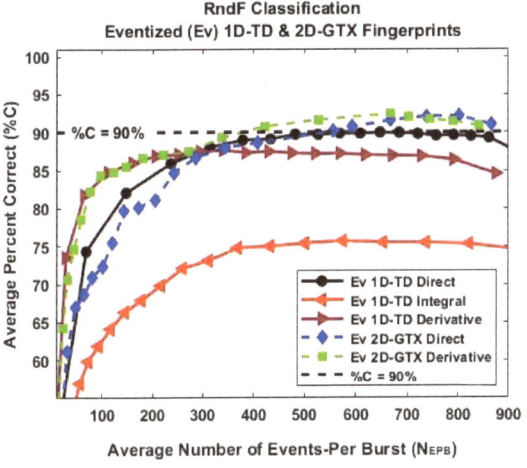

Figure 8. RndF classification performance for SINR = 16.5 dB channel conditions using 1D-TD and 2D-GTX eventized fingerprints. Results presented for a varying number of average Events-Per-Burst (N_{EPB}) that were generated by varying the method-dependent threshold values.

One operational implementation objective includes minimizing the number of fingerprint features, which in turn minimizes the required computational resources. This is supported by (1) setting eventization thresholds to minimize the number of detected Events-Per-Burst (N_{EPB}), while (2) achieving a given %C classification performance. As indicated in Table 3, the resultant N_{EPB} and %C classification performance is a function of the eventization method employed (method-dependent) and the specific threshold value (threshold-dependent) used with a given method. The results in Table 3 are a selected subset of results obtained for a range of empirical thresholds spanning $0.01 < \text{Thr} < 0.75$ in $\Delta \text{Thr} = 0.05$ increments. Table 3 shows the threshold value yielding the minimum N_{EPB} to achieve the maximum asymptotic %C for that threshold. As first used in Table 3, the color coded symbology for the various eventization methods is retained throughout the remainder of the paper in tables and figures to assist in tracking per-method performances.

Table 3. Method-dependent threshold values and resultant minimum N_{EPB} required to achieve the indicated maximum %C classification performance.

Eventization Method		Threshold (Thr)	Minimum N_{EPB}	Maximum %C
1D-TD	Direct (●)	0.60	482	89.71
	Integral (◄)	0.40	368	74.77
	Derivative (▶)	0.60	295	87.43
2D-GTX	Direct (◆)	0.05	738	92.06
	Derivative (■)	0.10	526	91.62

The %C classification results in Figure 8 were obtained for the indicated average (cross-burst, cross-device) number of N_{EPB} using a given eventization method. Relative to achieving a minimum N_{EPB} demonstration objective, there are several conclusions drawn from Figure 8.

- The 1D-TD Integral (◄) fingerprinting performance does not achieve the arbitrary performance benchmark of %C = 90% for all 50 < N_{EPB} < 900 considered—integral-based eventization was henceforth removed from further consideration;
- The 1D-TD Direct (●) fingerprinting performance is second-best and approaches the %C = 90% benchmark for N_{EPB} > 480. Of note here is the poorer performing 1D-TD Derivative (▶) fingerprints, which do achieve the %C = 90% benchmark for any N_{EPB} considered. Thus, the penalty in classification performance for taking a time domain derivative is severe, and the computational cost for doing so should be avoided;
- The 2D-GTX Derivative (■) fingerprinting performance is best overall across the entire range of 50 < N_{EPB} < 850 considered. This includes the %C = 90% benchmark being achieved for N_{EPB} > 380 detected events—the previously demonstrated benefit of eventized over non-eventized fingerprinting in Figure 7 is retained.

The classification performance for both 1D-TD (▶) and 2D-GTX (■) Derivative fingerprints is statistically equivalent over an approximate range of 50 < N_{EPB} < 215. For N_{EPB} > 215, the 2D-GTX (■) derivative fingerprints provide a clear benefit that includes an increase in performance of 1% < %C_Δ < 12.2%. This benefit is realized at the expense of performing one additional processing step that included taking a row-wise derivative of the input GTX matrix before performing eventization. There are other processing alternatives that may be considered (e.g., a column-wise derivative) to achieve similar or even greater benefit—alternatives remain on the list for future consideration.

4.2. Non-Eventized vs. Eventized Classification

To directly highlight the effect of eventization on device classification performance, a comparison of Non-Eventized (NonEv) vs. Eventized (Ev) RndF classification was performed. As presented in Figure 9, the comparisons were performed for SINR = 16.5 dB channel conditions using eventized fingerprints based on N_{EPB} = 600 events. A selection of these values is based on RndF achieving the %C = 85% benchmark, with SINR selected from Figure 7 and N_{EPB} selected from Figure 8 where all fingerprinting methods achieved the benchmark. The performance degradation resulting from eventization is reflected in the eventization difference metric %$C_{\Delta Ev} \equiv$ %C_{Ev} − %C_{NonEv} shown in Table 4 where %C_{Ev} is the performance using eventized fingerprints and %C_{NonEv} is the performance using non-eventized fingerprints.

As indicated in Table 4 %$C_{\Delta Ev}$ results, all eventization methods degraded performance to some degree with the GTX-based eventization methods degrading performance by %$C^{2D-Dir}_{\Delta Ev} \approx -4.96$% and %$C^{2D-Der}_{\Delta Ev} \approx -2.88$%. It is not unreasonable to expect that this level of degradation will ultimately be acceptable when considering other benefits provided by eventization. For example, achieving the %C = 90% benchmark in Figure 8 requires $N_{EPB} \approx 380$ events using the 2D-GTX derivative $\mathbf{F}^{\to 2D-GTX}_{Der-Ev}$ fingerprints that include an average of $(N_S = 1250) - (N_{EPB} \approx 380) = 870$ zero-valued elements. It is envisioned that a form of dimensional reduction (compression) may be possible for the derivative $F_{Der-Ev}^{\rightarrow 2D-GTX}$ fingerprints by removing zero-valued elements that are common across the pool of device fingerprints being used. If this can be done, there will be an approximate $[(1250 - 380)/1250] \times 100 \approx 70\%$ reduction in the number of eventized fingerprint features—this would decrease the computational processing requirements and may sufficiently offset the $-3\% < \%C_{\Delta Ev}^{2D-Der} < -4\%$ loss observed here with the RndF classifier. Furthermore, this level of degradation is considered marginal in light of the energy savings demonstrated by using SNN classification architectures [15].

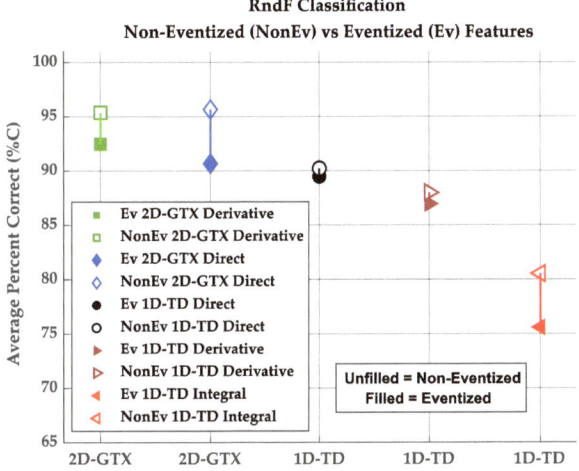

Figure 9. Summary of RndF classification performance using 1D-TD and 2D-GTX Non-Eventized (NonEv) and Eventized (Ev) fingerprints. Results are shown for SINR = 16.5 dB channel conditions and an average of Number of Events-Per-Burst of $N_{EPB} = 600$ for all eventization methods. The effect of eventization is reflected in the $\%C_{\Delta Ev}$ metrics with a negative ($-$) sign denoting degradation and a positive ($+$) sign denoting improvement.

Table 4. Comparison of non-eventized and eventized classification performance is shown in Figure 9 for each of the eventization methods considered.

Eventization Method		$\%C_{NonEv}$	$\%C_{Ev}$	$\%C_{\Delta Ev}$
1D-TD	Direct (●)	90.22	89.50	−0.72
	Integral (◀)	80.55	75.60	−4.95
	Derivative (▶)	88.02	87.00	−1.02
2D-GTX	Direct (◆)	90.70	95.66	−4.96
	Derivative (■)	92.50	95.38	−2.88

5. Summary and Conclusions

This paper summarizes first-step research activity aimed at realizing an envisioned "event radio" capability that mimics neuromorphic event-based camera processing. The energy efficiency of neuromorphic processing is orders of magnitude higher than traditional von Neumann-based processing architectures ($10\times$ to $1000\times$ [17–19]) and realized through synergistic design of brain-inspired software [47] and hardware [22] computing elements. The development and availability of event-based hardware devices supporting Radio Frequency (RF) applications are severely lagging when compared with activity in the event-based camera arena. Despite this lag, there remains considerable interest across the technical community in obtaining neuromorphic efficiency benefits through event-based

RF signal processing. For the Operational Technology (OT) protection needs addressed here, this processing includes efficient software computing capability to provide reliable device classification.

While not envisioned as the final solution, a Random Forest (RndF) classifier is first considered here as a reliable precursor to using a Convolutional Neural Network (CNN) classifier, which is, in turn, a precursor to eventual demonstrations using the desired neuromorphic-friendly Spiking Neural Network (SNN) classifier. The event-based RF fingerprints used for RndF device classification were generated from experimentally collected signals for N_{Dev} = 8 WirelessHART devices. The signals were pre-conditioned prior to fingerprint generation to induce the generally adverse effects of Cross-Channel Interference (CCI). The induced CCI is a result of the N_{Dev} = 8 devices simultaneously operating in N_{Ch} = 8 different FH-TDMA network channels [14]. For completeness, classification performance for non-eventized fingerprints was performed as well and used as a baseline for characterizing the effects of eventization on classifier performance. The average %C correct classification results show the following:

1. 2D-GTX non-eventized fingerprinting performs best with an MDA/ML classifier across the range of the 6.5 < SINR < 26.5 dB considered;
2. 2D-GTX eventized derivative fingerprinting performs best with a RndF classifier under SINR = 16.5 dB channel conditions across the range of the 50 < N_{EPB} < 900 considered;
3. 2D-GTX derivative-based eventization suffers a marginal performance penalty of %$C_{\Delta Ev}^{2D-GTX} \approx -2.88\%$ when compared with non-eventized performance;
4. An arbitrary performance benchmark of %$C \geq 90\%$ is achieved under SINR = 16.5 dB channel conditions using an average of $N_{EPB} \geq 400$ detected events per burst.

While RndF is not envisioned as being the final solution for achieving neuromorphic benefits in an event radio, the 2D-GTX eventized classification performance is promising and sufficiently motivates next-step demonstration activity, which includes (1) fingerprint dimensional reduction (compression) supported using rank-ordered RndF Gini feature relevance values such as illustrated in Figure 6—such a reduction minimizes the required computational generation, storage, transfer, complexity, etc.; and (2) transitioning to a Convolutional Neural Network (CNN) classifier as a precursor to near-end-game demonstrations using the desired neuromorphic-friendly Spiking Neural Network (SNN) classifier—RndF performance has historically provided a reliable baseline for highlighting the classification improvement that is achievable using a CNN classifier with similarly structured signals. It is believed that classification performance will remain favorable and represent one significant step toward identifying the software computing elements required to implement an event radio. It is also hoped that this will motivate researchers and designers to develop complementary hardware computing elements so that the full benefits of neuromorphic processing may be realized in the envisioned event radio.

Author Contributions: Conceptualization, M.J.S. and M.A.T.; Data curation, M.A.T.; Formal analysis, M.J.S., M.A.T. and J.W.D.; Investigation, M.J.S.; Methodology, M.J.S., M.A.T. and J.W.D.; Project administration, M.A.T. and J.W.D.; resources, M.A.T. and J.W.D.; supervision, M.A.T.; graphic visualization, M.J.S. and M.A.T.; writing—original draft, M.A.T.; writing—review and editing, M.J.S., M.A.T. and J.W.D. All authors have read and agreed to the published version of the manuscript.

Funding: This research was funded in part by support funding received from the Spectrum Warfare Division, Sensors Directorate, U.S. Air Force Research Laboratory, Wright-Patterson AFB, Dayton OH, during U.S. Government Fiscal Years 2021–2023.

Data Availability Statement: The experimentally collected WirelessHART data used to obtain results were not approved for public release at the time of paper submission. Requests for release of these data to a third party should be directed to the corresponding author. Data distribution to a third party will be made on a request-by-request basis and are subject to public affairs approval.

Acknowledgments: The views and conclusions contained in this document are those of the authors and should not be interpreted as representing the official policies, either expressed or implied, of the United States Air Force or the U.S. Government. This paper is approved for public release, Case Number 88ABW-2024-0175.

Conflicts of Interest: The authors declare no conflicts of interest. The funders had no role in the design of the study; in the collection, analyses, or interpretation of data; in the writing of the manuscript, or in the decision to publish the results.

Abbreviations

The following abbreviations are used throughout the manuscript:

1D-TD	One-Dimensional Time Domain
2D-GTX	Two-Dimensional Gabor Transform
CCI	Cross-Channel Interference
CNN	Convolutional Neural Network
DNA	Distinct Native Attribute
FH-TDMA	Frequency-Hopped, Time Division Multiple Access
FS	Frequency Slot
GTX	Gabor Transform
HART	Highway Addressable Remote Transducer
ID	Identity/Identification
IoT	Internet of Things
IIoT	Industrial Internet of Things
MDA	Multiple Discriminant Analysis
MHz	Megahertz
ML	Maximum Likelihood
MSps	Mega-Samples Per Second
RndF	Random Forest
SDR	Software Defined Radio
SNN	Spiking Neural Network
TS	Time Slot

References

1. FieldComm Group. WirelessHART: Proven and Growing Technology with a Promising Future, Global Control. 29 March 2018. Available online: https://tinyurl.com/fcgwirelesshartglobalcontrol (accessed on 11 May 2024).
2. Neshenko, N.; Bou-Harb, E.; Crichigno, J.; Kaddoum, G.; Ghani, N. Demystifying IoT Security: An Exhaustive Survey on IoT Vulnerabilities and a First Empirical Look on Internet-Scale IoT Exploitations. *IEEE Commun. Surv. Tutor.* **2019**, *21*, 2702–2733. [CrossRef]
3. Liu, X.; Qian, C.; Hatcher, W.G.; Xu, H.; Liao, W.; Yu, W. Secure Internet of Things (IoT)-based Smart-World Critical Infrastructures: Case Studies and Research Opportunities. *IEEE Access* **2019**, *7*, 79523–79544. [CrossRef]
4. Cyber Security and Infrastructure Agency (CISA). Cybersecurity and Physical Security Convergence. 2021. Available online: https://www.cisa.gov/cybersecurity-and-physical-security-convergence (accessed on 11 May 2024).
5. Stouffer, K.; Pease, M.; Tang, C.; Zimmerman, T.; Pillitteri, V.; Lightman, S. *Guide to Operational Technology (OT) Security*; Special Publication, NIST.SP.800-82r3; National Institute of Standards (NIST): Gaithersburg, MD, USA, 2023.
6. Soltanieh, N.; Norouzi, Y.; Yang, Y.; Karmakar, N.C. A Review of Radio Frequency Fingerprinting Techniques. *IEEE J. Radio Freq. Identif.* **2020**, *4*, 222–233. Available online: https://ieeexplore.ieee.org/document/8970312. (accessed on 11 May 2024). [CrossRef]
7. Fadul, M.K.M.; Reising, D.R.; Weerasena, L.P.; Loveless, T.D.; Sartipi, M.; Tyler, J.H. Improving RF-DNA Fingerprinting Performance in an Indoor Multipath Environment Using Semi-Supervised Learning. *IEEE Trans. Inf. Forensics Secur.* **2024**, *19*, 3194–3209. Available online: https://ieeexplore.ieee.org/document/10417054 (accessed on 11 May 2024). [CrossRef]
8. Mims, W.H.; Temple, M.A.; Mills, R.F. Active 2D-DNA Fingerprinting of WirelessHART Adapters to Ensure Operational Integrity in Industrial Systems. *Sensors* **2022**, *22*, 4906. Available online: https://www.mdpi.com/1424-8220/22/13/4906 (accessed on 11 May 2024). [CrossRef]
9. Long, J.D.; Temple, M.A.; Rondeau, C.M. Discriminating WirelessHART Communication Devices Using Sub-Nyquist Stimulared Responses. *Electronics* **2023**, *12*, 1973. [CrossRef]
10. Devan, P.A.M.; Hussin, F.A.; Ibrahim, R.; Bingi, K.; Khanday, F.A. A Survey on the Application of WirelessHART for Industrial Process Monitoring and Control. *Sensors* **2021**, *21*, 4951. [CrossRef]
11. Rondeau, C.M.; Temple, M.A.; Betances, J.A.; Schubert Kabban, C.M. Extending Critical Infrastructure Element Longevity Using Constellation-Based ID Verification. *J. Comput. Secur.* **2020**, *100*, 102073. [CrossRef]
12. Rondeau, C.M.; Temple, M.A.; Schubert Kabban, C.M. TD-DNA Feature Selection for Discriminating WirelessHART IIoT Devices. In Proceedings of the 53rd Hawaii International Conference on System Sciences (HICSS), Maui, HI, USA, 7–10 January

13. Gutierrez, J.A.; Borghetti, B.J.; Temple, M.A. Considerations for RF Fingerprinting across Multiple Frequency Channels. *Sensors* **2022**, *22*, 2111. [CrossRef]
14. Maier, M.J.; Hayden, H.S.; Temple, M.A.; Fickus, M.C. Ensuring the Longevity of WirelessHART Devices in Industrial Automation and Control Systems Using Distinct Native Attribute Fingerprinting. *Int. J. Crit. Infrastruct. Prot.* **2023**, *43*, 100641. [CrossRef]
15. Christensen, D.V.; Dittmann, R.; Linares-Barranco, B.; Sebastian, A.; Le Gallo, M.; Redaelli, A.; Slesazeck, S.; Mikolajick, T.; Spiga, S.; Menzel, S.; et al. 2022 Roadmap on Neuromorphic Computing and Engineering. *Neuromorphic Comput. Eng.* **2022**, *2*, 022501. [CrossRef]
16. Nunes, J.D.; Carvalho, M.; Carneiro, D.; Cardoso, J.S. Spiking Neural Networks: A Survey. *IEEE Access* **2022**, *10*, 60738–60764. Available online: https://ieeexplore.ieee.org/document/9787485 (accessed on 11 May 2024). [CrossRef]
17. Cai, F.; Kumar, S.; Van Vaerenbergh, T.; Sheng, X.; Liu, R.; Li, C.; Liu, Z.; Foltin, M.; Yu, S.; Xia, Q.; et al. Power-Efficient Combinatorial Optimization using Intrinsic Noise in Memristor Hopfield Neural Networks. *Nat. Electron.* **2020**, *3*, 409–418. Available online: https://www.nature.com/articles/s41928-020-0436-6 (accessed on 11 May 2024). [CrossRef]
18. Chatterjee, B.; Panda, P.; Maity, S.; Roy, K.; Sen, S. An Energy-Efficient Mixed-Signal Neuron for Inherently Error-Resilient Neuromorphic Systems. In Proceedings of the 2017 IEEE International Conference on Rebooting Computing (ICRC), Washington, DC, USA, 8–9 November 2017.
19. Clark, K.; Wu, Y. Survey of Neuromorphic Computing: A Data Science Perspective. In Proceedings of the IEEE 3rd International Conference on Computer Communication and Artificial Intelligence (CCAI), Taiyuan, China, 26–28 May 2023.
20. Gallego, G.; Delbrück, T.; Orchard, G.; Bartolozzi, C.; Taba, B.; Censi, A.; Leutenegger, S.; Davison, A.J.; Conradt, J.; Daniilidis, K.; et al. Event-Based Vision: A Survey. *IEEE Trans. Pattern Anal. Mach. Intell.* **2022**, *44*, 154–180. Available online: https://ieeexplore.ieee.org/document/9138762 (accessed on 11 May 2024). [CrossRef]
21. Davies, M.; Srinivasa, N.; Lin, T.H.; Chinya, G.; Cao, Y.; Choday, S.H.; Dimou, G.; Joshi, P.; Imam, N.; Jain, S.; et al. Loihi: A Neuromorphic Manycore Processor with On-chip Learning. *IEEE Micro* **2018**, *38*, 82–99. Available online: https://ieeexplore.ieee.org/document/8259423 (accessed on 11 May 2024). [CrossRef]
22. Farr, P.; Jones, A.M.; Bihl, T.; Boubin, J.; DeMange, A. Waveform Design Implemented on Neuromorphic Hardware. In Proceedings of the 2020 IEEE International Radar Conference (RADAR), Washington, DC, USA, 28–30 April 2020; pp. 934–939. Available online: https://ieeexplore.ieee.org/document/9114635 (accessed on 11 May 2024).
23. Ammar, M.A.; Abdel-Latif, M.S.; Badran, K.M.; Hassan, H.A. Deep Learning Achievements and Opportunities in Domain of EW Applications. In Proceedings of the 2021 Tenth International Conference on Intelligent Computing and Information Systems (ICICIS), Cairo, Egypt, 5–7 December 2021; Available online: https://ieeexplore.ieee.org/document/9694245/metrics#metrics (accessed on 11 May 2024).
24. Zhu, B.; Perez, A.; Hernandez, J. Event-based Time Series Data Preprocessing: Application to Traffic Flow Time Series. In Proceedings of the ITISE 2014: International Work-Conference on Time Series, Granada, Spain, 25–27 June 2014; Available online: https://oa.upm.es/36830/ (accessed on 11 May 2024).
25. Guralnik, V.; Srivastava, J. Event Detection From Time Series Data. In Proceedings of the Fifth ACM SIGKDD International Conference on Knowledge Discovery and Data Mining (KDD-99), San Diego, CA, USA, 15–18 August 1999; Available online: https://dl.acm.org/doi/pdf/10.1145/312129.312190 (accessed on 11 May 2024).
26. Benko, Z.; Babel, T.; Somogyvari, Z. Model-free Detection of Unique Events in Time Series. *Sci. Rep.* **2022**, *12*, 227. [CrossRef]
27. Auge, D.; Hille, J.; Mueller, E.; Knoll, A. A Survey of Encoding Techniques for Signal Processing in Spiking Neural Networks. *Neural Process Lett.* **2021**, *53*, 4693–4710. [CrossRef]
28. Petro, B.; Kasabov, N.; Kiss, R. Selection and Optimization of Temporal Spike Encoding Methods for Spiking Neural Networks. *IEEE Trans. Neural Netw.* **2020**, *31*, 358–370. Available online: https://ieeexplore.ieee.org/document/8689349 (accessed on 11 May 2024). [CrossRef] [PubMed]
29. Yarga, S.; Rouat, J.; Wood, S. Efficient Spike Encoding Algorithms for Neuromorphic Speech Recognition. In Proceedings of the International Conference on Neuromorphic Systems (ICONS), Knoxville, TN, USA, 27–29 July 2022; Available online: https://arxiv.org/pdf/2207.07073 (accessed on 11 May 2024).
30. Forno, E.; Fra, V.; Pignari, R.; Macii, E.; Urgese, G. Spike Encoding Techniques for IoT Time-Varying Signals Benchmarked on a Neuromorphic Classification Task. *Front. Neurosci.* **2022**, *16*, 999029. [CrossRef]
31. Siemens. WirelessHART Adapter, SITRANS AW210, 7MP3111. *User Manual*. November 2012. Available online: https://tinyurl.com/yyjbgybm (accessed on 11 May 2024).
32. Pepperl+Fuchs. WHA-BLT-F9D0-N-A0-*, WirelessHART Adapter. Manual. Available online: https://tinyurl.com/pepplusfucwirelesshart (accessed on 11 May 2024).
33. Emerson. Emerson Wireless 1410 Gateway. *Reference Manual 00809-0200-4410, Rev CA*, Sep 2020. Available online: https://www.emerson.com/documents/automation/manual-emerson-smart-wireless-gateway-1410-en-77632.pdf (accessed on 16 May 2024).
34. Ettus Research. USRP X300 and X300 X Series. *Specification Sheet*. Available online: https://www.ettus.com/wp-content/uploads/2024/01/X300_X310_Spec_Sheet_2024-01-23.pdf (accessed on 11 May 2024).
35. IEEE Standards Association. *Part 15.4: Low-Rate Wireless Personal Area Networks (LR-WPANs)*; IEEE Std 802.15.4™-2011; IEEE: New York, NY, USA, 2011.

36. HART Communication Foundation. Co-Existence of WirelessHART with Other Wireless Technologies. Tech Note: HCF_LIT-122, Rev. 1.0. 2010. Available online: https://www.emerson.com/documents/automation/white-paper-co-existence-of-wirelesshart-other-wireless-technologies-by-hcf-en-42582.pdf. (accessed on 11 April 2024).
37. Bastiaans, M.J. Gabor's Expansion of a Signal into Gaussian Elementary Signals. *Proc. IEEE* **1980**, *68*, 538–539. Available online: https://ieeexplore.ieee.org/document/1455955 (accessed on 11 May 2024). [CrossRef]
38. Bastiaans, M.J.; Geilen, M.C.W. On the Discrete Gabor Transform and the Discrete Zak Transform. *Signal Process.* **1996**, *49*, 151–166.
39. Qian, S.; Chen, D. Discrete Gabor Transform. *IEEE Trans. Signal Process.* **1993**, *41*, 2429–2438. Available online: https://ieeexplore.ieee.org/document/224251 (accessed on 11 May 2024). [CrossRef]
40. Rathinaraj, J.D.J.; McKinley, G.H. Gaborheometry: Applications of the Discrete Gabor Transform for Time Resolved Oscillatory Rheometry. *J. Rheol.* **2023**, *67*, 479–497. [CrossRef]
41. Lindeberg, T. A Time-causal and Time-recursive Analogue of the Gabor transform. *arXiv* **2023**, arXiv:2308.14512. [CrossRef]
42. Reising, D.R.; Temple, M.A.; Jackson, J.A. Authorized and Rogue Device Discrimination Using Dimensionally Reduced RF-DNA Fingerprints. *IEEE Trans. Inf. Forensics Secur.* **2015**, *10*, 1180–1192. Available online: https://ieeexplore.ieee.org/document/7031931 (accessed on 11 May 2024). [CrossRef]
43. Reising, D.R.; Temple, M.A. WiMAX Mobile Subscriber Verification Using Gabor-Based RF-DNA Fingerprints. In Proceedings of the 2024 IEEE International Conference on Communications (ICC), Ottawa, ON, Canada, 10–15 June 2012; Available online: https://ieeexplore.ieee.org/document/6364039 (accessed on 11 May 2024).
44. Tharwat, A. Classification Assessment Methods. *Appl. Comput. Inform.* **2020**, *17*, 168–192. [CrossRef]
45. Duda, R.O.; Hart, P.E.; Stork, D.G. *Pattern Classification*, 2nd ed.; John Wiley & Sons: New York, NY, USA, 2001.
46. Breiman, L. Random Forests. *Mach. Learn.* **2001**, *45*, 5–32. [CrossRef]
47. Eliasmith, C.; Anderson, C.H. *Neural Engineering: Computation, Representation, and Dynamics in Neurobiological Systems*; The MIT Press: Cambridge, MA, USA, 2004; ISBN 9780262550604.

Disclaimer/Publisher's Note: The statements, opinions and data contained in all publications are solely those of the individual author(s) and contributor(s) and not of MDPI and/or the editor(s). MDPI and/or the editor(s) disclaim responsibility for any injury to people or property resulting from any ideas, methods, instructions or products referred to in the content.

Opinion

Towards an AI-Enhanced Cyber Threat Intelligence Processing Pipeline

Lampis Alevizos [1,*] and Martijn Dekker [2]

1 School of Engineering and Computer Science, University of Central Lancashire (UCLan), Preston PR1 2HE, UK
2 Faculty of Economics and Business, Amsterdam Business School, University of Amsterdam (UvA), 1018 TV Amsterdam, The Netherlands; m.dekker4@uva.nl
* Correspondence: lampis@redisni.org

Abstract: Cyber threats continue to evolve in complexity, thereby traditional cyber threat intelligence (CTI) methods struggle to keep pace. AI offers a potential solution, automating and enhancing various tasks, from data ingestion to resilience verification. This paper explores the potential of integrating artificial intelligence (AI) into CTI. We provide a blueprint of an AI-enhanced CTI processing pipeline and detail its components and functionalities. The pipeline highlights the collaboration between AI and human expertise, which is necessary to produce timely and high-fidelity cyber threat intelligence. We also explore the automated generation of mitigation recommendations, harnessing AI's capabilities to provide real-time, contextual, and predictive insights. However, the integration of AI into CTI is not without its challenges. Thereby, we discuss the ethical dilemmas, potential biases, and the imperative for transparency in AI-driven decisions. We address the need for data privacy, consent mechanisms, and the potential misuse of technology. Moreover, we highlight the importance of addressing biases both during CTI analysis and within AI models, warranting their transparency and interpretability. Lastly, our work points out future research directions, such as the exploration of advanced AI models to augment cyber defenses, and human–AI collaboration optimization. Ultimately, the fusion of AI with CTI appears to hold significant potential in the cybersecurity domain.

Keywords: artificial intelligence; cyber threat intelligence; cyber resilience; ethical considerations; CTI and AI biases

Citation: Alevizos, L.; Dekker, M. Towards an AI-Enhanced Cyber Threat Intelligence Processing Pipeline. *Electronics* **2024**, *13*, 2021. https://doi.org/10.3390/electronics13112021

Academic Editors: Juan-Carlos Cano and Aryya Gangopadhyay

Received: 31 March 2024
Revised: 12 May 2024
Accepted: 16 May 2024
Published: 22 May 2024

Copyright: © 2024 by the authors. Licensee MDPI, Basel, Switzerland. This article is an open access article distributed under the terms and conditions of the Creative Commons Attribution (CC BY) license (https://creativecommons.org/licenses/by/4.0/).

1. Introduction and Motivation

Cyber threats are continuously growing in complexity and frequency, therefore the ability to rapidly process and act upon cyber threat intelligence (CTI) can mean the difference between a mitigated threat and a breach. CTI, as defined by the National Institute of Standards and Technology (NIST), includes information that allows organizations to understand the latest threats and to proactively defend against them [1]. However, the vast volume of CTI, coupled with its dynamic nature, poses significant challenges for timely processing and action.

Traditional CTI processing methodologies involve manual efforts, where analysts examine large amounts of data, attempting to recognize patterns, validate intelligence, and recommend actions [2]. Namely, analysts are trying to produce actionable and valuable intelligence by contextualizing information. This manual approach, while valuable, is increasingly becoming unsustainable given the scale and speed of modern cyber threats. The need for automation and enhanced analytical capabilities, therefore, has become evident.

AI, with its ability to handle large datasets and its capability to learn and adapt, offers a promising direction to augment the CTI processing pipeline. Preliminary research, such as the work by Buczak and Guven [3] have has already highlighted the potential of AI in cybersecurity, particularly in areas like anomaly detection and malware classification.

However, the integration of AI into the CTI processing pipeline, especially in a manner that highlights the collaboration with human expertise, remains an area of research.

This paper seeks to bridge this gap, presenting a comprehensive approach to harnessing AI for CTI processing. Our focus goes beyond automation, creating a collaborative framework where AI and human analysts work together, to produce rapid, accurate, and actionable CTI. By streamlining this pipeline, we aim to reduce the time from intelligence ingestion to the implementation of mitigating measures and, subsequent, resilience verification. We believe that combining AI with CTI offers a proactive and adaptable cybersecurity approach, rather than a reactive one. Our goal is to connect AI's capabilities with cybersecurity requirements, advancing future innovations in the field. The contributions by this paper can be summarized as follows:

(1) A blueprint of the AI-enhanced CTI processing pipeline: We present a comprehensive framework that integrates AI techniques at various stages of a threat-informed defense, starting with CTI data ingestion and progressing to resilience verification. We detail the components and functionalities, as well as highlighting the imperative collaboration between AI and human expertise.

(2) Innovation in real-time and predictive threat mitigation: Our research pioneers the use of AI for generating real-time, contextual, and predictive mitigation strategies. We explore the application of advanced AI algorithms that can swiftly analyze CTI data and suggest security measures, thereby enhancing the organizational responsiveness to cyber threats.

(3) Ethical and bias considerations: We perform a thorough examination of the ethical implications of using AI in CTI and strategies to address potential biases in both the CTI domain and AI models. We also propose methods to ensure unbiased and transparent AI-driven insights.

(4) Introduction of a cyber resilience index: We propose a novel cyber resilience index that serves as a barometer for an organization's defensive capabilities against cyber threats. Analogous to financial market indices, this metric offers a quick overview of an organization's cyber health, informing strategic defense decisions.

(5) Challenges in AI-driven CTI: We critically discuss the hurdles to embedding AI into CTI, starting with the ethical dilemmas, data bias, and the need for transparency in AI-driven decisions, presenting a roadmap for addressing these issues.

(6) Future research directions: We provide future research directions emerging from our findings, thus underlining areas of potential growth and innovation.

The structure of this paper is as follows. We begin with the background and a literature review. Next, we detail the components of the AI-enhanced CTI processing pipeline, namely (A) intelligence ingestion, (B) collaborative analysis, (C) automated mitigation, and (D) resilience verification. In the next section, we discuss the challenges and considerations for the AI-enhanced CTI processing pipeline. Lastly, we summarize the conclusions and propose future research directions.

2. Background and Literature Review

The convergence of CTI and AI is a relatively emerging field, but one that has gained significant attention due to the potential benefits it promises. CTI has evolved from basic threat feeds to sophisticated intelligence platforms that provide contextual information about threats [4]. The primary goal of a mature CTI capability should be to offer actionable and valuable insights that can guide defensive measures, essentially extracting the right signals throughout the vast "noise" within the cyber landscape [5]. The works of Chen et al. [6] provide a comprehensive overview of the CTI landscape, highlighting the challenges associated with intelligence validation and relevance determination.

The application of AI in cybersecurity is not completely new. Machine learning models have been employed for tasks like spam detection and network intrusion detection for years, as detailed in the work of Sarker et al. [7]. However, the integration of AI with CTI is a more recent endeavor and lacks research output. The potential of AI to process vast amounts of data rapidly makes it a natural fit for CTI processing. For instance, Ring et al. [8] explored the use of AI for threat hunting, highlighting the potential to uncover hidden threats in vast datasets.

Despite the advancements, several challenges persist in regard to CTI processing. The dynamic nature of the cyber threat landscape, coupled with the large volume of data, often leads to information overload [9]. Additionally, false positives, outdated intelligence, and a lack of context can hamper the effectiveness of CTI. Sauerwein et al. [10] researched these challenges, offering insights into potential mitigation strategies.

The collaboration between AI and human expertise is also an important topic of considerable interest. While AI excels at processing large datasets, human intuition and expertise remain irreplaceable for nuanced threat analysis [11]. The challenge lies in creating a framework or pipeline where AI augments human capabilities, without overwhelming them with data. Brundage et al. [12] discussed the potential pitfalls and best practices for AI–human collaboration in decision-making contexts. In our work, we aim to bridge this gap by establishing strong collaborative bonds between the CTI analyst and AI, where both complement each other's strengths and counter each other's weaknesses.

The work of Varma et al. [13] work primarily targets small and medium-sized enterprises (SMEs), providing them with a roadmap to integrate AI into their CTI processes. Although the work offers valuable insights for SMEs, its scope is limited when considering larger organizations, or more complex cybersecurity infrastructure. The researchers used AI only to enhance CTI, rather than create a comprehensive solution for cybersecurity. Our paper on the other hand, presents a comprehensive AI-enhanced CTI processing pipeline that is adaptable to organizations of any size. Additionally, the roadmap in the referenced work can be seen as a practical application, while our proposed pipeline offers a broader perspective, including all stages of CTI processing, making it more universally applicable, while also allowing organizations to select the components that best suit their needs. Suryotrisongko et al. [14] underlined the importance of trust in CTI sharing, advocating for the blending of explainable AI (XAI) with open-source intelligence (OSINT). While the work underlines the significance of transparency in AI-driven CTI, it primarily focuses on botnet detection, which is a specific subset of the broader CTI landscape. Our focus on transparency and interpretability in AI-driven insights aligns with the principles of XAI. However, our goal is to provide a more holistic view of the CTI landscape, addressing various challenges and stages of CTI processing, beyond just botnet detection. Ranade et al. [15] studied a niche, but crucial, challenge in regard to CTI, namely the generation of fake CTI descriptions using advanced AI models. Although the work highlights an emerging threat in the CTI domain, its primary focus was on the generation aspect rather than mitigation or validation. In our work we highlight the importance of validation and relevance determination in CTI. The challenge of fake CTI generation further highlights the need for robust validation mechanisms, which our paper addresses in detail, offering solutions and strategies to counter such threats.

Moraliyag et al. [16] proposed a proactive approach to CTI by classifying onion services based on content. Onion services, also known as hidden services, are services that are hosted on the Tor network (https://www.torproject.org/ (accessed on 4 January 2024)). Unlike traditional websites with public IP addresses, onion services use the Tor network's anonymizing technology to protect both users and service operators. Although this work provides valuable insights into dark web intelligence, its primary focus remains on classification techniques, potentially overlooking other crucial aspects of CTI processing. Our intelligence ingestion phase, in the proposed pipeline, can benefit from such classification techniques. Nonetheless, we propose a more comprehensive view, detailing various stages of CTI processing and addressing challenges beyond just classification. The research by Mitra et al. [17] research focuses on enhancing cybersecurity knowledge graphs with intelligence provenance, which is a novel approach to combat fake CTI. Nevertheless, relying solely on provenance might not address all the challenges associated with fake CTI, especially when considering sophisticated adversarial attacks. The integration of provenance information aligns with our pipeline's intelligence ingestion and collaborative analysis phases. However, in this work we provide a multi-faceted approach to CTI validation, which enables a more robust form of defense against fake intelligence. Mittal et al. [18] discussed the potential of AI in CTI, highlighting its role in uncovering hidden threats. Whilst this work offers valuable insights, it does not provide a detailed roadmap or framework for integrating AI into CTI processing. Our work builds on this premise but goes a step further by detailing how AI can be systematically integrated into various stages of CTI processing, offering a more structured approach.

3. The AI-Enhanced CTI Processing Pipeline

The fusion of AI capabilities with CTI processing has the potential to significantly enhance the speed, accuracy, and efficiency of threat intelligence operations. This section outlines a structured pipeline that integrates AI at various stages of CTI processing to enable the abovementioned attributes. Figure 1 visualizes the individual components comprising the AI-enhanced CTI processing pipeline, as a blueprint.

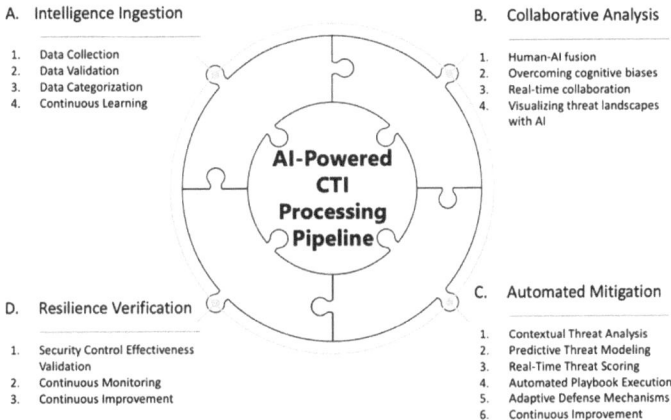

Figure 1. AI-enhanced CTI processing pipeline blueprint.

The central theme of Figure 1 is represented by the middle circle labelled "AI-enhanced CTI Processing Pipeline". Radiating outward from this central theme are four main components in a puzzle shape, namely:

A. Intelligence ingestion, which focuses on the initial stages of data collection, data validation, and data categorization using AI;
B. Collaborative analysis, which focuses on the collaborative analysis between human intelligence and artificial intelligence. We detail the concept of human–AI fusion, ways of overcoming cognitive biases, real-time collaboration, and visualizing threat landscapes with AI;
C. Automated mitigation, which focuses on analyzing threats in context using AI. It contains predictive threat modelling, real-time threat scoring, automated playbook execution, adaptive defense mechanisms, and a feedback loop for continuous improvement;
D. Resilience verification, which comprises of a proactive approach to security by simulating cyber-attacks. The focus is on continuous monitoring and continuous improvement led by AI, ultimately leading to a single cyber resilience metric, the cyber resilience index.

Each of these main components breaks down further into subcomponents, which we detail in the corresponding sections.

3.1. Intelligence Ingestion

The initial phase of the CTI processing pipeline is the ingestion of raw information or intelligence data. This means collecting, validating, and categorizing vast amounts of data from various sources, such as threat feeds, logs, and other intelligence or information repositories. The threat landscape is dynamic, with new threats emerging regularly; thereby, it is imperative for a CTI ingestion process, empowered by AI, to continuously learn from new data and adapt to the evolving threat environment. Figure 2 outlines the intelligence ingestion steps.

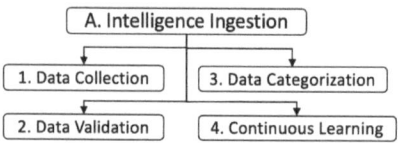

Figure 2. Intelligence ingestion steps.

3.1.1. Data Collection

Given the vast amount of CTI data that needs to be collected, manual collection, although feasible, is extremely time consuming and the added value of such an exercise is highly dependent on the analysts experience and expertise alone. AI-driven tools orchestrate multiple APIs (Application Programming Interfacesapplication programming interfaces) and, therefore, automate the collection process; thus, data is gathered in real-time and from a wide range of sources, such as open-source intelligence (OSINT), human intelligence (HUMINT), dark web monitoring, commercial threat feeds, internal organization threat data, vendor reports, social media monitoring, threat intelligence platforms (TIPs), and industry-specific threat reports.

3.1.2. Data Validation

Not all collected data are relevant. AI algorithms can quickly examine the data, discarding irrelevant information and highlighting potential threats using decision trees, as shown by Kotsiantis [19]. Then, such algorithms cross-reference data from multiple sources to verify its accuracy, using graph analytics to map the relationships between sources. For instance, if two independent sources report the same threat, it is more likely to be credible. However, the challenge is not just about collecting data, but producing meaningful and actionable intelligence from the overwhelming amount of "noise". Transforming information into actionable, valuable intelligence should be the goal. The total volume of data, combined with its dynamic nature, makes manual validation and analysis a challenging task. This is where AI plays a transformative role and can be implemented based on the following elements.

Signal extraction, using convolutional neural networks (CNNs), provides proven pattern recognition within large datasets, enabling such networks to detect indirect signs of malicious activity that may signify cyber threats. This capability allows AI to effectively extract the "signal", the meaningful, actionable intelligence, from the "noise", irrelevant or redundant information. Goodfellow et al. [20] provided the foundational theory on the capabilities of neural networks in terms of pattern recognition and anomaly detection. TensorFlow v2.16.0 (https://www.tensorflow.org/tutorials/generative/autoencoder, accessed on 12 December 2023) serves as a practical implementation of that theory, being a popular open-source machine learning framework that offers long short-term memory (LSTM) autoencoders, which can be used for anomaly detection in time-series data [21].

Cross-referencing and identifying correlations utilizing AI tools from multiple data sources automatically, enhances the validation process. For instance, if two independent sources report the same threat, the AI assigns a higher credibility score to that piece of intelligence. Advanced algorithms can also correlate seemingly unrelated pieces of information, uncovering hidden threats or tactics used by adversaries. A prime theoretical example, provided by Landwehr et al. [22] introduced logistic model trees, which can be used for correlating and cross-referencing data from multiple sources. Elastic Stack (https://www.elastic.co/, accessed on 19 December 2023) (Elasticsearch, Logstash, Kibana) is a practical implementation of this, which is widely used in cybersecurity for its capabilities in terms of data ingestion, indexing, and visualization. It can correlate logs from various sources to provide a unified view of events.

Transforming information into intelligence: AI bridges this gap by analyzing the context in which data are generated, determining its relevance to the organization's threat landscape, and providing actionable recommendations based on the analyzed intelligence [23]. However, it is important to note that the effectiveness of AI depends upon the accuracy and quality of the underlying data, e.g., data within the configuration management database (CMDB) of a company. The quality and integrity of the data sources are therefore crucial, as they directly impact the reliability of the intelligence generated.

Addressing human limitations: Traditional CTI relies on human analysts who, despite their expertise, have limitations in terms of processing capacity and speed. AI complements human analysts by managing vast datasets, ensuring that no potential threat goes unnoticed [3]. This collaboration between human intuition and AI's computational aptitude provides for comprehensive threat intelligence.

A feedback mechanism exists, where false positives or irrelevant data flagged by AI are reviewed and fed back into the system for continuous learning and improvement. This iterative process improves the AI model's accuracy over time. Figure 3 demonstrates the process of AI-driven validation, and the transformation of raw data into actionable intelligence.

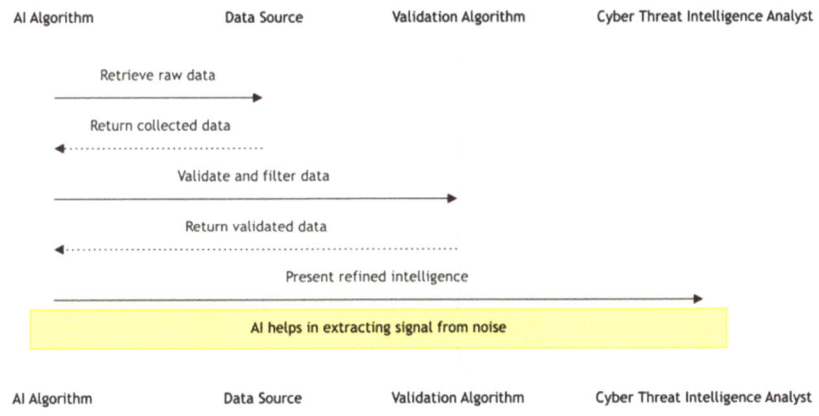

Figure 3. Extracting the signal from the noise using AI.

The AI algorithm retrieves raw data from the data source. The collected data is returned to the AI algorithm. Next, the AI algorithm sends this data to the validation algorithm to validate and filter out irrelevant or noisy data. The validated data is then returned to the AI algorithm. Finally, the refined intelligence is presented to the CTI analyst.

3.1.3. Data Categorization

Effective CTI requires the ingested data to be categorized into meaningful segments, which can guide the subsequent analysis and action as a result of the human–AI collaboration. The Latent Dirichlet Allocation (LDA) algorithm can be used to identify the underlying themes or topics within large text datasets, helping to categorize the content by subject matter [24]. For instance, threat actors (TAs), threat events, TTPs (tactics, techniques, and procedures), indicators of compromise (IoCs), the goals and motivations of TAs, and geopolitical trends. Moreover, Schonlau et al. showed how to use the random forest algorithm in this regard [25], while Sarker showed how to use neural networks to classify data into predefined categories based on training datasets [26], where the classification criteria are already known. Thus, using either method, AI takes unstructured data and organizes it into meaningful segments, ready for further analysis and action in the CTI pipeline.

3.1.4. Continuous Learning

The cyber threat landscape is not static; it evolves continuously with new vulnerabilities, TTPs, and threat actors emerging regularly. For an AI-enhanced CTI pipeline to remain effective, it must adapt to these changes. To successfully enable the AI-enhanced CTI pipeline to continuous learning, several methods can be used.

Machine learning (ML) for continuous relevance in CTI can be achieved utilizing the online learning algorithm [27]. This algorithm incrementally updates parameters in response to each new data point, thus providing adaptability to emerging threats without full retraining. This approach keeps the predictive model current, according to the evolving threat landscape.

Adversarial machine learning (AML) is used to anticipate potential evasion techniques that adversaries might employ. Red teams can either perform traditional attack simulations or use AML to simulate advanced evasion tactics. This will result in collecting data on novel attack vectors and improving defenses before they are exploited in the wild. Red teams should create adversarial examples led by cyber threat intelligence to assess an organization's defenses. Any successful evasion that is logged and analyzed, is in turn used to improve the data collection mechanisms. As the AI system processes new threat intelligence and interacts with human analysts, it will inevitably encounter false positives or misclassifications. Therefore, incorporating feedback from these interactions will allow the system to refine its algorithms, reducing errors over time.

Defensive distillation should be used to make machine learning models more robust against adversarial attacks [28]. Therefore, the data being collected will not be polluted by adversarial noise. Leveraging this technique to train the model on a "softened" version of the data, where the output probabilities are smoothed or "distilled" will make it harder for adversaries to find the precise distresses needed to deceive the model, thus data collection is cleaner.

Incorporating the broader context is important, which means that the AI system should continuously learn and incorporate insights from broader geopolitical, technological, and socio-economic contexts, enhancing its threat predictions. Political tensions often correlate with targeted cyber-attacks on government and critical infrastructure. Monitoring such geopolitical developments will allow the pipeline to anticipate increased risk levels and identify potential aggressors. Moreover, technological trends impact the nature and prevalence of cyber threats, if an innovative technology becomes widespread (e.g., blockchain), the AI system will prioritize threats targeting that technology. Lastly, economic challenges often lead to increased cybercrime activity. Tracking socio-economic indicators will help anticipate a rise in certain threat types. For instance, during economic downturns, there is typically a rise in financial fraud schemes.

Active learning is a specialized form of machine learning, where the model actively queries the human analyst for inputs on specific predictions [29]. For instance, if the AI system encounters a piece of data it is uncertain about, it seeks confirmation from a human expert. Over time, these interactions reduce the system's uncertainty and improve its accuracy.

3.2. Collaborative Analysis

Traditional analysis methods relying heavily on human expertise are oftentimes slow, potentially biased, and prone to errors. This is where artificial intelligence and, more specifically, machine learning, can play a crucial role [30]. Mishra's work [31] showed that gradient boosting machines (GBMs), trained on historical threat data, can provide insights into potential threats, their patterns, and possible implications. Human analysts collaborate with these AI insights, leveraging their expertise to understand the nuances and context behind each threat. Figure 4 outlines the collaborative analysis steps.

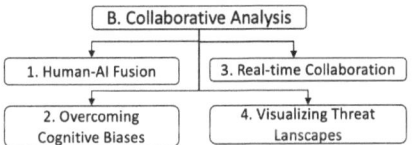

Figure 4. Collaborative analysis steps.

3.2.1. The Human–AI fusion

The collaboration between human analysts and AI is not about replacing one with the other, but about amplifying the strengths and mitigating the weaknesses of both. A summary of this phase is visualized in Figure 5.

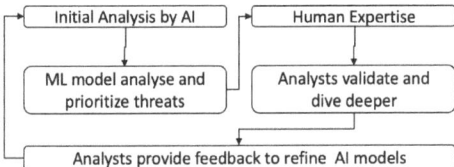

Figure 5. Human–AI Fusion.

The initial analysis by AI provides speed and scale. AI processes vast amounts of data at speeds incomprehensible to humans. For instance, Desmond et al. [32] showed that a model trained on extensive datasets can analyze millions of logs within minutes to detect anomalies, as opposed to humans. Chen's work [33] showed that AI can also provide pattern recognition through a deep learning algorithm, which is able to recognize patterns in data. Goodfellow et al. [20] proposed the use of neural networks that can identify patterns associated with malware traffic in network logs, even if the malware is a zero-day variant. In addition, AI offers fast prioritization. Based on detected patterns and historical data, AI can prioritize threats; therefore, the most imminent and dangerous threats can be addressed first. Existing AI-driven tools, like Cisco's Threat Grid v2.15 (https://www.cisco.com/c/en/us/products/security/threat-grid/index.html, accessed on 4 January 2024), analyze millions of samples daily, providing automated threat scores, based on the potential impact and prevalence of the detected threats [34].

Human expertise provides a broader contextual understanding. Although AI can recognize patterns, human analysts understand the broader context. For instance, in the SolarWinds attack (https://www.wired.com/story/the-untold-story-of-solarwinds-the-boldest-supply-chain-hack-ever/, accessed on 4 January 2024), while AI tools detected anomalies, it was the human analysts who pieced together the broader campaign, understanding the implications and the actors behind it. Human analysts bring intuition and experience. Analysts, with years of professional experience, can subjectively understand when something does not seem right, even if it passes AI checks. Their expertise allows them to focus on complex threats specific to the IT landscape, forming hypothesis and, therefore, uncovering potential hidden connections. Lastly, human analysts play a crucial role in validating AI findings. Although AI might flag a potential phishing email based on certain patterns, a human analyst can validate it by considering the sender's context, the email's content, and other information. Analysts interacting with the AI-enhanced pipeline must provide feedback for refining the AI model, thus its predictions become more accurate over time. However, a challenge to using AI in cybersecurity is the potential for false positives [10]. Human analysts should flag something as a false positive, so that eventually, the AI learns from it, thereby reducing similar false alarms in the future. False positives in AI cybersecurity systems are problematic because they waste the analyst's time and resources, leading to alert fatigue and potential oversight of genuine threats. By learning from flagged false positives, AI models can adapt and improve their detection accuracy over time. Ultimately, this will reduce unnecessary alerts and allow analysts to remain focused on real threats, while the AI-enhanced CTI pipeline increases its trustworthiness over time.

3.2.2. Overcoming Cognitive Bias

Cognitive biases are systematic patterns of deviation from the norm or rational judgment, thus leading analysts to create their own subjective reality from their perception of the input [35]. Such biases can significantly impact the decisions of CTI analysts, thereby potentially leading to disregarded threats or data misinterpretations.

One of the major ransomware-related attacks (https://www.csoonline.com/article/563017/wannacry-explained-a-perfect-ransomware-storm.html, accessed on 4 January 2024) happened in 2017. In the aftermath of this case, many organizations focused heavily on protecting themselves against similar ransomware threats. Although this is a valid concern, an overemphasis on one type of threat due to its recent occurrence (availability heuristic) can lead to neglecting other potential threats. The AI-driven CTI pipeline provides for a balanced focus on all the relevant threats, not just those that are currently in the spotlight. The integration of AI into the CTI process offers a unique opportunity to counteract these biases, ultimately allowing for more objective and comprehensive analysis.

CTI analysts may be subject to the following biases:

A. Confirmation bias: the analyst might prioritize data that aligns with their existing threat models, potentially overlooking new or unexpected threats. The analyst's perspective is unintentionally influencing the collection, analysis, and interpretation of CTI data. This bias (also known as observer bias) can lead analysts to favor information that confirms their presumptions or to overlook data that contradicts their beliefs, impacting the accuracy and objectivity of their analysis;
B. Availability heuristic: the analyst might give undue weight to a recent high-profile cyber-attack, neglecting other potential threats;
C. Anchoring bias: the analyst relies too heavily on the first piece of information encountered (the "anchor") when making decisions. Oftentimes, CTI analysts anchor their analysis directly to initial findings, therefore missing the broader threat landscape;
D. Status quo bias: a preference for the current situation, resisting change, leading to an over reliance on established threat models and an inability to adapt to the evolving cyber threat landscape.

AI's role in mitigating biases:

A. AI algorithms are inherently decoupled from emotions and preconceived notions, thereby providing an objective analysis of data. They treat each piece of information based on its merits and relevance, not on any external influence or bias [36];
B. AI models apply consistent criteria when analyzing data, thus using the same standards across all data, contrary to the human analyst, who might unconsciously alter their criteria based on their biases [37];
C. AI analyses data for decision making based on comprehensive data, rather than anecdotal evidence or recent events [37].

In conclusion, to successfully overcome cognitive bias, the goal is not to replace human analysts with AI, but to have them work together. AI provides objective analysis, empowered by human analysts, who bring contextual understanding and intuition. By working together, they can counteract the biases inherent in both human judgment and AI models, therefore leading to more balanced and comprehensive threat analysis.

In exploring how humans and AI work together to counteract bias and optimize collaboration, Dell'Acqua et al. [38] highlighted two main ways: the 'Cyborg' and the

'Centaur.' The 'Cyborg' mode mixes human and AI efforts closely, using AI for its fast processing and humans for their deep understanding and moral judgment. On the other hand, the 'Centaur' mode is about humans and AI working side by side, with each taking on tasks that suit their strengths. This division helps make the most of both AI's data handling abilities and human creativity and ethical insights. These dimensions can serve as guardrails and show how combining human and AI strengths can enhance strategic decision making (centaur) and improve efficiency and accuracy (cyborg).

3.2.3. Real-Time Collaboration

Real-time machine learning models bring a change in thinking in regard to how threat intelligence is processed and acted upon. Consider a zero-day vulnerability that has just been discovered, traditional threat intelligence systems might take hours, if not days, to update their databases and provide recommendations. However, a real-time AI-driven system can pick up discussions about this vulnerability from sources like social media, forums, commercial tools, TIPs, or dark web marketplaces within minutes. Such a system can then assess the potential impact of this vulnerability, generate alerts for human analysts, and even recommend immediate countermeasures [39]. The successful real-time collaboration between the CTI analyst and AI is based on the following elements:

A. Dynamic data ingestion, as cyber threat data is generated by all of the above-described sources continuously, it is imperative to have a system that can ingest this data in real-time. AI-driven models, especially those built on streaming data platforms, can process data as it flows in, without waiting for batch updates [40];
B. Instantaneous analysis, once the data are ingested. Real-time machine learning models analyze data instantaneously [41]. This means that as soon as a new threat indicator is detected, the AI-enhanced CTI pipeline can assess its severity, potential impact, and relevance;
C. Real-time alerts based on instantaneous analysis. The AI-enabled CTI pipeline generates real-time alerts for human analysts. These alerts can be prioritized based on the potential impact and, thereby, the analysts can focus on the most pressing threats first;
D. Human–AI interaction; real-time collaboration should not be just one way. Human analysts, upon receiving alerts, can interact with the AI-enhanced CTI pipeline, asking follow-up questions or clarifications;
E. Adaptive learning; one of the differentiating factors of real-time machine learning models is their ability to learn on-the-fly. As new data is processed, the model can update its understanding, hence its predictions and recommendations are always based on the latest threat intelligence [42].

3.2.4. Visualizing Threat Landscapes with AI

The ability to visualize and quickly comprehend threat models is paramount. As an example, one could think of a scenario where a CTI analyst comes across an image detailing the flow of a sophisticated malware attack. Instead of spending hours, or even days, deciphering the image, the analyst can use an AI tool to quickly understand the malware's propagation, its potential targets, and its behavior. The AI tool can also cross-reference the malware's signature with a database, providing insights into its origin, past variants, and potential countermeasures [43]. There are two prime examples where AI can supercharge collaborative analysis and empower the CTI analyst, as follows:

(i). Automated threat modelling:

Traditional threat modelling is a time-consuming exercise, requiring analysts to manually map out threats, vulnerabilities, and potential attack vectors. Moreover, CTI analysts and relevant stakeholders may lack the technical knowledge to perform threat modelling. Akhtar et al. [44] showed how AI automates this process, rapidly creating threat models based on the available data. Moreover, as new threat intelligence is ingested, AI can dynamically update the threat model, thus always reflecting the current threat landscape.

Lastly, AI algorithms and especially deep learning models can be used to identify difficult correlations and potential threats that might be overlooked in manual analysis [45].

(ii). Image recognition and explanation:

CTI analysts can drop images depicting complex IT landscapes, or complex threat actor flows, into an AI-powered tool. Iqbal et al. showed how AI can instantly analyze the image, recognizing various components, connections, and potential vulnerabilities [7]. Furthermore, in the same work, it was proven that AI can go beyond simple recognition. Advanced AI models can provide contextual explanations for the elements in the image [7]. For instance, if an analyst drops an image of a network topology, the AI can identify servers, firewalls, potential choke points, and even suggest potential attack vectors based on the layout. Furthermore, by cross-referencing the elements in the image with historical threat data, AI can provide insights into past vulnerabilities, attacks, or breaches associated with similar setups [45].

3.3. Automated Mitigation

Based on the combined intelligence from AI and human analysis, the pipeline integrates with organizational tools like configuration management databases (CMDBs), or taps into IT infrastructure data, to understand the environment and adapt recommendations. The recommendations range from technical solutions, like updating firewall rules or patching vulnerabilities, to strategic actions, such as user awareness campaigns or policy changes. In this section, we outline how AI can be harnessed to provide automated mitigation recommendations based on analyzed intelligence, and to visualize the integrated approach of AI-driven mitigation recommendations, a flowchart is presented in Figure 6.

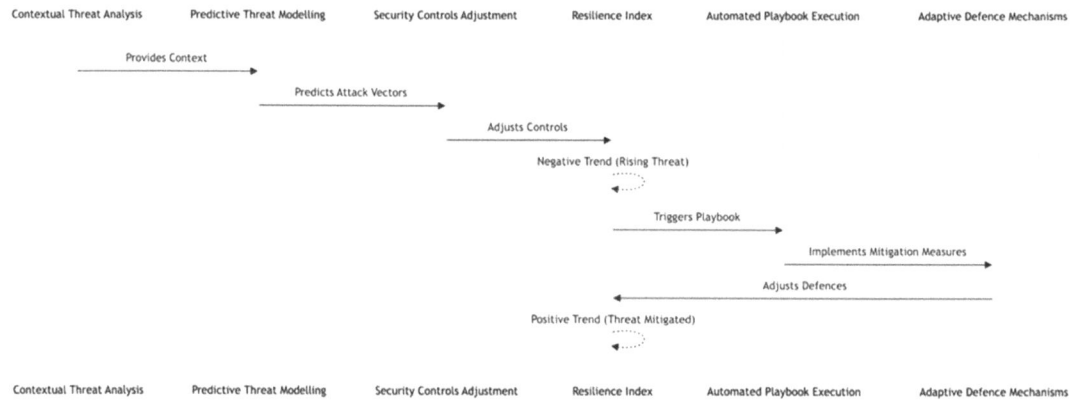

Figure 6. AI-driven CTI pipeline security control steering.

3.3.1. Contextual Threat Analysis

Before the AI-enhanced CTI pipeline recommends any mitigation strategies, it is crucial to understand the context of the threat against the operating IT landscape. Therefore, analysis of the threat in relation to the organization's infrastructure, assets, and previous incidents is necessary. This is achieved using natural language processing (NLP) to extract contextual information from threat intelligence reports [46].

Suppose an organization receives a threat intelligence report about a new ransomware strain targeting financial institutions. Using NLP, the AI system extracts keywords like "ransomware," "financial institutions," and cross-references these with the organization's IT and security landscape to determine the relevance and potential impact, thereby adjusting the relevant security controls accordingly.

3.3.2. Predictive Threat Modelling

AI trained with current intelligence and historical data can predict the likely progression of a threat based on patterns observed in past incidents [47]. As a result, we can pre-emptively strengthen defenses in vulnerable areas. For example, if historical data indicates that every time there is a spike in traffic from a particular region, a DDoS attack follows, the AI can predict a potential DDoS attack when it observes a similar traffic pattern in the future and, therefore, proactively adjust the relevant security controls accordingly.

3.3.3. Real-Time Threat Scoring

Not all threats have the same level of severity or relevance to an organization. AI provides a real-time threat score based on the specific IT landscape and organizational information, helping to prioritize mitigation efforts. As a result, it can score threats faster based on factors like the potential impact, exploitability, and the organization's vulnerability faster [48]. As a result, the most critical threats are addressed first. For instance, an organization might receive thousands of alerts daily. An AI system can score a detected phishing attempt as "high" risk if it is linked to a known APT group, while a generic malware detection might be scored as "medium" risk.

3.3.4. Automated Playbook Execution

For known threats or attack patterns, AI automatically executes predefined mitigation playbooks, reducing the response time subject to integration with threat intelligence systems and security orchestration, automation, and response (SOAR) platforms. Upon detecting a recognized threat pattern, the pipeline triggers the corresponding playbook, for immediate action. For example, if the AI pipeline detects patterns consistent with the "Emotet" malware, it can trigger a predefined playbook that isolates affected systems, blocks associated IPs, and sends notifications to the incident response team. Although the AI-enhanced CTI pipeline speeds up the threat response by running preset mitigation strategies, it is imperative to involve humans in key decision making to handle high-risk situations. For instance, cybersecurity experts should review and authorize actions chosen by AI in high-risk scenarios, combining the speed of AI with human insight to minimize the risk of automated responses.

3.3.5. Adaptive Defense Mechanisms

AI dynamically adjusts defenses based on ongoing threat analysis, using reinforcement learning (RL) models that can adapt security configurations in real-time [49]. For instance, if the AI pipeline detects increased traffic from a specific IP range associated with malicious activities, it can dynamically adjust the firewall rules to block or throttle that traffic. Or if the AI pipeline observes that every Friday evening there is an attempt to exfiltrate data, it can dynamically adjust the egress firewall rules during that time to add an additional layer of scrutiny.

3.3.6. Continuous Improvement

After implementing mitigation measures, it is essential to assess their effectiveness and refine the strategies accordingly. For instance, after blocking a suspected malicious IP address, the AI system can monitor for any subsequent attempts or changes in attack patterns from that IP address, refining the threat profile over time.

3.4. Resilience Verification

The simulation or emulation of attack scenarios, based on the received cyber threat intelligence, to assess the resilience of the implemented measures, is imperative at this stage. A security control effectiveness evaluation measures the resilience of organizations against potential cyber threats, eventually producing an alternative to a stock market index, but for cybersecurity. Much like financial indices, which provide traders with a snapshot of the market's health or trends, for e.g., the S&P 500 index, a cyber resilience index can offer decision makers a quick overview of their organization's cybersecurity posture. This index, updated by the AI in real-time, can serve as a barometer of an organization's cyber health and, therefore, can be used by decision makers to steer their defenses and resources accordingly. As a result, organizations are not just reactive, but proactive, in their cybersecurity approach.

To achieve this, we define the following three steps for resilience verification leading to the formation of a cyber resilience index: (1) automated penetration testing, (2) continuous monitoring, and (3) continuous improvement, as illustrated in Figure 7.

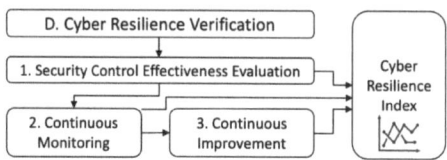

Figure 7. Cyber Resilience verification steps.

3.4.1. Security

AI provides a governance layer in regard to processes or tools to simulate or emulate cyber-attacks on a system, to identify vulnerabilities and assess the effectiveness of the implemented mitigation measures. As a result, AI simulates and emulates advanced persistent threats (APTs) to assess how well a system can withstand prolonged, targeted attacks. Moreover, since AI governs the CTI pipeline, it can adapt the strategies of APTs based on the received CTI and counter to the system's responses, mimicking the behavior of real-world adversaries [50]. Therefore, AI can provide a factual security control effectiveness validation rather than a checklist-based theoretical assessment.

3.4.2. Continuous Monitoring

AI continuously monitors the network traffic, system logs, and other relevant data sources throughout the IT landscape in which it is deployed. As a result, it provides real-time detection of suspicious activity that might indicate a breach or vulnerability exploitation. It can also be trained on historical network traffic data to recognize patterns related to known cyber threats, thus serving as an AI-powered intrusion detection system (IDS). Once deployed, it can monitor network traffic in real-time, flagging any deviations from the norm for further investigation.

3.4.3. Continuous Improvement

One of the key advantages of integrating AI into resilience verification is its ability to continuously self-improve. As an example, AI has the ability to observe and evaluate the ecosystem in which it operates and can gain knowledge from any newly identified threats or weaknesses, improving its algorithms for the next evaluation. For instance, when an intrusion detection system identifies a new form of malware on the network, it can adjust the AI algorithms to identify this threat in the future. This continuous learning helps to keep the pipeline updated with the latest threat intelligence, coupled with the latest data from the IT landscape it is deployed on.

4. Challenges and Considerations

Although an AI-enhanced CTI processing pipeline may offer significant capabilities, it comes with several challenges and considerations that we discuss in this section. Given that we are in the nascent stages of building AI tools for corporate use, it is imperative for organizations starting to work with AI technologies, and especially within CTI, that they should follow guidelines set by NIST AI RMF 1.0 [51], the EU AI Act [52,53], and ISO 5338 [54]. Adhering to these standards will help organizations develop AI systems that are effective, efficient, ethically responsible, and compliant with global regulations.

4.1. Ethical Consideration in AI-Enhanced CTI Processing

4.1.1. Data Privacy and Confidentiality

AI-enhanced CTI processing requires access to vast amounts of data, some of which may be sensitive or confidential. Such datasets must be managed ethically, with respect to privacy laws and regulations. Organizations, therefore, should consider anonymizing the data used for training AI models to ensure that personally identifiable information (PII) is sufficiently protected. Moreover, data breaches or misuse can lead to significant reputational damage and legal consequences. It is, therefore, essential to implement strict data anonymization techniques, using differential privacy, while ensuring that data are stored securely and encrypted [55].

4.1.2. Consent, Surveillance, and Proportionality

While hunting for threat vectors, evidence, and indicators of compromise, either manually or with the use of AI, oftentimes the lines between legitimate surveillance and invasion of privacy are blurred. Hence, any data collection or surveillance activities should be executed with the necessary prior consent and in accordance with legal and ethical standards. Ultimately, the relevant individuals should be aware of and agree to the monitoring and data collection processes, respecting their autonomy and rights. Nonetheless, the implementation of clear consent mechanisms, regularly updating terms of service, and transparency in regard to data collection practices can solve this challenge [56]. The scale of the surveillance must also be proportionate, thus preventing overreach and the potential misuse of data. Solving this challenge requires regular audits and setting clear boundaries on data collection [57].

4.1.3. Technology Misuse

Like any innovative tool, the AI-enhanced CTI pipeline can potentially be misused. There is a potential risk of being used for malicious purposes, for instance, spreading misinformation or running unauthorized surveillance. Organizations should implement strict access controls and behavioral monitoring to prevent misuse. NIST recently provided a thorough AI risk management framework describing this challenge and potential solutions [51].

4.2. Addressing Potential Bias in AI-Enhanced CTI Models

4.2.1. Training Data Scrutiny

AI models are only as good as the data they are trained on. If the training data contains biases, the AI model will likely inherit those biases. It is crucial to use training datasets that are diverse and representative to avoid unintentional bias in AI-driven insights. Biased training data will lead to skewed AI predictions, ultimately leading to unfair outcomes. To address this challenge, we need to consider diverse datasets, employing fairness-enhancing interventions, and regular bias audits. However, even with unbiased training data, algorithms themselves can introduce biases [58]. It is, therefore, evident that regular audits and evaluations of AI models can help to identify and rectify any inherent biases in AI algorithms [59].

4.2.2. Continuous Model Evaluation

Continuous model evaluation will provide reasonable assurance that the AI-enabled pipeline will remain relevant and unbiased as new CTI data emerges. Moreover, biases in the AI model may create loops where the model's predictions reinforce existing biases. For instance, if an AI model incorrectly flags certain types of network traffic as malicious due to bias, it may lead to increased scrutiny of similar traffic in the future, reinforcing the bias. It is, therefore, imperative to implement real-time evaluation metrics and periodic retraining of the models [59]. If the AI-driven CTI pipeline makes a wrong decision, it is crucial to have a reliable rollback mechanism in place to restore the system to its previous working state. This can be done by creating epochs or checkpoints before carrying out any playbook actions [60]. Therefore, if a decision is found to be incorrect or harmful, the system can easily go back to a state before the action was taken, reducing the likelihood of disruptions.

4.2.3. Systematic Bias Detection

The AI-enhanced CTI pipeline may be prone to observer bias, which appears when the subjective predispositions of individuals involved in the AI's development or operational phases influence the selection of training data or the interpretation of the system's outputs. Such biases can inadvertently lead to misrepresentation in the AI model, both in the CTI, as well as in regard to its threat detection capabilities. This will potentially result in the disproportionate identification or neglect of specific threats. To protect the precision and impartiality of the AI-enhanced CTI pipeline, it is imperative to, firstly, acknowledge and address the presence of observer bias. Moreover, training data bias, a form of availability bias, originates from the initial CTI collection phase of the pipeline. This may occur when the training data predominantly consists of easily accessible information, or when over or under sampling leads to a training dataset that does not accurately represent real-world scenarios. Nonetheless, mitigations for both these biases have been proposed by [61] and other scholars [62] such as the utilization of heterogeneous data sources, and the implementation of systematic bias detection and correction mechanisms throughout the lifecycle of the AI model. Additionally, organizations should consider regularly testing the AI system's performance using blind or double-blind methods, where neither the testers nor the AI system has information that might influence the outcome of the test. This would provide a method of assessing the AI system's ability to identify threats without bias.

4.3. Transparency and Interpretability within the AI-Enhanced CTI Pipeline

4.3.1. Explainable AI (XAI) and Stakeholder Trust

Many advanced AI models, especially deep learning models, are oftentimes seen as "black boxes", where their decision-making processes are not easily interpretable. This poses a significant challenge in regard to CTI, where understanding the rationale behind insights is crucial for decision making. XAI enables AI model predictions to become understandable by humans, fostering trust, and facilitating better decision making. To address the black box issue, there is a growing amount of stress on model explainability. SHAP (SHapley Additive exPlanation) or LIME (Local Interpretable Model-Agnostic Explanation) techniques have emerged as a means to provide insights into how AI models arrive at their decisions. Therefore, using interpretable models, employing post-hoc explanation techniques, and visualizing model decisions are demonstrated ways forward to solve this challenge effectively [63].

Lastly, it is of utmost importance for AI-driven CTI systems to be effective, that stakeholders must trust the insights with which they are provided. Guaranteeing transparency and interpretability is key to building and maintaining this trust. Regular communication about how the AI models work, their limitations, and the steps taken to address inaccuracy can help foster trust among stakeholders.

4.3.2. Feedback within the Human—AI fusion

Feedback from CTI analysts can refine AI models and, therefore, verify their continued relevance and accuracy. This can be achieved through the implementation of an iterative refinement model [64], and by fostering a collaborative AI–human environment. For instance, a CTI analyst provides real-world insights and corrections, which can be used to fine tune AI algorithms. Or, if an AI model misclassifies a type of malware, feedback from analysts can correct this mistake, leading to improved future detection capabilities. Additionally, comprehensive documentation warrants that all stakeholders understand the workings, limitations, and scope of AI models, thus maintaining detailed model logs, providing clear documentation on algorithms and training data, and confirming transparency in model updates are all important [65].

4.4. AI Model Robustness and Adversarial Attacks

Prior to adopting an AI-driven CTI pipeline, a prime consideration is that the pipeline itself might become a potential target for cyber adversaries. Similar to traditional software systems that can be exploited, AI models have their own set of vulnerabilities, especially to adversarial attacks. These attacks involve feeding the model specially crafted input data designed to deceive it, leading to incorrect outputs or predictions [66]. Adversarial attacks can be categorized on a high level, into two types. White-box attacks, where the attacker has complete knowledge of the AI model, the architecture, and its weights, and black-box attacks, where the attacker has no knowledge of the model's internal aspects and only has access to its inputs and outputs. Common adversarial attack techniques include adding imperceptible noise to input data, generating adversarial examples, or exploiting model transferability, where an adversarial example crafted for one model affects another [67]. An adversarial attack could lead to several adverse outcomes, for instance, the misclassification of benign network traffic as malicious, or vice versa, incorrect threat scoring, leading to missed prioritization of threats, or even deceptive insights that could mislead incident response teams or decision-makers [67].

To protect the AI-driven CTI pipeline against adversarial attacks, several strategies can be followed. For instance, training the AI model on adversarial examples, making it more robust to such attacks. Input filtering or normalization can detect and mitigate adversarial input data [68]. The use of collaborative models will increase its robustness, as an attacker would need to deceive multiple models simultaneously. Moreover, continuously updating and retraining the AI model confirms that it is equipped to manage new adversarial techniques. Organizations should also consider adopting frameworks to perform red teaming against generative AI models, such as PyRIT v0.2.1 (https://www.microsoft.com/en-us/security/blog/2024/02/22/announcing-microsofts-open-automation-framework-to-red-team-generative-ai-systems/, accessed on 14 January 2024), to assess and improve the security posture of AI models, making them more resilient to adversarial threats and countering bias within the models. Lastly, implementing real-time monitoring to detect unusual patterns in the model's predictions can flag potential adversarial attacks.

5. Conclusions and Future Research

In this work, we defined an AI-enhanced CTI processing pipeline and presented how the integration of AI into CTI processing has the potential to revolutionize the cybersecurity landscape. AI can automate, enhance, and expedite numerous CTI tasks, starting with augmenting the human analyst in separating the "signal" from the "noise", namely actionable cyber intelligence. Moving on to data ingestion, collaborative analysis, automated mitigation and, lastly, to cyber resilience verification. Therefore, organizations can achieve a more proactive and adaptive cybersecurity posture, staying a step ahead of the evolving threats, as opposed to adopting a reactive approach. Successful implementation would signal the beginning of an AI-based end-to-end cyber defense system. Currently, human CTI analysts and AI are connected by strong collaborative bonds, where both complement each other's strengths and counter each other's weaknesses.

However, the integration of AI into CTI brings implementation challenges, ethical considerations, potential biases, and the need for transparency and interpretability that require attention. Nonetheless, with a balanced approach that combines the strengths of AI with human expertise, these challenges can be addressed effectively.

With this work, we set the foundational framework for an AI-enhanced CTI processing pipeline, and we identify several research routes. Some potential directions for future research may involve advanced AI models. As AI continues to evolve, exploring newer models and architectures tailored for CTI tasks could yield even more accurate and efficient results. Another angle is the ethical use of AI in cybersecurity. Thorough research into the ethical implications of AI use in cybersecurity would be useful, alongside the development of guidelines and best practices for responsible deployment.

Moreover, on the Human–AI collaboration aspect, further research on optimizing the collaboration between AI systems and human analysts could demonstrate how each complements the other's strength. Researching the integration of more diverse and unconventional data sources into the CTI pipeline, such as social media, could potentially allow for a more accurate prediction of the threat actor's next attacks. Another research route is to study ways to mitigate bias in the pipeline and explore ways to make risk decisions about which runbacks can be automated and which cannot. Lastly, a more technical angle would be to investigate the feasibility and methodologies for real-time threat intelligence processing using AI, enabling an instantaneous response to emerging threats, while also achieving automated compliance with security policies, which is our next focus.

Author Contributions: Conceptualization, L.A.; Investigation, L.A.; Writing—original draft, L.A.; Writing—review & editing, M.D.; Supervision, M.D. All authors have read and agreed to the published version of the manuscript.

Funding: This research received no external funding.

Data Availability Statement: No new data were created or analyzed in this study. Data sharing is not applicable to this article.

Conflicts of Interest: The authors declare no conflict of interest.

References

1. Johnson, C.; Badger, L.; Waltermire, D.; Snyder, J.; Skorupka, C. Guide to Cyber Threat Information Sharing. October 2016. Available online: https://nvlpubs.nist.gov/nistpubs/SpecialPublications/NIST.SP.800-150.pdf (accessed on 22 September 2023).
2. Phythian, M. Studies in Intelligence. In *Understanding the Intelligence Cycle*; Routledge Taylor & Francis Group: London, UK; New York, NY, USA, 2013; pp. 21–43.
3. Buczak, A.; Guven, E. A Survey of Data Mining and Machine Learning Methods for Cyber Security Intrusion Detection. *IEEE Commun. Surv.* **2016**, *18*, 1153–1176. [CrossRef]
4. Strom, B.E.; Applebaum, A.; Miller, D.P.; Nickels, K.C.; Pennington, A.G.; Thomas, C.B. The MITRE Corporation. March 2020. Available online: https://attack.mitre.org/docs/ATTACK_Design_and_Philosophy_March_2020.pdf (accessed on 27 September 2023).
5. Dekker, M.; Alevizos, L. A threat-intelligence driven methodology to incorporate uncertainty in cyber risk analysis and enhance decision-making. *Wiley Secur. Priv.* **2023**, *7*, e333. [CrossRef]
6. Chen, T.M.; Abu-Nimeh, S. Lessons from Stuxnet. *Computer* **2011**, *44*, 91–93. [CrossRef]
7. Sarker, I.H.; Furhad, H.M.; Nowrozy, R. AI-Driven Cybersecurity: An Overview, Security Intelligence Modeling and Research Directions. *SN Comput. Sci.* **2021**, *2*, 173. [CrossRef]
8. Ring, M.; Wunderlich, S.; Grudl, D. Flow-based benchmark data sets for intrusion detection. In Proceedings of the 16th European Conference on Cyber Warfare and Security, Dublin, Ireland, 29–30 June 2017; p. 361.
9. Brown, R.; Nickels, K. *SANS 2023 CTI Survey: Keeping up with a Changing Threat Landscape*; SANS Institute: Boston, MA, USA, 2023.
10. Sauerwein, C.; Sillaber, C.; Mussmann, A.; Breu, R. Threat Intelligence Sharing Platforms: An Exploratory Study of Software Vendors and Research Perspectives. In Proceedings of the 13 Internationalen Tagung Wirtschaftsinformatik (WI 2017), St. Gallen, Switzerland, 12–15 February 2017.
11. Sundar, S.S. Rise of Machine Agency: A Framework for Studying the Psychology of Human–AI Interaction (HAII). *J. Comput.-Mediat. Commun.* **2020**, *25*, 74–88. [CrossRef]
12. Brundage, M.; Avin, S.; Wang, J.; Belfield, H.; Krueger, G.; Hadfield, G.; Khlaaf, H. arXiv—Computer Science—Computers and Society. 20 April 2020. Available online: https://arxiv.org/abs/2004.07213 (accessed on 30 September 2023).
13. Varma, A.J.; Taleb, N.; Said, R.A.; Ghazal, T.M.; Ahmad, M.; Alzoubi, H.M.; Alshurideh, M. A Roadmap for SMEs to Adopt an AI Based Cyber Threat Intelligence. In *The Effect of Information Technology on Business and Marketing Intelligence Systems*; Springer: Cham, Switzerland, 2023; pp. 1903–1926.
14. Suryotrisongko, H.; Musashi, Y.; Tsuneda, A.; Sugitani, K. Robust Botnet DGA Detection: Blending XAI and OSINT for Cyber Threat Intelligence Sharing. *IEEE Access* **2022**, *10*, 34613–34624. [CrossRef]
15. Ranade, P.; Piplai, A.; Mittal, S.; Joshi, A.; Finin, T. arXiv—Generating Fake Cyber Threat Intelligence Using Transformer-Based Models. 18 June 2021. Available online: https://arxiv.org/abs/2102.04351 (accessed on 1 October 2023).
16. Moraliyage, H.; Sumanasena, V.; De Silva, D.; Nawaratne, R.; Sun, L.; Alahakoon, D. Multimodal Classification of Onion Services for Proactive Cyber Threat Intelligence Using Explainable Deep Learning. *IEEE Access* **2022**, *10*, 56044–56056. [CrossRef]
17. Mitra, S.; Piplai, A.; Mittal, S.; Joshi, A. Combating Fake Cyber Threat Intelligence using Provenance in Cybersecurity Knowledge Graphs. In Proceedings of the IEEE International Conference on Big Data (Big Data), Orlando, FL, USA, 15–18 December 2021.

18. Mittal, S.; Joshi, A.; Finin, T. Cyber-All-Intel: An AI for Security Related Threat Intelligence. 7 May 2019. Available online: https://arxiv.org/pdf/1905.02895.pdf (accessed on 3 October 2023).
19. Kotsiantis, S. Decision trees: A recent overview. *Artif. Intell. Rev.* **2011**, *39*, 261–283. [CrossRef]
20. Goodfellow, I.; Bengio, Y.; Courville, A. *Deep Learning*; The MIT Press: London, UK, 2016.
21. Nguyen, H.; Tran, K.; Thomassey, S.; Hamad, M. Forecasting and Anomaly Detection approaches using LSTM and LSTM Autoencoder techniques with the applications in supply chain management. *Int. J. Inf. Manag.* **2021**, *57*, 102282. [CrossRef]
22. Landwehr, N.; Hall, M.; Frank, E. Logistic Model Trees. *Mach. Learn.* **2005**, *59*, 161–205. [CrossRef]
23. Chen, H.; Chiang, R.H.L.; Storey, V.C. Business Intelligence and Analytics: From Big Data to Big Impact. *MIS Q.* **2012**, *36*, 1165–1188. [CrossRef]
24. Tian, K.; Revelle, M.; Poshyvanyk, D. Using Latent Dirichlet Allocation for automatic categorization of software. In Proceedings of the 6th IEEE International Working Conference on Mining Software Repositories, Vancouver, BC, Canada, 16–17 May 2009.
25. Schonlau, M.; Zou, R.Y. The random forest algorithm for statistical learning. *Stata J. Promot. Commun. Stat. Stata* **2024**, *20*, 3–29. [CrossRef]
26. Sarker, I.H. Deep Learning: A Comprehensive Overview on Techniques, Taxonomy, Applications and Research Directions. *SN Comput. Sci.* **2021**, *2*, 420. [CrossRef] [PubMed]
27. Marschall, O.; Cho, K.; Savin, C. A Unified Framework of Online Learning Algorithms for Training Recurrent Neural Networks. *J. Mach. Learn. Res.* **2020**, *21*, 1–34.
28. Catak, F.O.; Kuzlu, M.; Catak, E.; Cali, U.; Guler, O. Defensive Distillation-Based Adversarial Attack Mitigation Method for Channel Estimation Using Deep Learning Models in Next-Generation Wireless Networks. *IEEE Access* **2022**, *10*, 98191–98203. [CrossRef]
29. Settles, B. Synthesis Lectures on Artificial Intelligence and Machine Learning (SLAIML). In *Active Learning*; Springer: Berlin/Heidelberg, Germany, 2012; pp. 1–114.
30. Kuhl, N.; Goutier, M.; Baier, L.; Wolf, C.; Martin, D. Human vs. supervised machine learning: Who Learn. Patterns Faster? *Cogn. Syst. Res.* **2022**, *76*, 78–92. [CrossRef]
31. Mishra, S. An Optimized Gradient Boost Decision Tree Using Enhanced African Buffalo Optimization Method for Cyber Security Intrusion Detection. *Appl. Sci.* **2023**, *12*, 12591. [CrossRef]
32. Desmond, M.; Muller, M.; Ashktorab, Z.; Dugan, C.; Duesterwald, E.; Brimijoin, K.; Finegan-Dollak, C.; Brachman, M.; Sharma, A. Increasing the Speed and Accuracy of Data Labeling Through an AI Assisted Interface. In Proceedings of the IUI '21: 26th International Conference on Intelligent User Interfaces, College Station, TX, USA, 13–17 April 2021.
33. Chen, C.P. Deep learning for pattern learning and recognition. In Proceedings of the IEEE 10th Jubilee International Symposium on Applied Computational Intelligence and Informatics, Timisoara, Romania, 21–23 May 2015.
34. Angelelli, M.; Arima, S.; Catalano, C.; Ciavolino, E. Cyber-Risk Perception and Prioritization for Decision-Making and Threat Intelligence. 1 August 2023. Available online: https://arxiv.org/abs/2302.08348 (accessed on 12 April 2023).
35. Lemay, A. Leblanc and Sylvain. Cognitive Biases in Cyber Decision-Making. In Proceedings of the ICCWS 2018 13th International Conference on Cyber Warfare and Security, Washington, DC, USA, 8–9 March 2018.
36. Kartal, E. A Comprehensive Study on Bias in Artificial Intelligence Systems: Biased or Unbiased AI, That's the Question! *Int. J. Intell. Inf. Technol. (IJIIT)* **2022**, *18*, 1–23. [CrossRef]
37. Lorente, A. Setting the goals for ethical, unbiased, and fair AI. In *AI Assurance*; Academic Press: Cambridge, MA, USA, 2023; pp. 13–54.
38. Dell'Acqua, F.; McFowland, E., III; Mollick, E.; Lifshitz-Assaf, H.; Kellogg, K.C.; Rajendran, S.; Krayer, L.; Candelon, F.; Lakhani, K.R. Navigating the Jagged Technological Frontier: Field Experimental Evidence of the Effects of AI on Knowledge Worker Productivity and Quality 22 September 2024. Available online: https://www.hbs.edu/ris/Publication%20Files/24-013 _d9b45b68-9e74-42d6-a1c6-c72fb70c7282.pdf (accessed on 2 February 2024).
39. Kaloudi, N.; Li, J. The AI-Based Cyber Threat Landscape: A Survey. *ACM Comput. Surv.* **2020**, *53*, 1–34. [CrossRef]
40. Sarker, I.H. AI-Based Modeling: Techniques, Applications and Research Issues towards Automation, Intelligent and Smart Systems. *SN Comput. Sci.* **2022**, *3*, 1–20.
41. Gupta, C.; Johri, I.; Srinivasan, K.; Hu, Y.-C.; Qaisar, S.M.; Huang, K.-Y. A Systematic Review on Machine Learning and Deep Learning Models for Electronic Information Security in Mobile Networks. *Sensors* **2022**, *22*, 2017. [CrossRef] [PubMed]
42. Sarker, I.H. Machine Learning: Algorithms, Real-World Applications and Research Directions. *SN Comput. Sci.* **2021**, *2*, 160. [CrossRef] [PubMed]
43. Djenna, A.; Bouridane, A.; Rubab, S.; Marou, I.M. Artificial Intelligence-Based Malware Detection, Analysis, and Mitigation. *Symmetry* **2023**, *15*, 677. [CrossRef]
44. Akhtar, M.S.; Feng, T. Malware Analysis and Detection Using Machine Learning Algorithms. *Symmetry* **2022**, *14*, 2304. [CrossRef]
45. Mohamed, N. Current trends in AI and ML for cybersecurity: A state-of-the-art survey. *Cogent Eng.* **2023**, *10*, 2. [CrossRef]
46. Jain, J. Artificial Intelligence in the Cyber Security Environment. In *Artificial Intelligence and Data Mining Approaches in Security Frameworks*; Wiley: Hoboken, NJ, USA, 2021.
47. Sree, S.V.; Koganti, S.C.; Kalyana, S.K.; Anudeep, P. Artificial Intelligence Based Predictive Threat Hunting in the Field of Cyber Security. In Proceedings of the 2nd Global Conference for Advancement in Technology (GCAT), Bangalore, India, 1–3 October 2021.

48. Gupta, I.; Gupta, R.; Singh, A.K.; Wen, X. An AI-Driven VM Threat Prediction Model for Multi-Risks Analysis-Based Cloud Cybersecurity. *Trans. Syst. Man Cybern. Syst.* **2023**, *53*, 6815–6827.
49. Deep Reinforcement Learning for Cyber Security. *Trans. Neural Netw. Learn. Syst.* **2023**, *34*, 3779–3795. [CrossRef]
50. Confido, A.; Ntagiou, E.V.; Wallum, M. Reinforcing Penetration Testing Using AI. In Proceedings of the 2022 IEEE Aerospace Conference (AERO), Big Sky, MT, USA, 5–12 March 2022.
51. NIST. Artificial Intelligence Risk Management Framework (AI RMF 1.0). January 2023. Available online: https://nvlpubs.nist.gov/nistpubs/ai/NIST.AI.100-1.pdf (accessed on 8 October 2023).
52. P. R. Committee. Council of the European Union. 25 November 2022. Available online: https://data.consilium.europa.eu/doc/document/ST-14954-2022-INIT/en/pdf (accessed on 12 January 2024).
53. Council of the European Union. 26 January 2024. Available online: https://data.consilium.europa.eu/doc/document/ST-5662-2024-INIT/en/pdf (accessed on 4 February 2024).
54. *ISO/IEC 5338:2023*; Information Technology—Artificial Intelligence—AI System Life Cycle Processes. International Standards Organization: Geneva, Switzerland, 2023. Available online: https://www.iso.org/standard/81118.html (accessed on 18 January 2024).
55. Sweeney, L. k-Anonymity: A Model for Protecting Privacy. *Int. J. Uncertain. Fuzziness Knowl.-Based Syst.* **2002**, *10*, 557–570. [CrossRef]
56. Solove, D.J. Privacy Self-Management and the Consent Dilemma. *Harv. Law Rev.* **2013**, *126*, 1880.
57. Tene, O.; Polonetsky, J. Big Data for All: Privacy and User Control in the Age of Analytics. *J. Technol. Intellect. Prop.* **2013**, *11*, 240–272.
58. Solon, B.; Selbst, D.A. Big Data's Disparate Impact. *Calif. Law Rev.* **2016**, *104*, 671–732.
59. Danks, D.; London, A.J. Algorithmic Bias in Autonomous Systems. In Proceedings of the 26th International Joint Conference on Artificial Intelligence (IJCAI 2017), Pittsburgh, PA, USA, 19–25 August 2017.
60. Abdullah, I.U.T. MLOps: A Step forward to Enterprise Machine Learning. 27 May 2023. Available online: https://arxiv.org/pdf/2305.19298.pdf (accessed on 6 October 2023).
61. Schwartz, R.; Vassilev, A.; Greene, K.K.; Perine, L. Towards a Standard for Identifying and Managing Bias in Artificial Intelligence. 15 March 2022. Available online: https://nvlpubs.nist.gov/nistpubs/SpecialPublications/NIST.SP.1270.pdf(accessed on 24 February 2024).
62. Ha, T.; Kim, S. Improving Trust in AI with Mitigating Confirmation Bias: Effects of Explanation Type and Debiasing Strategy for Decision-Making with Explainable AI. *Int. J. Hum.-Comput. Interact.* **2023**, 1–12. [CrossRef]
63. Ribeiro, M.T.; Singh, S.; Guestrin, C. "Why Should I Trust You?": Explaining the Predictions of Any Classifier. In Proceedings of the KDD '16: 22nd ACM SIGKDD International Conference on Knowledge Discovery and Data Mining, San Francisco, CA, USA, 13–17 August 2016.
64. Holstein, K.; Vaughan, J.W.; Daume, H.; Dudik, M.; Wallach, H. Improving Fairness in Machine Learning Systems: What Do Industry Practitioners Need? In Proceedings of the CHI '19: Proceedings of the 2019 CHI Conference on Human Factors in Computing Systems, Glasgow, UK, 4–9 May 2019.
65. Gebru, T.; Morgenstern, J.; Vecchione, B.; Vaughan, J.W.; Wallach, H.; Daumé, H.; Crawford, K. Datasheets for Datasets. 1 December 2021. Available online: https://arxiv.org/abs/1803.09010 (accessed on 14 October 2023).
66. Vassilev, A.; Oprea, A.; Fordyce, A.; Anderson, H. Adversarial Machine Learning: A Taxonomy and Terminology of Attacks and Mitigations. January 2024. Available online: https://nvlpubs.nist.gov/nistpubs/ai/NIST.AI.100-2e2023.pdf (accessed on 25 January 2024).
67. Adversarial Machine Learning Attacks and Defense Methods in the Cyber Security Domain. *ACM Comput. Surv.* **2021**, *54*, 5.
68. Kaur, R.; Gabrijelčič, D.; Klobučar, T. Artificial intelligence for cybersecurity: Literature review and future research directions. *Inf. Fusion* **2023**, *97*, 101804. [CrossRef]

Disclaimer/Publisher's Note: The statements, opinions and data contained in all publications are solely those of the individual author(s) and contributor(s) and not of MDPI and/or the editor(s). MDPI and/or the editor(s) disclaim responsibility for any injury to people or property resulting from any ideas, methods, instructions or products referred to in the content.

Article

HotCFuzz: Enhancing Vulnerability Detection through Fuzzing and Hotspot Code Coverage Analysis

Chunlai Du [1], Yanhui Guo [2,*], Yifan Feng [1] and Shijie Zheng [1]

[1] School of Information Science and Technology, North China University of Technology, Beijing 100144, China; duchunlai@ncut.edu.cn (C.D.); 180630@mail.ncut.edu.cn (Y.F.); heyralap@mail.ncut.edu.cn (S.Z.)
[2] Department of Computer Science, University of Illinois Springfield, Springfield, IL 62703, USA
* Correspondence: yguo56@uis.edu

Abstract: Software vulnerabilities present a significant cybersecurity threat, particularly as software code grows in size and complexity. Traditional vulnerability-mining techniques face challenges in keeping pace with this complexity. Fuzzing, a key automated vulnerability-mining approach, typically focuses on code branch coverage, overlooking syntactic and semantic elements of the code. In this paper, we introduce HotCFuzz, a novel vulnerability-mining model centered on the coverage of hot code blocks. Leveraging vulnerability syntactic features to identify these hot code blocks, we devise a seed selection algorithm based on their coverage and integrate it into the established fuzzing test framework AFL. Experimental results demonstrate that HotCFuzz surpasses AFL, AFLGo, Beacon, and FairFuzz in terms of efficiency and time savings.

Keywords: vulnerability mining; hotspot code; fuzzing; AFL

Citation: Du, C.; Guo, Y.; Feng, Y.; Zheng, S. HotCFuzz: Enhancing Vulnerability Detection through Fuzzing and Hotspot Code Coverage Analysis. *Electronics* **2024**, *13*, 1909. https://doi.org/10.3390/electronics13101909

Academic Editors: Abdussalam Elhanashi and Pierpaolo Dini

Received: 13 April 2024
Revised: 8 May 2024
Accepted: 10 May 2024
Published: 13 May 2024

Copyright: © 2024 by the authors. Licensee MDPI, Basel, Switzerland. This article is an open access article distributed under the terms and conditions of the Creative Commons Attribution (CC BY) license (https://creativecommons.org/licenses/by/4.0/).

1. Introduction

In recent years, a surge in security incidents, including hacker attacks and user information leaks, has underscored the pivotal role of software vulnerabilities in such events. As reported by the CVE Detail website, the steady increase in vulnerabilities demonstrates a significant upward trajectory. These vulnerabilities stem from weaknesses in software algorithm models and flaws in code implementation. As software code expands and logic grows more intricate, the task of uncovering hidden vulnerabilities becomes increasingly daunting. Coverage-guided fuzzing has emerged as a highly effective method for automated vulnerability mining. This approach entails the automated or semi-automated generation of unexpected, random, abnormal, or invalid data, which are then inputted into the target software. By monitoring the software's response to these inputs, potential vulnerabilities are unearthed if the program crashes or behaves abnormally. Moreover, the integration of deep-learning techniques into vulnerability-mining research has led to advancements in source code vulnerability mining, as exemplified by projects like SySeVR [1]. These initiatives extract code representations capturing both syntax and semantic information, enabling vulnerability mining at the source code level.

Path-coverage-based fuzzing test techniques, exemplified by AFL [2], emphasize the exploration of new paths and the minimization of test inputs for seed selection. However, these techniques lack consideration for semantic code analysis. Conversely, AI-based vulnerability-mining techniques centered on the program source code leverage feature learning from code blocks harboring vulnerabilities. In practical vulnerability-mining scenarios, these techniques can identify the location of vulnerability blocks but often struggle to precisely pinpoint the exact vulnerability point. Moreover, they encounter difficulties in uncovering unknown vulnerabilities.

Therefore, we explore a vulnerability-mining method based on hotspot code coverage. During the static analysis phase, deep-learning techniques are employed to identify and fine-tune suspicious hotspot codes. Subsequently, the research pivots to dynamic fuzzing, where vulnerability mining is guided by hotspot code coverage. Our contributions can be summarized as follows:

(1) The introduction of a test input seed-filtering algorithm designed for hotspot code coverage, facilitating the priority-oriented testing of hotspot code blocks.
(2) The proposal of HotCFuzz, a vulnerability-mining model guided by hotspot code coverage. Leveraging the calibration of hotspot codes identified during the static analysis phase, we prioritize vulnerability mining within hotspot code regions. This approach ensures the effective mining not only for hotspot codes but also for non-hotspot codes.
(3) Experimental results demonstrate the superior performance of our proposed HotC-Fuzz model compared to AFL, AFLGo, Beacon, and FairFuzz.

The rest of the paper is organized as follows: Section 2 introduces related work. In Section 3, the proposed MemConFuzz model is described. In Section 4, the experimental process and the results are discussed. Finally, we conclude the paper in Section 5.

2. Related Work

In terms of fuzzing tests, AFLgo [3] identifies the source code during compilation and computes the distance to the target basic block based on edges in the Control Flow Graph (CFG) of the Program Under Test (PUT). During runtime, it aggregates the distance values for each basic block and computes the average to assess seeds. The prioritization of seeds is based on distances, favoring those closer to the target. Hawkeye [4] measures the similarity between the execution traces of seeds and the target at the function level, incorporating the basic block trace distance with the coverage function similarity for seed prioritization and power scheduling. LOLLY [5] employs a user-specified sequence of program statements as the target, evaluating seeds based on their ability to cover the target sequence (i.e., sequence coverage). Berry [6] enhances LOLLY by considering the execution context of the target sequence, upgrading target sequences with "essential nodes," and using the similarity between the target execution trace and the enhanced target sequence to differentiate seed priority. DrillerGo [7] utilizes semantic information from logs and CVE descriptions to guide fuzzing, whereas SAVIOR [8] leverages Sanitizer information to prioritize seeds with collaborative executions and verifies all vulnerable program locations on the execution program path. UAFuzz [9] employs a sequence-aware target similarity metric to gauge the similarity between seed executions and error tracing after the target is freed. RDFuzz [10] combines the basic block execution frequency and distance to the target path for seed prioritization, aiming to identify frequently executed code regions close to the target path to effectively trigger and discover potential vulnerabilities. Beacon [11] steers the execution path and employs lightweight static analysis to compute abstract preconditions to the target, thereby discarding numerous infeasible execution paths at runtime to enhance execution efficiency. FairFuzz [12] automatically identifies branches covered by a small number of inputs and biases the mutation towards these branches, leading to a higher coverage. AFLChurn [13] assigns numerical weights to basic blocks based on recent changes or the change frequency to identify and prioritize code regions more likely to introduce new vulnerabilities. WindRanger [14] considers deviations from basic blocks, meaning blocks deviating from the target location starting from the execution trace when calculating the "distance" to the target path. This approach aims to identify paths potentially concealing vulnerabilities by analyzing changes in execution paths.

In the realm of AI-based techniques, Hin et al. [15] introduced LineVD, a deep-learning framework that approaches statement-level vulnerability detection as a node classification task. It captures control and data dependency relationships between statements within functions using graph neural networks, while encoding original source code tokens using a transformer-based model. Li et al. [16] pioneered the VulDeePecker vulnerability detection method, which, first, segments the program forward and backward based on the Program Dependency Graph (PDG) of the source code. Subsequently, it employs a bidirectional long short-term memory network to ascertain whether program slices contain vulnerabilities.

To overcome the limitations of the grammar features in VulDeePecker's program-slicing dependencies, Li et al. [1] proposed the SySeVR vulnerability detection method. This method broadens the grammar features of the slicing dependencies into four categories, extracting vulnerability code blocks containing both semantic and syntactic information from program slices. Zhou et al. [17] introduced Devign, a methodology that extracts valuable features from learned rich node representations through a novel convolutional module, utilizing them for graph-level classification. Additionally, Wu et al. [18] developed VulCNN, a model inspired by deep-learning-based image classification techniques. Vul-CNN transforms the source code of functions into an image representation, parsing and tokenizing the source code while preserving crucial syntactic and semantic information. This approach enables the model to effectively learn the internal representation of the source code.

3. Methodology

3.1. Motivation

The dynamic vulnerability-mining method based on fuzzing tests hinges on two pivotal factors: (1) the selection of input seeds for generating distorted test inputs, and (2) the distortion strategies employed on these input seeds. It can be argued that the quality of the chosen input seeds significantly influences the efficacy of vulnerability mining through fuzz testing. To elucidate the concepts presented in this paper, the following definitions are provided:

Definition 1. *Dangerous Instruction Code: Instructions within a program's codebase that operate on operands influenced by untrustworthy input data from the external environment. These instructions have the potential to induce program exceptions or alter the normal execution flow, posing a risk to the program's integrity and security.*

Definition 2. *Dangerous Function Code: Functions within a software system designed to process input data originating from untrusted external sources. These functions, however, fail to implement rigorous boundary checks on the received input data, thereby exposing the system to potential security vulnerabilities.*

Definition 3. *Hotspot Code Region: A set of code segments within a program characterized by their propensity to trigger program crashes or other undesirable behaviors. Hotspot code regions encompass both dangerous functions, which handle untrusted input data without stringent boundary checks, and dangerous instruction codes, which operate on operands susceptible to manipulation by untrusted external sources.*

If the target program under test contains hotspot code regions, the likelihood of triggering program crashes during the testing process is significantly higher compared to other code regions. Building upon this logical assumption, we prioritize hotspot code region coverage as a fundamental factor for orienting fuzzing tests. Figure 1 illustrates the core concept of vulnerability mining presented in this paper.

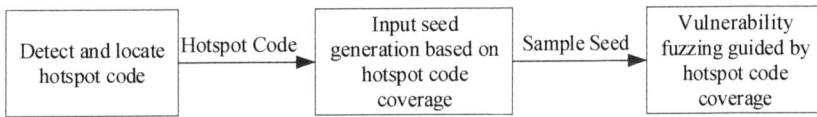

Figure 1. The idea behind this article.

In the static analysis phase, hotspot code regions are discerned from the program's source code employing a deep-learning framework. Subsequently, during the dynamic analysis phase, the coverage of each test input concerning the identified hotspot codes during the execution of the target program is logged. An evaluation method for test inputs is then formulated to allocate greater emphasis on input seed selection within the fuzzing test framework. This ensures that test inputs exhibiting high coverage of hotspot code regions are prioritized for selection as input seeds. Finally, a hotspot-code-coverage-guided fuzzing framework is constructed around the chosen seed inputs.

In summary, our proposed model meticulously analyzes the semantics of code behavior by examining dependencies in static analysis. Furthermore, leveraging dynamic fuzzing, it promptly retains context information during program crashes along with the corresponding test inputs responsible for triggering the crashes. This capability allows for precise pinpointing of the crash location, thereby facilitating subsequent analysis of the underlying cause of the vulnerability.

3.2. HotCFuzz Framework

According to the concept of prioritizing testing of hotspot code as described in Section 3.1, we have developed a vulnerability-mining framework named HotCFuzz, which operates under the guidance of hotspot code coverage. The components of HotCFuzz are depicted in Figure 2 below:

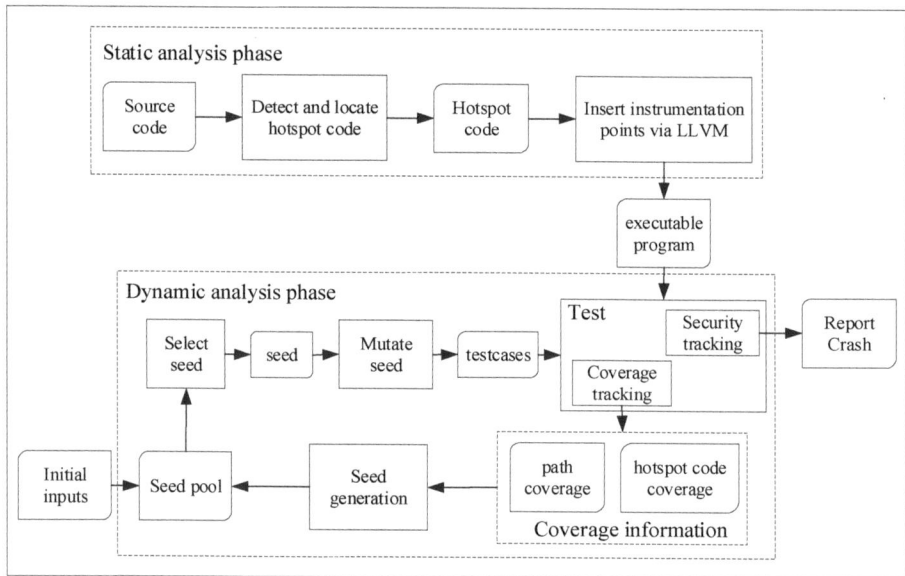

Figure 2. HotCFuzz model.

HotCFuzz comprises two distinct phases: (1) Static Analysis Phase: This phase involves scrutinizing the source code to pinpoint and annotate hotspot code. The phase yields location information pertaining to the identified hotspot code. Utilizing LLVM, instrumentation code is inserted at the outset of all identified hotspot codes within the source code of the target program. This instrumentation code records the execution information of the instrumented hotspot code during runtime. (2) Dynamic Testing Phase: This phase encompasses two primary facets. Firstly, it monitors the execution coverage of the hotspot code for each test input processed by the program. Secondly, it evaluates the quality of the test inputs based on the coverage of the hotspot code. Leveraging this evaluation, superior test inputs are selected as seed inputs to generate a new round of test input sets.

3.2.1. Detect and Locate Hotspot Codes

The program comprises a sequence of functions interconnected by sequential control flow relationships, with each function capable of being converted into an Abstract Syntax

Tree (AST) representation. To achieve this, we employ Joern, which operates by generating a Code Property Graph (CPG) initially. This CPG amalgamates the AST, Control Flow Graph (CFG), and Program Dependence Graph (PDG) of the program's code. Subsequently, Joern produces the corresponding AST, CFG, and PDG from the CPG, tailored to meet the task requirements. For example, let us consider the source code depicted in Figure 3, which encompasses two functions, namely, Value and func, with a calling relationship established between them.

```
1:   int Value(z)
2:   {
3:       int w = z;
4:       w = w + 3;
5:   }
6:   void func()
7:   {
8:       int x = UserInput();
9:       if ( x < MaxValue)
10:      {
11:          int y = x * 5;
12:          Value(y);
13:      }
14:  }
```

Figure 3. Sample code.

The corresponding ASTs for the functions *Value* and *func* generated by Joern from CPG are shown in Figures 4 and 5:

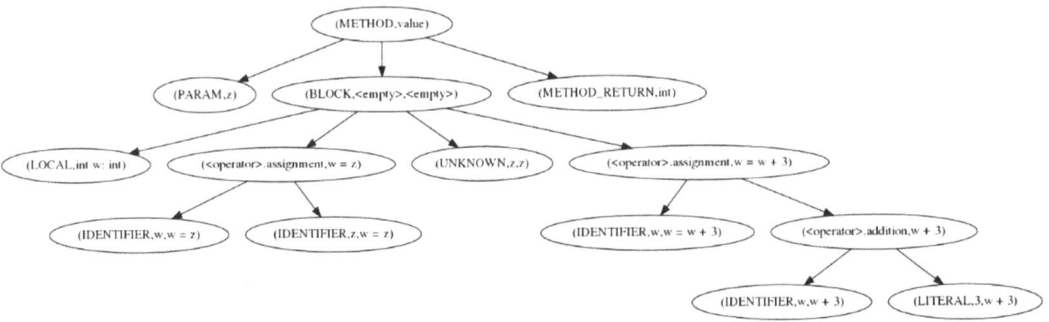

Figure 4. AST for the function *Value*.

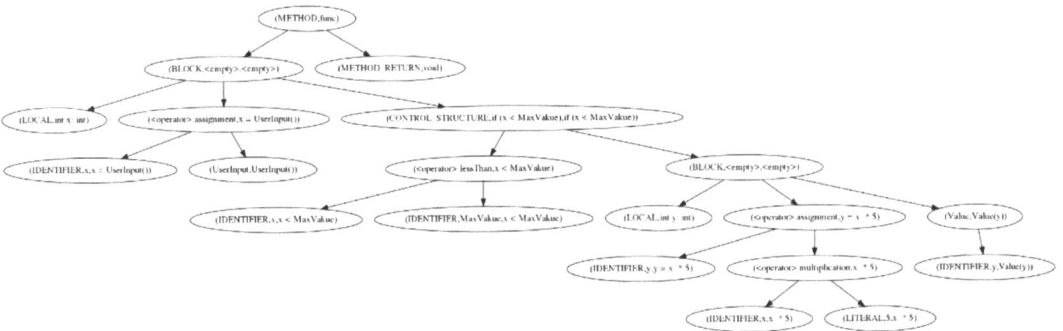

Figure 5. AST for the function *func*.

In order to identify hotspots within programs, we first detect potentially hazardous elements within the AST of the target program. These hazardous elements encompass API functions, array variables, pointer variables, etc. Due to variations in variable naming conventions across different programs, direct comparisons are often infeasible. Therefore, our AST-based hazardous element detection focuses solely on hazardous API functions. Subsequently, we consider semantic information of API functions to find association relations hidden in instructions. For example, hazardous instruction sequence of heap operation on the same memory block can be found through assignment propagation dependencies among heap pointer variables in the DDG. To realize the detection of hazardous API functions within AST elements, we necessitate the construction of vulnerability syntax characteristics tailored for the search, comparison, and identification of AST elements. We employed an outstanding static auditing tool, Checkmarx, to identify snippets of vulnerable code within known vulnerability programs. This strategic utilization of Checkmarx significantly enhances our ability to scrutinize the source code and pinpoint hotspot areas effectively. By leveraging Checkmarx's advanced capabilities, we ensure a thorough examination of critical code segments, thereby augmenting the precision of our hotspot code annotation process. The incorporation of Checkmarx's analysis results into our methodology adds an additional layer of assurance regarding the effectiveness of our hotspot code identification and annotation process.

Subsequently, we extract hazardous API functions from the identified vulnerable code snippets to represent the vulnerability syntax characteristics. In our research, we selected 5891 snippets which encompass C language vulnerabilities in the SARD dataset [19]. Utilizing the represented vulnerability syntax characteristics, we performed a comprehensive traversal search on the AST encompassing all functions within the target program. Since the code elements for vulnerability may reside as either leaf nodes or intermediate nodes within the AST, upon encountering a node that meets a vulnerability syntax characteristic, we meticulously record the location of corresponding code area. Subsequently, through this analysis, we discern the locations of hotspot codes within the target program.

In certain scenarios, specific sequences of function abnormalities are necessary in order to pose a vulnerability threat. For instance, in the case of a heap vulnerability such as double *free*, the occurrence of two *free* operations on memory allocated by *malloc* is required. Hence, it is imperative to apply further filtering to the identified hotspot functions in the target program to accommodate such cases.

The PDG delineates the data dependencies and control dependencies among instructions within a function. Figures 6 and 7, respectively, illustrate the PDGs of the functions *Value* and *func* from Figure 2.

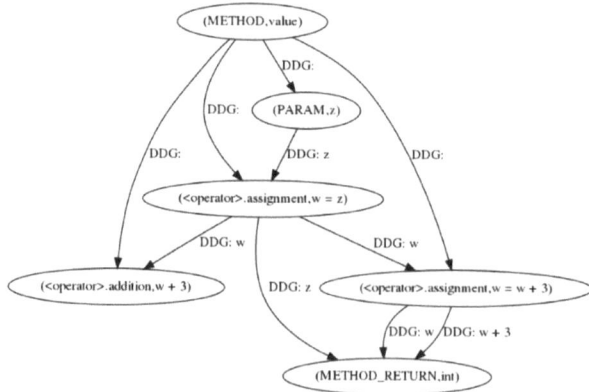

Figure 6. PDG for the function *Value*.

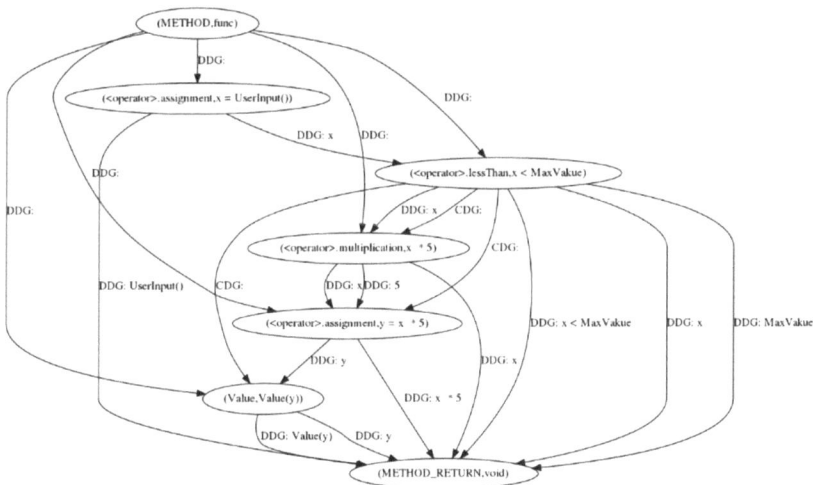

Figure 7. PDG for the function *func*.

Since the Program Dependence Graphs (PDGs) generated by Joern are function-oriented, it becomes necessary to merge the PDGs of related functions based on their calling relationships. This is accomplished by establishing an equivalence mapping between the data within the calling function and the parameters of the called function, taking into account the relationship of parameter passing. As a result, we establish semantic correlations between functions or instructions. For instance, when dealing with operations on the same heap memory, despite potential differences in the names of pointer variables between the calling function and the called function, these disparities can be reconciled through equivalence mapping. Ultimately, this process enables us to discern the hotspot code within the target program (Algorithm 1).

Algorithm 1: Detect and Locate Hotspot Functions in Target Programs
Input: Vulnerability syntax characteristic
Output: Hotspot code information
1:    generate the AST of the target program function
2:    traversal search of function AST based on vulnerability syntax characteristics
3:    for AST not traversed, do
4:       if AST grammar element matching the characteristics, then
5:          grammatical elements incorporated into a collection of hotspot codes
6:       endif
7:    endfor
8:    using Joern to generate the PDG of the target program function
9:    for function call not processed, do
10:       constructing mapping equivalence tables for variable names
11:   endfor
12:   traverse the function PDG vertically and comb through the sequence of anomalies between functions based on data dependencies.
13:   update collection set of hotspot code regions

3.2.2. Input Seed Filtering based on Hotspot Code Coverage

Based on the identification of the hotspot code within the target program, we undertook two tasks. Firstly, we defined a bitmap matrix named *hotspot_shm*, which shares the same dimensions as the AFL bitmap. This matrix serves to track and record coverage information of executed hotspot code functions, specifically capturing the frequency of triggering each hotspot function, denoted as *hotspot_funcs*. Secondly, as illustrated in Figure 2, during the static phase, we utilized LLVM tools to inject instrumentation code ahead of all hotspot code segments. The objective was to capture the execution of hotspot code. Throughout the fuzzing process, when the target program encounters a hotspot code, the corresponding position in the hotspot code bitmap matrix is set to 1.

To obtain input seeds prioritizing the coverage of hotspot code, we employ hotspot_funcs as a metric. Each input is evaluated based on the number of hotspot code functions it activates, with inputs triggering more hotspot code functions assigned a higher coefficient. Consequently, such inputs receive a higher score and are accorded higher priority as seeds. We adhere to the following two principles:

Principle 1. *The coefficient increases proportionally with the recorded coverage of hotspot functions during execution. Consequently, the seed's score and energy are augmented.*

Principle 2. *AFL's original scoring strategy is also factored into our approach. We ensure that the final score of the seed does not become excessively large, as this could potentially lead to the seed being trapped in local code blocks during execution.*

Therefore, the input seed preference evaluation formula proposed in this paper is as follows:

$$\text{Priority score}(\text{seed}_i) = \begin{cases} P_{afl}(\text{seed}_i) * \left(a - \frac{e^{-\text{hotspot\_funcs}}}{b}\right), & \text{hotspot\_funcs} \\ P_{afl}(\text{seed}_i), & \text{otherwise} \end{cases} \quad (1)$$

where a is the deviation factor, which is a consideration for implementing principle 2 and takes the value of 1.3 in the experimental test. b is adjustment to ensure $\left(a - \frac{e^{-\text{hotspot\_funcs}}}{b}\right) > 1$ whenever hotspot codes are executed. When hotspot codes are found but the input seeds concerning hotspot code have been decreasing, b should be increased. In our experiment, b is set to 6 based on experiment statistics. The hotspot_funcs denotes the total number of hotspot code functions triggered by a test seed during the fuzzing test loop. seed_i refers to the current test input in the fuzzing process. $P_{afl}(\text{seed}_i)$ represents AFL's original seed selection strategy. Equation (1) not only considers the prioritization of input seeds that explore hotspot code regions but also ensures that other excellent input seeds are not discarded.

3.2.3. HotCFuzz Model

The HotCFuzz model is constructed based on the AFL framework. As depicted in Algorithm 2, during the static analysis phase, hotspot code information is first obtained from the target program (line 1). Subsequently, instrumentation codes are inserted at the heads of hotspot codes to record the execution of these hotspot codes during the fuzzing of the target program. These recorded data serve to evaluate the quality of test inputs, thereby aiding in the selection of superior input seeds (line 2). During the fuzzing phase, HotCFuzz initially provides a seed set S, from which inputs s are chosen and mutated for fuzzing in a continuous loop until a timeout is reached or fuzzing is aborted. Within this loop, AFL typically selects seeds using ChooseNext from a circular queue in the order they were added in the original framework. However, we modify ChooseNext to prioritize seeds based on their importance (line 5). AssignEnergy determines the number of inputs generated from s (line 6). The generated inputs, denoted as s', are produced by randomly mutating s (line 8). If the generated input s' triggers a crash, it is added to the CrashSet. Otherwise, if s' covers a new branch (line 12–14) or multiple hotspot codes (line 15–17), it is appended to the seed set S.

Algorithm 2: HotCFuzz
Input: Test Input Set S Output: CrashSet
1:    Static analysis phase to obtain hotspot code information 2:   Insert codes for recording hotspot code executed by LLVM 3:   CrashSet ← ∅ 4:   repeat 5:   select seed s from Set S 6:   determine the number p of abnormal inputs according to s by power schedule 7:   for i from 1 to p, do 8:   mutate s to create new test input s' 9:   if s' triggers crash, then 10:   add s' to CrashSet 11:   else 12:   if new paths are found, then 13:   add s' to S 14:   endif 15:   if hotspot code areas are found, then 16:   evaluating s' and add s' to S according to score 17:   endif 18:   endif 19:   endfor 20:   until timeout reached or abort-signal

4. Experiments

4.1. Experiment Setup

In this experiment, our objective is to assess the performance of the HotCFuzz model and compare it with other fuzzing test techniques. To achieve this, we have selected a publicly available test input set comprising programs such as cxxfilt v2.30, readelf v2.25, objdump v2.25, and openjpeg v2.3.0. It is noteworthy that some of these programs have known vulnerabilities, duly identified by CVE numbers, thereby ensuring the authenticity and reliability of the experiment.

For comparison with HotCFuzz, we have selected AFL, AFLGo, Beacon, and FairFuzz as the fuzzing test tools. These tools represent advanced techniques in the field of fuzzing and are extensively utilized for vulnerability-mining purposes.

4.2. Evaluation Indicators

To assess the performance of the HotCFuzz model, we have identified the following three critical evaluation metrics:

(1) Vulnerability Discovery Count: This metric involves comparing the number of vulnerabilities detected by different fuzzing tools within the same duration of execution. It serves to validate the efficacy of vulnerability mining across different testers.

(2) Same Vulnerability Trigger Time: This metric entails comparing the time required by different fuzzing tools to reproduce known vulnerabilities. It validates the effectiveness of various testers in triggering known vulnerabilities.

(3) Seed Execution Path Gain Count: This metric involves comparing the number of gained execution paths for seeds among different fuzzing tools within the same duration of execution. This metric visualizes code coverage, revealing testing depth and breadth, and also reflects the efficiency of the fuzzing tester in generating test cases.

4.3. Experimental Results

We evaluated the performance of the HotCFuzz model using the above evaluation metrics and compared it with AFL, AFLGo, Beacon, and FairFuzz. The experimental results are shown in Tables 1–3.

Table 1. Number of vulnerability mining for the same testing time (24 h) for different fuzzing testers.

Program	AFL	AFLGo	Beacon	FairFuzz	HotCFuzz
cxxfilt	373	6	401	39	493
readelf	86	65	54	181	91
objdump	3	2	4	10	18
openjpeg	8	10	0	7	21

Table 2. Trigger time for real-world exploits (min).

Program	Vulnerability	AFL	AFLGo	Beacon	FairFuzz	HotCFuzz
cxxfilt	CVE-2016-4492	17	9	6	8	5
readelf	CVE-2017-9039	6	12	6	6	9
openjpeg	CVE-2017-12982	768	960	Timeout	930	673

Table 3. Number of seed execution paths.

Program	AFL	AFLGo	Beacon	FairFuzz	HotCFuzz
cxxfilt	2783	3433	4106	3826	7948
readelf	3049	3988	3125	5895	4013
objdump	2876	3934	3877	4671	4832
openjpeg	3946	3771	1285	3016	4011

4.4. Discussion

To validate the efficiency of HotCFuzz in vulnerability mining, we assessed its performance by comparing the number of vulnerabilities detected within the same timeframe. As illustrated in Table 1, across various test programs, HotCFuzz consistently outperforms AFL, AFLGo, Beacon, and FairFuzz in terms of vulnerability discovery. Specifically, in the cxxfilt test program, HotCFuzz discovers a greater number of vulnerabilities compared to its counterparts. Similarly, in the readelf and objdump test programs, HotCFuzz excels, surpassing AFL, AFLGo, and FairFuzz in vulnerability mining. Even in the openjpeg test program, HotCFuzz outshines other fuzzing testers by identifying more vulnerabilities. These findings underscore the efficiency, superiority, and effectiveness of HotCFuzz in vulnerability mining. The experimental result in Table 1 highlights the superiority of our approach over established methods such as AFL, AFLGo, Beacon, and FairFuzzy.

HotCFuzz aims to expedite the triggering of exploits located at the target location. The experimental findings, outlined in Table 2, demonstrate HotCFuzz's superior performance in this aspect across most scenarios. With the exception of a longer vulnerability reproduction time observed in CVE-2017-9039, HotCFuzz consistently outperforms AFL, AFLGo, Beacon, and FairFuzz in terms of vulnerability triggering speed. For instance, in the vulnerability reproduction of CVE-2016-4492, HotCFuzz showcases greater efficiency compared to the latest Beacon tool, owing to its well-crafted seed selection strategy. Conversely, in the vulnerability reproduction of CVE-2017-9039, HotCFuzz experiences a longer reproduction time due to the specific or intricate triggering conditions associated with the vulnerability. In cases where the target vulnerability relies on a particular execution environment state or necessitates a specific input sequence for triggering, HotCFuzz may encounter challenges in the absence of adequate dynamic execution information feedback. The outcomes of these experiments presented in Table 2 demonstrate the shortest time consumption achieved by our model and underscore its effectiveness in vulnerability detection.

Table 3 presents the results of experiments on the number of execution paths covered, showcasing HotCFuzz's exceptional performance compared to other fuzzing testers. However, it is worth noting that, in the case of testing readelf, HotCFuzz's results are slightly inferior to those of FairFuzz. This outstanding experimental performance can be attributed to HotCFuzz's innovative approach, which integrates both static analysis and dynamic execution information to guide the seed selection and mutation process. Through meticulous analysis of the program's data flow and control flow, HotCFuzz efficiently identifies potential vulnerability trigger paths, thereby optimizing test cases to explore a greater number of execution paths. Consequently, HotCFuzz outshines similar tools in seed path coverage experiments, underscoring its effectiveness and superiority in vulnerability mining.

5. Conclusions

Vulnerability mining has long been a focal point in software security research, with fuzzing remaining a prominent technology in this domain. While fuzzing has seen significant success, it often overlooks the syntax and semantics of the target program. To address this gap, we aimed to identify suspicious hotspot codes within the target program by leveraging syntax and semantic features extracted from vulnerability code fragments. Based on this concept, we introduced HotCFuzz, a vulnerability fuzzing model centered on hotspot code coverage. In the HotCFuzz model, we introduced an algorithm to detect and pinpoint hotspot codes during the static analysis stage. Instrumentation codes are inserted at the beginning of hotspot codes to record their execution, facilitating the selection of superior input seeds. Experimental results demonstrate that HotCFuzz surpasses AFL, AFLGo, Beacon, and FairFuzz in terms of performance and effectiveness.

Author Contributions: Conceptualization, C.D. and Y.G.; methodology, C.D. and Y.G.; software, Y.F.; validation, S.Z.; investigation, Y.F.; writing—original draft preparation, Y.F. and S.Z.; writing—review and editing, Y.G.; visualization, Y.F. and S.Z.; project administration, C.D.; funding acquisition, C.D. All authors have read and agreed to the published version of the manuscript.

Funding: This research was funded by the National Natural Science Foundation of China grant number 62172006.

Data Availability Statement: The test results data presented in this study are available upon request. The dataset can be found in public web sites.

Conflicts of Interest: The authors declare no conflicts of interest.

References

1. Li, Z.; Zou, D.; Xu, S.; Jin, H.; Zhu, Y.; Chen, Z. SySeVR: A Framework for Using Deep Learning to Detect Software Vulnerabilities. *IEEE Trans. Dependable Secur. Comput.* **2022**, *19*, 2244–2258. [CrossRef]
2. American Fuzzing Lop. Available online: https://lcamtuf.coredump.cx/afl/ (accessed on 8 April 2024).
3. Böhme, M.; Pham, V.T.; Nguyen, M.D.; Roychoudhury, A. Directed Greybox Fuzzing. In Proceedings of the 2017 ACM SIGSAC Conference on Computer and Communications Security, Dallas, TX, USA, 30 October–3 November 2017; pp. 2329–2344.
4. Chen, H.; Xue, Y.; Li, Y.; Chen, B.; Xie, X.; Wu, X.; Liu, Y. Hawkeye: Towards a Desired Directed Grey-Box Fuzzer. In Proceedings of the 2018 ACM SIGSAC Conference on Computer and Communications Security, Toronto, ON, Canada, 15–19 October 2018; pp. 2095–2108.
5. Liang, H.; Zhang, Y.; Yu, Y.; Xie, Z.; Jiang, L. Sequence Coverage Directed Greybox Fuzzing. In Proceedings of the 2019 IEEE/ACM 27th International Conference on Program Comprehension (ICPC), Montreal, QC, Canada, 25–26 May 2019; IEEE Computer Society; pp. 249–259.
6. Liang, H.; Jiang, L.; Ai, L.; Wei, J. Sequence Directed Hybrid Fuzzing. In Proceedings of the 2020 IEEE 27th International Conference on Software Analysis, Evolution and Reengineering (SANER), London, ON, Canada, 18–21 February 2020; pp. 127–137.
7. Kim, J.; Yun, J. Poster: Directed Hybrid Fuzzing on Binary Code. In Proceedings of the 2019 ACM SIGSAC Conference on Computer and Communications Security, London, UK, 11–15 November 2019; pp. 2637–2639.
8. Chen, Y.; Li, P.; Xu, J.; Guo, S.; Zhou, R.; Zhang, Y.; Wei, T.; Lu, L. Savior: Towards Bug-Driven Hybrid Testing. In Proceedings of the 2020 IEEE Symposium on Security and Privacy (SP), San Francisco, CA, USA, 18–21 May 2020; pp. 1580–1596.
9. Nguyen, M.D.; Bardin, S.; Bonichon, R.; Groz, R.; Lemerre, M. Binary-Level Directed Fuzzing for {use-after-Free} Vulnerabilities. In Proceedings of the 23rd International Symposium on Research in Attacks, Intrusions and Defenses (RAID 2020), San Sebastian, Spain, 14–16 October 2020; pp. 47–62.
10. Ye, J.; Li, R.; Zhang, B. RDFuzz: Accelerating Directed Fuzzing with Intertwined Schedule and Optimized Mutation. *Math. Probl. Eng.* **2020**, *2020*, 7698916. [CrossRef]
11. Huang, H.; Guo, Y.; Shi, Q.; Yao, P.; Wu, R.; Zhang, C. Beacon: Directed Grey-Box Fuzzing with Provable Path Pruning. In Proceedings of the 2022 IEEE Symposium on Security and Privacy (SP), San Francisco, CA, USA, 23–25 May 2022; pp. 36–50.
12. Lemieux, C.; Sen, K. FairFuzz: A Targeted Mutation Strategy for Increasing Greybox Fuzz Testing Coverage. In Proceedings of the 33rd ACM/IEEE International Conference on Automated Software Engineering, Montpellier, France, 3–7 September 2018; pp. 475–485.

13. Zhu, X.; Böhme, M. Regression Greybox Fuzzing. In Proceedings of the 2021 ACM SIGSAC Conference on Computer and Communications Security, Virtual Event, Republic of Korea, 15–19 November 2021; pp. 2169–2182.
14. Du, Z.; Li, Y.; Liu, Y.; Mao, B. WindRanger: A Directed Greybox Fuzzer Driven by Deviation Basic Blocks. In Proceedings of the 44th International Conference on Software Engineering, Pittsburgh, PA, USA, 21–29 May 2022; pp. 2440–2451.
15. Hin, D.; Kan, A.; Chen, H.; Babar, M.A. LineVD: Statement-Level Vulnerability Detection Using Graph Neural Networks. In Proceedings of the 19th International Conference on Mining Software Repositories, Pittsburgh, PA, USA, 23–24 May 2022; pp. 596–607.
16. Li, Z.; Zou, D.; Xu, S.; Ou, X.; Jin, H.; Wang, S.; Deng, Z.; Zhong, Y. VulDeePecker: A Deep Learning-Based System for Vulnerability Detection. In Proceedings of the 2018 Network and Distributed System Security Symposium, San Diego, CA, USA, 18–21 February 2018.
17. Zhou, Y.; Liu, S.; Siow, J.; Du, X.; Liu, Y. Devign: Effective Vulnerability Identification by Learning Comprehensive Program Semantics via Graph Neural Networks. *Adv. Neural Inf. Process. Syst.* **2019**, *32*.
18. Wu, Y.; Zou, D.; Dou, S.; Yang, W.; Xu, D.; Jin, H. VulCNN: An Image-Inspired Scalable Vulnerability Detection System. In Proceedings of the 44th International Conference on Software Engineering, Pittsburgh, PA, USA, 21–29 May 2022; pp. 2365–2376.
19. Nist Sofware Assurance Reference Dataset. Available online: https://samate.nist.gov/SARD/ (accessed on 8 April 2024).

Disclaimer/Publisher's Note: The statements, opinions and data contained in all publications are solely those of the individual author(s) and contributor(s) and not of MDPI and/or the editor(s). MDPI and/or the editor(s) disclaim responsibility for any injury to people or property resulting from any ideas, methods, instructions or products referred to in the content.

Article

Sampling-Based Machine Learning Models for Intrusion Detection in Imbalanced Dataset

Zongwen Fan [1,†], **Shaleeza Sohail** [2,†], **Fariza Sabrina** [3,*,†] **and Xin Gu** [4,†]

1. College of Computer Science and Technology, Huaqiao University, Xiamen 361021, China; zongwen.fan@hqu.edu.cn
2. College of Engineering, Science and Environment, The University of Newcastle, Callaghan, NSW 2308, Australia; shaleeza.sohail@newcastle.edu.au
3. School of Engineering and Technology, Central Queensland University, Rockhampton, QLD 4701, Australia
4. Lincoln Institute of Higher Education, Sydney, NSW 2000, Australia; xin.gu@lincolnau.nsw.edu.au
* Correspondence: f.sabrina@cqu.edu.au
† These authors contributed equally to this work.

Abstract: Cybersecurity is one of the important considerations when adopting IoT devices in smart applications. Even though a huge volume of data is available, data related to attacks are generally in a significantly smaller proportion. Although machine learning models have been successfully applied for detecting security attacks on smart applications, their performance is affected by the problem of such data imbalance. In this case, the prediction model is preferable to the majority class, while the performance for predicting the minority class is poor. To address such problems, we apply two oversampling techniques and two undersampling techniques to balance the data in different categories. To verify their performance, five machine learning models, namely the decision tree, multi-layer perception, random forest, XGBoost, and CatBoost, are used in the experiments based on the grid search with 10-fold cross-validation for parameter tuning. The results show that both the oversampling and undersampling techniques can improve the performance of the prediction models used. Based on the results, the XGBoost model based on the SMOTE has the best performance in terms of accuracy at 75%, weighted average precision at 82%, weighted average recall at 75%, weighted average F1 score at 78%, and Matthews correlation coefficient at 72%. This indicates that this oversampling technique is effective for multi-attack prediction under a data imbalance scenario.

Keywords: cybersecurity; oversampling technique; undersampling technique; multi-class classification; machine learning

1. Introduction

The Internet of Things (IoT) is used in a huge number of smart applications that exchange data and perform tasks autonomously like smart agriculture, smart homes, and smart cities [1]. IoT devices produce a huge volume of data that is utilized by smart applications. Due to their large number, distributed nature, and diverse functionalities, IoT networks always face a huge number of cyberattacks [2]. Therefore, securing IoT networks from attackers is critical to ensure the safe and reliable use of these devices and applications. The heterogeneity of the devices and protocols, and the limited processing power, memory, and energy of many IoT devices, are some of the major challenges in choosing security measures [3].

Recently machine learning has been used to detect cyberattacks on smart applications. These machine learning algorithms create security models for detecting attacks based on the training data that consists of data regarding normal and malicious traffic [4]. In most of the datasets used by these algorithms, the proportion of data related to normal and attack classes may be imbalanced. The imbalanced dataset is a big obstacle in detecting attacks accurately, especially for minority classes [5].

In machine learning, it is common to encounter datasets that are imbalanced, where one class has significantly more or fewer instances than other classes. An imbalanced dataset can play a role in creating a biased model favoring the majority classes and performing poorly on the minority classes [6]. Common approaches that have been used to overcome this challenge are using oversampling and undersampling techniques. By randomly duplicating records in smaller or minority classes, oversampling tries to match the number of records to the bigger or majority classes. This can also be accomplished by artificially creating new records or instances. Some of the oversampling approaches include synthetic minority oversampling technique (SMOTE) [7], distributed random oversampling [8], BorderlineSMOTE [9], borderline oversampling with support vector machine (SVM) [10], adaptive synthetic sampling (ADASYN) [11], etc.

On the other hand, undersampling is based on the idea that the number of records in majority classes is to be reduced in order to keep them the same in number as the instances in minority classes. This can be achieved by randomly removing instances from the majority class or selecting a representative subset of instances based on certain criteria. Some of the undersampling approaches are condensed nearest-neighbor undersampling [12], Tomek Links method [13], edited nearest neighbors [14], one-sided selection [15], and instance hardness threshold [16].

Both oversampling and undersampling techniques have their advantages and disadvantages [17,18]. Oversampling can be effective in increasing the number of instances in the minority class and reducing the bias toward the majority class. By providing more instances in the minority class, oversampling can help the model learn the patterns and characteristics of the minority class and improve its ability to generalize to new instances. However, it can also lead to overfitting and generate synthetic instances that do not accurately represent the minority class. Undersampling can be effective in reducing the number of instances in the majority class and focusing on the most informative instances. Undersampling can help to focus on the most informative instances and reduce the noise and redundancy in the dataset as well. However, it can also lead to information loss and remove instances that are important for the model's performance.

It is important to point out that oversampling and undersampling techniques should be used with caution and in conjunction with different preprocessing techniques, such as feature selection and normalization [18]. Furthermore, the choice of oversampling or undersampling technique should be based on the specific characteristics of the dataset and the goals of the machine learning task [19]. In this paper, we focus on exploring the effect of multiple oversampling and undersampling approaches for attack detection for different machine learning algorithms. The IoT dataset we chose shows very poor detection performance for most of the minority classes if no oversampling or undersampling approach is used. The main contributions of the paper are as follows:

- We identify that the traditional machine learning algorithms may not detect minority attack classes when no sampling technique is used. Therefore, in reality, for network attack detection systems, such a limitation may result in failure to detect some of the attacks completely, which emphasizes the importance of using sampling techniques for imbalanced datasets.
- We thoroughly investigate the effect of different oversampling and undersampling techniques on the performance of multiple traditional and ensemble machine learning algorithms.
- We identify the best sampling approach for network attack detection using one of the latest IoT datasets.

Section 2 provides a discussion and analysis of the relevant literature in two areas: machine learning models and relevant research using the IOTID20 dataset. Section 3 provides the details of machine learning models and sampling techniques. Section 4 provides an in-depth analysis of the collected results for the intrusion detection system. A summary of the paper and some options about the possible future work are provided in Section 5.

2. Related Work

In this section, we provide an overview of the machine learning approaches used in the domain of cybersecurity. We also present recent efforts on the use of oversampling and undersampling approaches to improve attack detection for imbalanced datasets. It is important to mention here that oversampling and undersampling approaches are used effectively in improving overall attack detection; however, most of the literature does not specify the detection accuracy for the minority classes. Due to a small proportion of records for the minority classes, the overall detection accuracy can be very high even when the system fails to detect smaller classes completely.

A random forest-based attack detection approach is proposed that uses smart feature selection to improve attack prediction performance [20]. The use of oversampling and some feature selection approaches is explored for imbalanced datasets in the area of cybersecurity, specifically intrusion detection. Decision trees were used for different binary and muti-class attack detection models, and the models performed reasonably [21]. The multilayer perceptron network showed very good anomaly detection abilities with a small number of features for multi-class problems [22]. A combination of random forest and optimization approaches produced very good results for classifying cyberattacks and reducing false alarm rates [23].

The overall accuracy of these proposed approaches was good, but the prediction accuracy for all minority classes was not investigated. Hence, the performance for all attack types cannot be analyzed and compared to this work. With a very small sample size for most minority classes, the overall attack detection accuracy can be very high, but the detection for some of the minority classes may be very low or even zero. This results in completely missing some cyberattacks, which may have a catastrophic effect on the security systems.

An intrusion detection system using an artificial neural network provides very high attack detection when the hyperparameters of the neural network are tuned [24]. Another artificial neural network-based intrusion detection system for binary classification showed promising results for a simulated IoT network [25]. An artificial neural network-based approach for three different levels of classification of attacks is proposed while tuning the hyperparameters for optimal performance [26]. With the proper tuning of hyperparameters, the neural network model showed very high accuracy for most of the cases.

The ANN approach provides a good option for intrusion detection systems and has high performance but faces the challenges of being complex, computationally expensive, and requiring the selection of hyperparameters. Also, the low detection rate for minority classes still exists in the above-mentioned ANN-based approaches, so the above-mentioned works did not consider that specifically.

Ensemble classifiers provide an excellent option for gaining the combined benefits of two different algorithms. Jabbar et al. [27] used an ensemble classifier for the binary detection problem of cyberattacks and combined random forest with another approach. The same research group also proposed ADTree and KNN ensemble classifiers for detecting cyberattacks [28]. To reduce the time of model building and training, a tree-based approach was combined with a bagging method for the classification of attacks [29]. Most of the above-mentioned approaches provide high accuracy for overall attack detection; however, when it comes to multi-class attack detection, the detection rate of smaller or minority attack classes is a great challenge.

Karthik and Krishnan [30] proposed a novel approach to detect IoT attacks using a combination of random forest techniques with a novel oversampling approach. The proposed method was evaluated on different datasets and compared with several approaches; it showed good results in terms of accuracy, precision, recall, and F1 score. Bej et al. [31] proposed a new oversampling technique for imbalanced datasets. The minority samples were scaled and stretched to create new samples for smaller classes. With extensive experiments and testing on numerous imbalanced datasets, the proposed approach showed very promising results.

We used the IOTID20 dataset for testing our approach. Qaddoura et al. [32] addressed the class imbalance issue in the IoTID20 dataset by considering clustering and oversampling techniques. Support vector machine (SVM) with an oversampling technique was investigated for classification and achieved good performance for attack detection at the binary level, where only attacks and normal classes were detected. Farah [33] compared the performance of multiple techniques to detect attacks in the IOTID20 dataset and detected some classes of attacks. However, subcategory attacks that were of minority classes were not detected. Krishnan. Nawaz and Lin [34] compared the attack detection considering random forest, XGBoost, and SVC approaches for detecting normal and attack classes only and achieved very high accuracy. However, the minority class detection was not targeted, as subcategory-based attack detection was not considered.

The main motivation of this work is to detect cyberattacks belonging to minority classes when imbalanced datasets are considered for attack detection, which is a significant concern in almost all datasets in this domain, as previously discussed. In the Section 4, we provide detection accuracy, precision, recall, and F1 score for all attack classes, including minority classes that have very few samples, to show that our approach significantly improves the detection of these small attack classes. In Table 1, we compare our work with other existing studies that use the same dataset to show that we succeeded in detecting minority classes, which were not considered by the other studies due to the small sample sizes for these classes.

Table 1. Comparison with existing approaches.

Research Work	Approaches Used	Detection	Results Provided
Qaddoura et al. [32]	SLFN, SVM	Binary Detection (anomaly/attack)	Only for normal attack
Farrah [33]	Multiple	Binary and Category	Only for normal attack and Categories (majority classes)
Krishnan, Nayaz and Liu [34]	SVC, XGBoost RF	Binary	Only for normal attack
Our work	DT, RF MLP, XGBoost CatBoost	Multi-class Detection	All normal attack, Categories (majority classes) and Subcategories (minority classes)

Undersampling is one efficient method to handle imbalanced datasets as it focuses on reducing the number of samples from the majority classes. An undersampling approach based on the theory of evidence is proposed in evidential undersampling [35]. This approach considers a very important factor, which is to avoid removing meaningful samples. The samples in the majority classes are assigned a soft evidential label after removing unclear samples. When tested with different ML algorithms, this approach outperformed some basic undersampling approaches. An undersampling approach based on consensus clustering is proposed to handle imbalanced learning [36]. The consensus clustering-based scheme used a different combination of clustering algorithms for the undersampling purpose. The results obtained with different ML algorithms showed that different combinations can produce very different results. A novel two-step undersampling approach is proposed [37]. Firstly, the majority class is considered for similar instances, which are grouped together into subclasses. Then, from those subclasses, unrepresentative data samples are removed. The proposed approach performed significantly better than other undersampling approaches.

3. Methods

In this section, we first introduce the machine learning algorithms. Next, we describe the oversampling techniques, followed by the undersampling techniques. Finally, we present the flowchart of the prediction model for the intrusion detection prediction. For researchers to duplicate the outcomes, we have shared our code with the GitHub repository [38]. All undersampling and oversampling approaches that we used are from imblearn library [39]. A snapshot of example source code is shown in Figure 1. All undersampling and oversampling source codes are provided in our GitHub repository [38]. Beside the ReadMe file, five Python Notebook files uploaded to the Github repository, which are ADASYN-sent to github.ipynb, Baseline.ipynb, InstaceHardnessThreshold-sent to github.ipynb, RandomUnderSampler-sent to github.ipynb, and SMOTE-sent to github.ipynb.

```python
def kFoldCV(model, data, n_fold=10):
    diff = int(len(data)/n_fold)
    results = np.zeros((1, 4))
    predictY = data[:,-1].astype('int')
    targetY = deepcopy(predictY).astype('int')
#    predictY = deepcopy(data[:,-1]).astype('int')
    cv = StratifiedKFold(n_splits=n_fold)
    X, y = data[:,:-1],data[:,-1].astype('int')
    begin = 0
    for fold, (train_index, test_index) in enumerate(cv.split(X, y)):
        X_train, y_train = X[train_index], y[train_index]
#        print(X_train.shape,y[test_index].shape)
        sc = StandardScaler()
        X_train = sc.fit_transform(X_train)
        X_train, y_train = InstanceHardnessThreshold().fit_resample(X_train, y_train)
        X_test = sc.transform(X[test_index])
        predictY[begin:begin+len(X_test)] = model.fit(X_train, y_train).predict(X_test)
        targetY[begin:begin+len(X_test)] = y[test_index]
        begin += len(X_test)
#        targetY[begin:end] = test[:,-1]
#        predictY[begin:end] = model.fit(X_train, y_train).predict(X_test)
    t = classification_report(targetY, predictY)
    print(t)
    print(matthews_corrcoef(targetY, predictY))
```

Figure 1. A snapshot of the code for InstanceHardnessThreshold undersampling method.

3.1. Machine Learning Algorithms

In this section, five frequently utilized machine learning algorithms are briefly explained, including the decision tree (DT), multilayer perceptron (MLP), random forest (RF), extreme gradient boosting (XGBoost), and category boosting (CatBoost).

3.1.1. DT

The DT [40] is a tree-based supervised algorithm that can be used for classification tasks. A DT algorithm is constructed by three types of nodes, which are the decision node, change node, and end node. Each type of node has its task. More specifically, the decision node indicates a choice that needs to be determined. Consequently, the chance node analyzes the probabilities of the results. The end node presents the ultimate result of a decision pathway. By calculating the value of each option in the tree, the DT is able to achieve promising results by minimizing the risk and maximizing the likelihood.

The primary purpose of the DT algorithm is to obtain the measure of information gain. Specifically, the DT model first evaluates the entropy, as given in Equation (1). After that, the conditional entropy is calculated using Equation (2). Finally, the information gain is obtained using Equation (3).

$$H(D) = -\sum_{k=1}^{K} p_k log_2 P_k, \qquad (1)$$

$$H(D|A) = -\sum_{i=1}^{n} p_i H(D|A = x_i), \qquad (2)$$

$$g(D, A) = H(D) - H(D|A), \qquad (3)$$

where D is a given dataset, K represents the count of categories, n stands for the number of features, p_k denotes the probability rate of the kth category, and p_i signifies the probability of the feature A in the ith subset.

3.1.2. MLP

The MLP is composed of three types of layers, which are the input layer, the hidden layer, and the output layer [41]. Every layer is linked to its neighboring layers. Similarly, every neuron within the hidden and output layers is connected to all neurons in the preceding layer via a weight vector. Each layer proceeds its own computation. The output of each layer is generated by passing the weighted sum of inputs and bias terms through a non-linear activation function, which then becomes the input for the subsequent layer. In the input layer, the number of neurons corresponds to the number of input features, while the output layer represents the model's output. For a binary classification, a single neuron will be generated as the result. The hidden layer neurons reside between the input and output layers, forming connections with both. These interconnected neurons enable communication and information exchange among themselves. Through adjusting weights in the connections between neurons, the MLP can mimic the information analysis and processes like a human brain.

For a binary classification problem, the MLP generates one single neuron in the output layer where its value can be obtained using Equation (4).

$$f(X) = W^2 g(W^1 X + b^1) + b^2, \qquad (4)$$

where W^1 represents the weights, and b^1 denotes a bias term in the transition from the input layer to the neighboring hidden layer. Likewise, W^2 represents the weights, and b^2 denotes a bias term when passing from the hidden layer to the output layer. $g(\cdot)$ signifies an activation function.

3.1.3. RF

The RF is an ensemble algorithm that leverages multiple decision trees [42]. By constructing numerous decision trees using bootstrap samples, the RF algorithm enhances prediction accuracy and stability. It effectively addresses overfitting issues by utilizing resampling and feature selection techniques. During the training process, the RF generates multiple sub-datasets, each containing the same number of samples as the original training set, through resampling. For each sub-dataset, individual decision trees are trained using a recursive partitioning approach. This involves searching for the best feature splits within the selected features. Ultimately, the RF algorithm combines the predictions from all decision trees by taking their average as the final output.

3.1.4. XGBoost

XGBoost is also an ensemble algorithm [43]. To achieve the final result, this algorithm employs gradient boosting to aggregate multiple outcomes from the decision tree-based algorithms. To scale down the impact of overfitting, this algorithm uses shrinkage and feature subsampling techniques. The XGBoost method is tailored to real-world applications that require high computation time and storage memory. Therefore, it is well suited for applications that necessitate parallelization, distributed computing, out-of-core computing, and cache optimization. Additionally, this ensemble algorithm enables parallel tree boosting, alternately referred to as gradient-boosted decision tree and gradient boosting machine.

Gradient boosting aims to discover the function that most effectively approximates the data by optimizing Equation (5).

$$Obj = \sum_i L(\hat{y}_i, y_i) + \sum_i \Omega(f_i), \qquad (5)$$

$$\Omega(f) = \gamma T + (1/2)\lambda ||\mathbf{w}||^2, \qquad (6)$$

where L represents a convex loss function that quantifies the dissimilarity between the target value y_i and the predicted value $\hat{y}$. The weight vector is denoted by $\mathbf{w}$, f_i refers to the ith function, T signifies the number of leaves in the tree, and $\Omega(f)$ penalizes model complexity, while γT and $\lambda||\mathbf{w}||^2$ impose constant penalties for each additional tree leaf and extreme weights, respectively.

3.1.5. CatBoost

CatBoost, based on categorical boosting, is an ensemble model that is effective for prediction tasks involving categorical features [44]. Distinguished from the other gradient boosting algorithms, an ordered boosting technique is employed to mitigate the issue of target leakage. Furthermore, it effectively resolves the issues with categorical features by replacing the original features with one or more numerical values. Constructed on the foundation of the traditional gradient boosting-based algorithms that can lead to the overfitting problem, CatBoost addresses this issue by utilizing random permutation for leaf value estimation to minimize the issue of overfitting. It can rapidly construct a model for big data projects with a high level of generalization. By combining many base estimators, this algorithm is able to build a strong competitive prediction model that achieves better performance than random selection.

3.2. Oversampling Techniques

One or more classes that are characterized with few examples are referred to as minority classes, while one or more classes with a significant number of examples are known as majority classes. When minority classes and majority classes are in the same dataset, they cause an imbalanced class distribution. As a common practice in dealing with a binary (two-class) classification problem, class 0 is recognized as the majority class, and class 1 signifies the minority class.

3.2.1. SMOTE

One of the most significant obstacles in classification problems is that there are far more majority classes than minority ones. To overcome this, an oversampling technique was adopted to balance the data before the prediction models were trained. The synthetic minority oversampling technique (SMOTE) was applied to oversample the data, and it was found that it effectively mitigates the overfitting problem of the prediction model to the majority class [7]. The SMOTE is an oversampling technique that generates synthetic samples for minority classes. To mitigate the overfitting problems, this algorithm generates new instances by utilizing the interpolation between the positive instances that are in proximity, with a focus on the feature space.

The SMOTE first randomly selects a minority class instance a. Then, the algorithm searches for its k nearest neighbors which are also minority classes. Consequently, one of the k nearest neighbors b is identified by chance. Connecting a and b to form a line segment in the feature space creates synthetic instances. As a result, the synthetic instances are created as a convex combination of the two chosen neighboring instances a and b.

3.2.2. ADASYN

ADASYN represents a generalized version of SMOTE. This algorithm also creates synthetic instances to oversample the minority classes. However, it considers the density distribution that determines the number of synthetic instances to be generated for samples, which is difficult to learn [11]. By doing so, this algorithm dynamically adjusts the decision boundaries based on the samples difficult to learn. This is where the fundamental distinction exists between ADASYN and SMOTE.

3.3. Undersampling Techniques

Undersampling techniques are tailored to address the issues of skewed distribution in classification datasets.

3.3.1. RandomUnderSampler

Random undersampling, also termed RandomUnderSampler, arbitrarily selects samples from the majority class and then deletes them from the training dataset [45]. This approach is regarded as the simplest undersampling technique. Although it is simple, it is effective. However, this algorithm is not without limitations. A drawback of this technique is that samples are eliminated without considering their potential usefulness or importance in determining the decision boundary between the classes. In random undersampling models, the instances in the majority class are deleted randomly to reach a balanced distribution. This potentially results in the removal of valuable information.

3.3.2. InstanceHardnessThreshold

Instance hardness threshold (InstanceHardnessThreshold) is an undersampling method that can be used to alleviate class imbalance by removing samples with the aim of balancing the dataset [16]. In other words, the samples classified with a low probability will be removed from the dataset. Consequently, the prediction model can be trained based on the simplified dataset. The probability of misclassification for each sample is defined by the hardness threshold, which is considered the core difference between the instance hardness threshold and other undersampling techniques.

3.4. Flowchart of a Prediction Model for the Intrusion Detection Prediction

A flowchart for intrusion detection prediction using the machine learning models, namely the DT, MLP, RF, XGBoost, and CatBoost, is presented in Figure 2. As illustrated in the figure, parameter tuning incorporated a grid search with 10-fold cross-validation. Specifically, the given dataset was preprocessed. After that, the grid search with 10-fold cross-validation tuned the model parameters with the use of different sampling techniques, including the oversampling techniques (SMOTE and ADASYN) and the undersampling techniques (RandomUnderSampler and InstanceHardnessThreshold). The dataset was divided into ten subsets, with each subset used once as the validation data while the model was trained on the remaining nine subsets. This process was repeated 10 times, with each subset serving as the validation data exactly once. The average performance across all folds was then computed to evaluate the model's performance under different hyperparameter configurations. Finally, with the best parameters obtained, the prediction model was trained and evaluated.

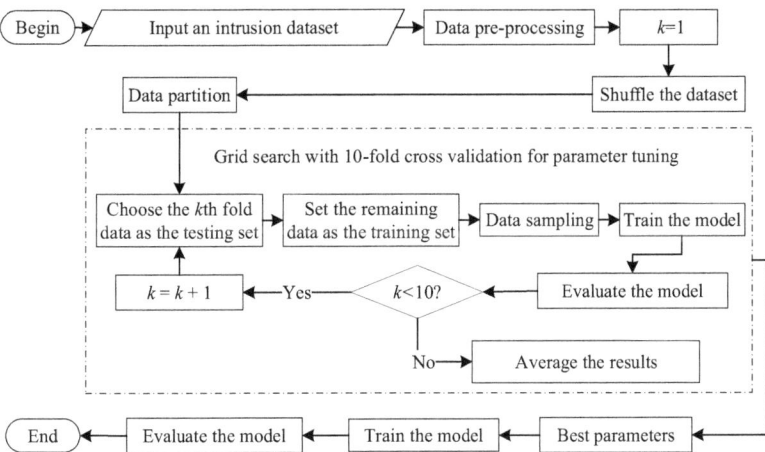

Figure 2. A flowchart showing the prediction procedure for the intrusion detection prediction.

4. Experimental Setup and Results

The main motivation behind this work is to evaluate the efficacy of undersampling and oversampling approaches to improve minority class detection for imbalanced datasets using different machine learning approaches. With extensive experimentation, we compared the detection accuracy of IoT attacks belonging to minority classes using matching learning models outlined in Section 3.

4.1. System Setup

The experiments were conducted and assessed using the Python programming language within Jupyter Notebook from the Anaconda distribution. We utilized the sklearn library in our implementation. Furthermore, the experiments were executed on a computer system featuring an i7-12700H CPU operating at 2.30 GHz and 64.0 GB of RAM.

4.2. Data Description

Datasets are required to train machine learning models for attack detection. Without a dataset, no training/testing of models can be carried out. In recent years, several datasets have been developed and made available by researchers, including UNSW-NB15 [46] and Bot-IoT [47]. Specifically, the IoTID20 dataset [48] was chosen for this research as this is one of the most recent datasets established in the IoT scenario for depicting realistic and up-to-date network traffic features [26]. A number of studies have been conducted based on this dataset for experimentation. Ullah and Mahmoud [48] used this dataset. Anomalous activities in attack detection were divided into binary, category, and subcategory levels. Five machine learning models (i.e., DT, MLP, RF, XGBoost, and CatBoost) were applied to detect IoT attacks. As minority attack classes are important in cybersecurity, we focused on the detection of subcategory attacks (nine-class classification). The IoTID20 dataset, developed by Ullah and Mahmoud [48], contains 86 attributes, including 3 categorical attributes. The categorical attribute is Sub_Cat. The data types in Sub_Cat data column are Mirai-Ackflooding, Mirai-Hostbruterforceg, Mirai-HTTP Flooding, Mirai-UDF Flooding, MITM ARP Spoofing, DoS-Synflooding, Scan Hostport, Scan Port OS, and Normal. The distributions of nine data types of Sub_Cat are presented in Figure 3.

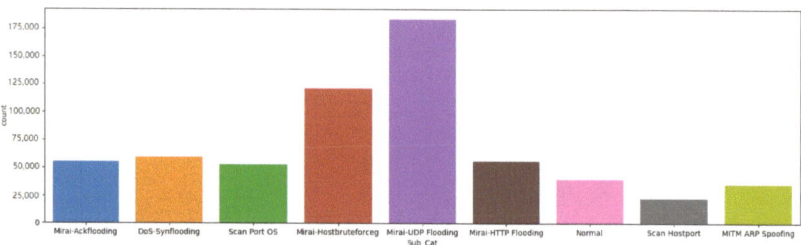

Figure 3. IoTID20 data distribution based on Sub_Cat.

4.3. Experimental Settings

Machine learning models are sensitive to the setting of hyperparameters; hence, we applied the grid search approach to select the best hyperparameters. For a given model, the grid search approach exhaustively generates several candidate values from a list of hyperparameter values to determine the optimal hyperparameters. In this paper, the grid search was used with 10-fold cross-validation in the machine learning models for parameter tuning. To be specific, for the DT, candidate values of [3, 4, 5] were considered for the maximum tree depth [49]. For the MLP, candidate values of [100, 200, 300] for the neuron number in the hidden layer and [100, 200, 300] for the maximum iterations were considered. For the ensemble models (RF, XGBoost, and CatBoost), candidate values of [10, 100, 200] for the number of estimators and [3, 4, 5] for the maximum tree depth in each estimator were considered [50]. The final results were obtained based on the models trained using the optimal hyperparameters. Take the RF as an example; nine pairs of hyperparameters ([10, 3], [10, 4], [10, 5], [100, 3], [100, 4], [100, 5], [200, 3], [200, 3], [200, 4], and [200, 5]) were used to trained and evaluate the model. The pair with the best results were utilized as the optimal hyperparameters for the final evaluation.

For the evaluation of different sampling-based machine learning models, macro-accuracy was used to calculate the accuracy for each class independently and then determine the unweighted average. It treats all classes equally regardless of their imbalance. Similar to macro-accuracy, these metrics account for class imbalance by considering precision, recall, and F1 score for each class and then taking the macro-average. Moreover, weighted accuracy was utilized to take the class imbalance by assigning higher weights to minority classes during calculation. It was calculated as the average accuracy of each class weighted by the number of instances in each class. Similar to weighted accuracy, these metrics account for class imbalance by considering precision, recall, and F1 score for each class and then taking the weighted average. Further, we introduced the Matthews correlation coefficient (MCC) [51] to evaluate the model performance.

4.4. Multi-Class Classification Based on Oversampling

In this section, we compare the results obtained from the oversampling-based machine learning models, namely the DT, MLP, RF, XGBoost, and CatBoost, as shown in Tables 2–6, based on SMOTE and ADASYN. As we can see from Table 2, without the use of oversampling techniques, some results of the DT have zero values for precision, recall, and F1 score. This is because the number of MITM ARP Spoofing (35,377), Mirai-Ackflooding (55,124), and Scan Hostport (22,192) is far less than that of Mirai-UDP Flooding (183,189), making the DT overfit to the majority class, while with the use of SMOTE and ADASYN, this problem can be alleviated. Specifically, MITM ARP Spoofing can be detected with the use of SMOTE, while MITM ARP Spoofing and Mirai-Ackflooding can be detected with the use of ADASYN. Among all the results, the baseline DT has the best MCC.

Table 2. Oversampling for multi-class classification using the DT.

Label	Baseline				SMOTE			ADASYN		
	Precision	Recall	F1 Score	Support	Precision	Recall	F1 Score	Precision	Recall	F1 Score
DoS-Synflooding	1.00	0.99	1.00	59,391	1.00	1.00	1.00	0.60	1.00	0.75
MITM ARP Spoofing	0.00	0.00	0.00	35,377	0.28	0.41	0.33	0.27	0.70	0.39
Mirai-Ackflooding	0.00	0.00	0.00	55,124	0.00	0.00	0.00	0.32	0.12	0.17
Mirai-HTTP Flooding	1.00	0.00	0.00	55,818	0.25	0.86	0.39	0.27	0.60	0.37
Mirai-Hostbruteforceg	0.70	0.59	0.64	121,178	0.61	0.30	0.40	0.99	0.19	0.32
Mirai-UDP Flooding	0.58	0.97	0.72	183,189	1.00	0.59	0.74	1.00	0.59	0.74
Normal	0.84	0.80	0.81	40,073	1.00	0.61	0.76	0.52	0.83	0.64
Scan Hostport	0.00	0.00	0.00	22,192	0.00	0.00	0.00	0.00	0.00	0.00
Scan Port OS	0.43	0.94	0.59	53,073	0.38	0.95	0.54	0.42	0.79	0.55
Accuracy			0.62	625,415			0.54			0.53
Macro avg	0.50	0.48	0.42	625,415	0.50	0.52	0.46	0.49	0.53	0.44
Weighted avg	0.58	0.62	0.53	625,415	0.64	0.54	0.54	0.68	0.53	0.51
MCC			0.55	625,415			0.50			0.48

Oversampling for multi-class classification using the MLP can be found in Table 3. As we can see from the table, similar to the results in Table 2, some results of the MLP have low values for precision, recall, and F1 score (e.g., Mirai-Ackflooding) without the use of oversampling techniques. However, with the use of SMOTE and ADASYN, the performance of the minority attack precision can be improved. For example, the F1 score of Mirai-Ackflooding attack is improved from 28% to 38% and 42% with the use of SMOTE and ADASYN, respectively. The F1 score of the Scan Hostport attack is improved from 11% to 34% and 33% with the use of SMOTE and ADASYN, respectively. In addition, with the use of the ADASYN technique, the performance of MLP can be improved from 63% macro-average accuracy to 68% and from 70% weighted average F1 score to 73%. However, the ADASYN-based MLP sacrifices the performance of accuracy and MCC.

Table 3. Oversampling for multi-class classification using the MLP.

Label	Baseline				SMOTE			ADASYN		
	Precision	Recall	F1 Score	Support	Precision	Recall	F1 Score	Precision	Recall	F1 Score
DoS-Synflooding	1.00	1.00	1.00	59,391	1.00	1.00	1.00	0.97	1.00	0.98
MITM ARP Spoofing	0.79	0.80	0.79	35,377	0.72	0.91	0.80	0.58	0.86	0.69
Mirai-Ackflooding	0.32	0.25	0.28	55,124	0.32	0.49	0.38	0.32	0.42	0.36
Mirai-HTTP Flooding	0.33	0.32	0.32	55,818	0.32	0.47	0.38	0.33	0.46	0.38
Mirai-Hostbruteforceg	0.72	0.84	0.77	121,178	0.91	0.73	0.81	0.81	0.66	0.73
Mirai-UDP Flooding	0.80	0.79	0.80	183,189	1.00	0.70	0.82	1.00	0.70	0.82
Normal	0.96	0.88	0.91	40,073	0.91	0.91	0.91	0.70	0.96	0.81
Scan Hostport	0.62	0.06	0.11	22,192	0.27	0.47	0.34	0.27	0.42	0.33
Scan Port OS	0.62	0.84	0.71	53,073	0.65	0.62	0.63	0.57	0.47	0.52
Accuracy			0.71	625,415			0.71			0.67
Macro avg	0.68	0.64	0.63	625,415	0.68	0.70	0.68	0.62	0.66	0.63
Weighted avg	0.71	0.71	0.70	625,415	0.78	0.71	0.73	0.74	0.67	0.69
MCC			0.66	625,415			0.66			0.62

Oversampling for multi-class classification using the RF can be found in Table 4. As we can see from the table, the F1 score results from the baseline are low, while the results from SMOTE and ADASYN are greatly improved. For example, for [Mirai-Ackflooding, Mirai-HTTP Flooding, Scan Hostport], their results are improved from [1%, 8%, 0%] to [44%, 41%, 16%] and [29%, 39%, 16%] using SMOTE and ADASYN, respectively. This is very important in real-world intrusion detection applications when the intrusion will severely affect user information security. According to the results, the SMOTE-based RF has the best performance in terms of macro-average F1 score, weighted average F1 score, and MCC.

Table 4. Oversampling for multi-class classification using the RF.

Label	Baseline				SMOTE			ADASYN		
	Precision	Recall	F1 Score	Support	Precision	Recall	F1 Score	Precision	Recall	F1 Score
DoS-Synflooding	1.00	0.99	1.00	59,391	1.00	1.00	1.00	0.95	1.00	0.97
MITM ARP Spoofing	0.92	0.16	0.27	35,377	0.32	0.88	0.47	0.31	0.58	0.40
Mirai-Ackflooding	0.41	0.00	0.01	55,124	0.30	0.44	0.36	0.30	0.28	0.29
Mirai-HTTP Flooding	0.37	0.05	0.08	55,818	0.31	0.41	0.35	0.30	0.53	0.39
Mirai-Hostbruteforceg	0.45	0.91	0.60	121,178	0.89	0.22	0.35	0.96	0.20	0.34
Mirai-UDP Flooding	0.66	0.90	0.76	183,189	0.98	0.71	0.82	0.99	0.70	0.82
Normal	0.85	0.81	0.83	40,073	0.80	0.85	0.82	0.52	0.95	0.67
Scan Hostport	0.91	0.00	0.00	22,192	0.18	0.14	0.16	0.14	0.18	0.16
Scan Port OS	0.73	0.24	0.36	53,073	0.47	0.83	0.60	0.32	0.55	0.40
Accuracy			0.62	625,415			0.60			0.56
Macro avg	0.70	0.45	0.43	625,415	0.58	0.61	0.55	0.53	0.55	0.49
Weighted avg	0.64	0.62	0.54	625,415	0.72	0.60	0.60	0.70	0.56	0.56
MCC			0.55	625,415			0.55			0.51

Table 5. Oversampling for multi-class classification using the XGBoost.

Label	Baseline				SMOTE			ADASYN		
	Precision	Recall	F1 Score	Support	Precision	Recall	F1 Score	Precision	Recall	F1 Score
DoS-Synflooding	1.00	1.00	1.00	59,391	1.00	1.00	1.00	1.00	1.00	1.00
MITM ARP Spoofing	0.96	0.96	0.96	35,377	0.92	0.98	0.95	0.86	0.99	0.92
Mirai-Ackflooding	0.30	0.31	0.30	55,124	0.29	0.55	0.38	0.30	0.43	0.36
Mirai-HTTP Flooding	0.31	0.29	0.30	55,818	0.28	0.31	0.29	0.31	0.44	0.36
Mirai-Hostbruteforceg	0.84	0.95	0.89	121,178	0.97	0.86	0.91	0.93	0.87	0.90
Mirai-UDP Flooding	0.81	0.78	0.79	183,189	1.00	0.70	0.83	1.00	0.70	0.83
Normal	0.94	0.92	0.93	40,073	0.91	0.94	0.92	0.74	0.98	0.84
Scan Hostport	0.96	0.42	0.59	22,192	0.49	0.60	0.54	0.44	0.65	0.53
Scan Port OS	0.79	0.88	0.83	53,073	0.80	0.83	0.82	0.82	0.59	0.69
Accuracy			0.76	625,415			0.75			0.74
Macro avg	0.77	0.72	0.73	625,415	0.74	0.75	0.74	0.71	0.74	0.71
Weighted avg	0.76	0.76	0.76	625,415	0.82	0.75	0.77	0.80	0.74	0.76
MCC			0.72	625,415			0.72			0.70

Table 6. Oversampling for multi-class classification using the CatBoost.

Label	Baseline				SMOTE			ADASYN		
	Precision	Recall	F1 Score	Support	Precision	Recall	F1 Score	Precision	Recall	F1 Score
DoS-Synflooding	1.00	1.00	1.00	59,391	1.00	1.00	1.00	0.98	1.00	0.99
MITM ARP Spoofing	0.90	0.90	0.90	35,377	0.83	0.97	0.89	0.74	0.98	0.84
Mirai-Ackflooding	0.31	0.29	0.30	55,124	0.31	0.60	0.41	0.32	0.45	0.37
Mirai-HTTP Flooding	0.32	0.27	0.29	55,818	0.31	0.33	0.32	0.32	0.45	0.38
Mirai-Hostbruteforceg	0.81	0.93	0.87	121,178	0.95	0.83	0.88	0.92	0.82	0.86
Mirai-UDP Flooding	0.79	0.80	0.79	183,189	1.00	0.70	0.83	1.00	0.70	0.82
Normal	0.94	0.91	0.92	40,073	0.89	0.93	0.91	0.72	0.97	0.83
Scan Hostport	0.92	0.39	0.55	22,192	0.47	0.59	0.52	0.43	0.64	0.52
Scan Port OS	0.77	0.85	0.81	53,073	0.80	0.81	0.81	0.81	0.58	0.68
Accuracy			0.75	625,415			0.75			0.73
Macro avg	0.75	0.70	0.71	625,415	0.73	0.75	0.73	0.69	0.73	0.70
Weighted avg	0.75	0.75	0.75	625,415	0.82	0.75	0.77	0.79	0.73	0.75
MCC			0.70	625,415			0.71			0.69

Oversampling for multi-class classification using the XGBoost can be found in Table 5. As we can see from the table, XGBoost has no zero results. The reason could be that it is an ensemble model that integrates multiple estimators for the final prediction. Although the baseline of XGBoost outperforms the DT, MLP, and RF in Tables 2–4, with the use of SMOTE and ADASYN, the performance of the XGBoost model can be further improved, from 76% weighted average precision to 82% and 80%, respectively. According to the results, the SMOTE-based XGBoost has the best weighted average F1 score and MCC.

Oversampling for multi-class classification using the CatBoost can be found in Table 6. As we can see from the table, similar to the results in Table 5, the CatBoost model has no zero results. The reason could be that CatBoost is also an ensemble model that can reduce the effect of the imbalance problem. Although the baseline of CatBoost outperforms the DT, MLP, and RF in Tables 2–4, its performance is slightly worse than the XGBoost model in Table 5. However, with the use of SMOTE, its performance can be further improved, from 75% weighted average F1 score to 77%. According to the results, the SMOTE-based model has the best performance for accuracy, precision, recall, F1 score, and MCC.

4.5. Multi-Class Classification Based on Undersampling

Although the oversampling techniques are able to improve machine learning models' performance in the dataset of imbalanced scenarios, in this study, we conducted experiments using undersampling techniques to see if the machine learning models could still be improved. We compare the results obtained from the undersampling-based machine learning models, namely the DT, MLP, RF, XGBoost, and CatBoost, as shown in Tables 7–11, based on the RandomUnderSampler and InstanceHardnessThreshold. It is worth noting that, with the use of undersampling techniques, fewer samples are used for training, which can improve the training efficiency. This is more practical in real-world applications, especially in big data scenarios. As we can see from Table 7, with the use of RandomUnderSampler and InstanceHardnessThreshold, the performance of DT can be improved. This is because the RandomUnderSampler is utilized to randomly remove the samples, which will remove some important information. However, with the use of InstanceHardnessThreshold, the performance of DT can be improved from [62%, 42%, 53%, 0.55] to [66%, 58%, 66%, 0.61] in terms of accuracy, macro-average F1 score, weighted average, and MCC, respectively.

Table 7. Undersampling for multi-class classification using DT.

Label	Baseline			RandomUnderSampler				InstanceHardnessThreshold		
	Precision	Recall	F1 Score	Support	Precision	Recall	F1 Score	Precision	Recall	F1 Score
DoS-Synflooding	1.00	0.99	1.00	59,391	1.00	1.00	1.00	1.00	0.99	0.99
MITM ARP Spoofing	0.00	0.00	0.00	35,377	0.33	0.53	0.41	0.65	0.56	0.60
Mirai-Ackflooding	0.00	0.00	0.00	55,124	0.30	0.00	0.01	0.31	0.56	0.40
Mirai-HTTP Flooding	1.00	0.00	0.00	55,818	0.25	0.85	0.39	0.34	0.10	0.16
Mirai-Hostbruteforceg	0.70	0.59	0.64	121,178	0.70	0.33	0.45	0.54	0.80	0.64
Mirai-UDP Flooding	0.58	0.97	0.72	183,189	1.00	0.59	0.74	0.98	0.71	0.82
Normal	0.84	0.80	0.81	40,073	0.95	0.67	0.79	0.82	0.81	0.81
Scan Hostport	0.00	0.00	0.00	22,192	0.17	0.17	0.17	0.86	0.04	0.08
Scan Port OS	0.43	0.94	0.59	53,073	0.44	0.89	0.59	0.64	0.75	0.69
Accuracy			0.62	625,415			0.56			0.66
Macro avg	0.50	0.48	0.42	625,415	0.57	0.56	0.50	0.68	0.59	0.58
Weighted avg	0.58	0.62	0.53	625,415	0.70	0.56	0.56	0.72	0.66	0.66
MCC			0.55	625,415			0.52			0.61

Undersampling for multi-class classification using the MLP model can be found in Table 8. As shown in the table, unlike the results in Table 2, although with fewer samples obtained using the InstanceHardnessThreshold model, the undersampling-based MLP can greatly outperform its baseline. Without the use of undersampling techniques, sample results have low values, while with the use of RandomUnderSampler and InstanceHardnessThreshold, this problem can be improved. In addition, comparing the two undersampling techniques, the performance of InstanceHardnessThreshold-based MLP outperforms RandomUnderSampler-based MLP. The macro-average F1 score of the baseline is 63%, while the macro-average F1 score of InstanceHardnessThreshold-based MLP is 66%; the weighted average F1 score of the baseline is 70%, while the weighted average F1 score of InstanceHardnessThreshold-based MLP is 72%. According to the results, the MLP has the best performance for MCC.

Table 8. Undersampling for multi-class classification using MLP.

Label	Baseline			RandomUnderSampler				InstanceHardnessThreshold		
	Precision	Recall	F1 Score	Support	Precision	Recall	F1 Score	Precision	Recall	F1 Score
DoS-Synflooding	1.00	1.00	1.00	59,391	1.00	1.00	1.00	1.00	1.00	1.00
MITM ARP Spoofing	0.79	0.80	0.79	35,377	0.59	0.93	0.72	0.79	0.78	0.78
Mirai-Ackflooding	0.32	0.25	0.28	55,124	0.31	0.56	0.40	0.30	0.49	0.37
Mirai-HTTP Flooding	0.33	0.32	0.32	55,818	0.31	0.33	0.32	0.31	0.43	0.36
Mirai-Hostbruteforceg	0.72	0.84	0.77	121,178	0.81	0.59	0.68	0.87	0.72	0.79
Mirai-UDP Flooding	0.80	0.79	0.80	183,189	1.00	0.70	0.82	1.00	0.70	0.82
Normal	0.96	0.88	0.91	40,073	0.90	0.89	0.90	0.98	0.85	0.91
Scan Hostport	0.62	0.06	0.11	22,192	0.24	0.47	0.31	0.19	0.46	0.27
Scan Port OS	0.62	0.84	0.71	53,073	0.59	0.55	0.57	0.67	0.60	0.63
Accuracy			0.71	625,415			0.67			0.69
Macro avg	0.68	0.64	0.63	625,415	0.64	0.67	0.64	0.68	0.67	0.66
Weighted avg	0.71	0.71	0.70	625,415	0.75	0.67	0.69	0.78	0.69	0.72
MCC			0.66	625,415			0.62			0.64

Table 9. Undersampling for multi-class classification using RF.

Label	Baseline			RandomUnderSampler				InstanceHardnessThreshold		
	Precision	Recall	F1 Score	Support	Precision	Recall	F1 Score	Precision	Recall	F1 Score
DoS-Synflooding	1.00	0.99	1.00	59,391	1.00	1.00	1.00	1.00	1.00	1.00
MITM ARP Spoofing	0.92	0.16	0.27	35,377	0.32	0.89	0.47	0.90	0.16	0.27
Mirai-Ackflooding	0.41	0.00	0.01	55,124	0.30	0.40	0.34	0.30	0.42	0.35
Mirai-HTTP Flooding	0.37	0.05	0.08	55,818	0.30	0.46	0.36	0.34	0.20	0.25
Mirai-Hostbruteforceg	0.45	0.91	0.60	121,178	0.90	0.22	0.35	0.43	0.92	0.59
Mirai-UDP Flooding	0.66	0.90	0.76	183,189	0.99	0.70	0.82	0.93	0.73	0.82
Normal	0.85	0.81	0.83	40,073	0.79	0.85	0.82	0.73	0.84	0.78
Scan Hostport	0.91	0.00	0.00	22,192	0.17	0.11	0.13	0.70	0.00	0.01
Scan Port OS	0.73	0.24	0.36	53,073	0.47	0.83	0.60	0.00	0.00	0.00
Accuracy			0.62	625,415			0.60			0.60
Macro avg	0.70	0.45	0.43	625,415	0.58	0.61	0.54	0.59	0.47	0.45
Weighted avg	0.64	0.62	0.54	625,415	0.73	0.60	0.60	0.63	0.60	0.57
MCC			0.55	625,415			0.55			0.54

Undersampling for multi-class classification using the RF can be found in Table 9. As we can see from the table, with the use of undersampling techniques, the performance of RF is improved. For the macro average F1 score and weighted average F1 score of RF, their results are improved from [43%, 54%] to [54%, 60%] and [45%, 57%] with the use of RandomUnderSampler and InstanceHardnessThreshold, respectively. According to all the results, the RF and RandomUnderSampler-based RF have the same MCC.

Table 10. Undersampling for multi-class classification using XGBoost.

Label	Baseline			RandomUnderSampler				InstanceHardnessThreshold		
	Precision	Recall	F1 Score	Support	Precision	Recall	F1 Score	Precision	Recall	F1 Score
DoS-Synflooding	1.00	1.00	1.00	59,391	1.00	1.00	1.00	1.00	1.00	1.00
MITM ARP Spoofing	0.96	0.96	0.96	35,377	0.89	0.98	0.93	0.94	0.87	0.90
Mirai-Ackflooding	0.30	0.31	0.30	55,124	0.31	0.53	0.39	0.28	0.49	0.35
Mirai-HTTP Flooding	0.31	0.29	0.30	55,818	0.31	0.41	0.35	0.28	0.39	0.32
Mirai-Hostbruteforceg	0.84	0.95	0.89	121,178	0.97	0.84	0.90	0.96	0.81	0.88
Mirai-UDP Flooding	0.81	0.78	0.79	183,189	1.00	0.70	0.83	1.00	0.70	0.83
Normal	0.94	0.92	0.93	40,073	0.90	0.94	0.92	0.99	0.86	0.92
Scan Hostport	0.96	0.42	0.59	22,192	0.48	0.61	0.54	0.31	0.85	0.45
Scan Port OS	0.79	0.88	0.83	53,073	0.81	0.82	0.81	0.88	0.53	0.66
Accuracy			0.76	625,415			0.75			0.71
Macro avg	0.77	0.72	0.73	625,415	0.74	0.76	0.74	0.74	0.72	0.70
Weighted avg	0.76	0.76	0.76	625,415	0.82	0.75	0.78	0.83	0.71	0.75
MCC			0.72	625,415			0.72			0.68

Table 11. Undersampling for multi-class classification using CatBoost.

Label	Baseline				RandomUnderSampler			InstanceHardnessThreshold		
	Precision	Recall	F1 Score	Support	Precision	Recall	F1 Score	Precision	Recall	F1 Score
DoS-Synflooding	1.00	1.00	1.00	59,391	1.00	1.00	1.00	1.00	1.00	1.00
MITM ARP Spoofing	0.90	0.90	0.90	35,377	0.80	0.97	0.87	0.89	0.83	0.86
Mirai-Ackflooding	0.31	0.29	0.30	55,124	0.31	0.52	0.39	0.28	0.49	0.36
Mirai-HTTP Flooding	0.32	0.27	0.29	55,818	0.32	0.43	0.37	0.28	0.38	0.32
Mirai-Hostbruteforceg	0.81	0.93	0.87	121,178	0.95	0.81	0.88	0.94	0.79	0.86
Mirai-UDP Flooding	0.79	0.80	0.79	183,189	1.00	0.70	0.82	1.00	0.70	0.83
Normal	0.94	0.91	0.92	40,073	0.89	0.93	0.91	0.99	0.85	0.92
Scan Hostport	0.92	0.39	0.55	22,192	0.46	0.60	0.52	0.29	0.83	0.43
Scan Port OS	0.77	0.85	0.81	53,073	0.80	0.79	0.80	0.87	0.53	0.66
Accuracy			0.75	625,415			0.75			0.71
Macro avg	0.75	0.70	0.71	625,415	0.73	0.75	0.73	0.73	0.71	0.69
Weighted avg	0.75	0.75	0.75	625,415	0.81	0.75	0.77	0.82	0.71	0.74
MCC			0.70	625,415			0.71			0.67

Undersampling for multi-class classification using the XGBoost can be found in Table 10. As shown in the table, similar to the results of RF in Table 9, with the use of undersampling techniques, the performance of baseline has similar results compared to the RandomUnderSampler-based XGBoost. The reason could be that although the RandomUnderSampler removes samples, the XGBoost model can still maintain its high performance by integrating multiple estimators for the final prediction. Although the baseline of XGBoost outperforms the DT, MLP and RF in Tables 7–9, with the use of RandomUnderSampler, the performance of the XGBoost model can be further improved, from 76% weighted average F1 score to 78%. From the results, it is evident that in terms of the macro-average F1 score, weighted average F1 score, and MCC, RandomUnderSampler-based XGBoost has the best performance.

Undersampling for multi-class classification using the CatBoost can be found in Table 11. As we can see from the table, although the baseline of CatBoost outperforms the DT, MLP and RF in Tables 7–9, its performance is slightly worse than the XGBoost model in Table 10. Based on all the results, RandomUnderSampler-based model has the best performance in terms of macro-average F1 score, weighted average F1 score, and MCC.

4.6. Comparison of Machine Learning Using Different Sampling Techniques for Multi-Class Classification

Based on the results of oversampling techniques in Tables 2–6, the ensemble models (XGBoost, and CatBoost) have better performance than the single models (DT and MLP). This is reasonable because the ensemble model is able to reduce the overfitting problem by aggregating multiple estimators for final prediction. In addition, with the use of SMOTE, all machine learning models used can be further improved. Based on all the results, the SMOTE-based XGBoost has the best performance for this multi-class classification task. Based on the results of undersampling techniques in Tables 7–11, the ensemble models (XGBoost and CatBoost) have better performance than the single models (DT and MLP).

We can also see from Table 12 that XGBoost and CatBoost have high performance with the use of oversampling or undersampling techniques. It is surprising to see that with the use of undersampling techniques, the performance of machine learning models is similar to the oversampling-based machine learning models. The reasons could be that (1) undersampling reduces the size of the majority class, which can simplify the learning task for the model; (2) oversampling techniques may introduce synthetic instances into the minority class, which could potentially add noise to the dataset; (3) undersampling ensures that the model focuses on the most relevant instances in the dataset by reducing the dominance of the majority class, which could lead to better discrimination between classes and improved model performance; and (4) oversampling techniques may generate synthetic instances that are outliers in feature space, potentially affecting the model's performance. However, the choice between undersampling and oversampling should focus on the specific problems. According to all the results, the SMOTE-based XGBoost and CatBoost have the best performance for this multi-class classification task, with an accuracy of 75%, a weighted average precision of 82%, a weighted average recall of 75%, and a weighted average F1 score of 77%, while the RandomUnderSampler-based CatBoost has similar performance with their results of [75%, 81%, 75%, 77%] for [accuracy, weighted average precision, weighted average recall, weighted average F1 score].

Table 12. Comparison of machine learning using different sampling techniques for multi-class classification.

Method	Baseline				SMOTE				ADASYN			
	Accuracy	Precision	Recall	F1 Score	Accuracy	Precision	Recall	F1 Score	Accuracy	Precision	Recall	F1 Score
DT	0.62	0.58	0.62	0.54	0.64	0.54	0.54	0.60	0.53	0.68	0.53	0.51
MLP	0.71	0.71	0.71	0.70	0.71	0.78	0.71	0.73	0.67	0.74	0.67	0.69
RF	0.62	0.64	0.62	0.54	0.60	0.72	0.60	0.60	0.56	0.70	0.56	0.56
XGBoost	0.76	0.76	0.76	0.76	0.75	0.82	0.75	0.77	0.74	0.80	0.74	0.76
CatBoot	0.75	0.75	0.75	0.75	0.75	0.82	0.75	0.77	0.73	0.79	0.73	0.75
Method	Baseline				RandomUnderSampler				InstanceHardnessThreshold			
DT	0.62	0.58	0.62	0.54	0.56	0.70	0.56	0.56	0.66	0.72	0.66	0.66
MLP	0.71	0.71	0.71	0.70	0.67	0.75	0.67	0.69	0.69	0.78	0.69	0.72
RF	0.62	0.64	0.62	0.54	0.60	0.73	0.60	0.60	0.60	0.63	0.60	0.57
XGBoost	0.76	0.76	0.76	0.76	0.75	0.82	0.75	0.78	0.71	0.83	0.71	0.75
CatBoot	0.75	0.75	0.75	0.75	0.75	0.81	0.75	0.77	0.71	0.82	0.71	0.74

5. Conclusions and Future Work

In the context of a smart home environment, one of the major challenges is the users' inability to understand and take the necessary security precautions. However, the data of attacks are imbalanced, making it difficult to accurately predict the right category. In this paper, five machine learning models were used for security attack detection on smart applications. In addition, oversampling and undersampling techniques were introduced to solve the imbalanced problem. The results show that the SMOTE-based XGBoost has the best performance with the best accuracy, weighted average precision, weighted average recall, weighted average F1 score, and MCC, with values of 75%, 82%, 75%, 77%, and 72%, respectively. This indicates that these sampling techniques are effective for multi-attack prediction. Further, without the use of sampling techniques, the traditional machine learning models could not detect minority attack classes in some cases, while the model with the sampling techniques used was able to address this problem. This indicates that this sampling-based model is effective for intrusion detection.

However, deploying machine learning models in the real-world IoT environment presents several challenges. The implementation of these models in IoT devices, which often have limited resources, introduces several considerations that can impact their effectiveness and feasibility. In addition, the computational requirements of the models directly impact their deployment on IoT devices. For model training, the computation time ranged from 295 s to 56,777 s depending on which machine learning model and sampling technique was used.

5.1. Threats to Validity

In terms of threats to validity, the first is data quality. The data used to train models or inform research should be accurate and reliable. This means ensuring that data collection methods are valid and that the data are free from errors or biases. In addition, given the class imbalance problem discussed in this paper, sampling techniques (oversampling or undersampling) should be applied to alleviate this problem. In addition, the security of the system is also important. The system should be designed to resist tampering and ensure that it operates as intended without being compromised. Finally, cognitive understanding of system results is crucial for allowing users to understand how decisions are made. By doing so, users can trust the system and understand its limitations.

5.2. Future Work

For future work, considering that the undersampling-based XGBoost model showed the best performance, we will investigate more of the RandomUnderSampler technique to further address the imbalanced problem. In addition, as the deep learning approaches have feature learning capabilities, we will investigate how to integrate feature generation into the ensemble models to further improve detection accuracy. Finally, considering the sampling techniques that can be used to improve the prediction performance for imbalanced data, more advanced techniques (e.g., generative adversarial networks [52–55]) will be investigated to generate synthetic data and handle imbalance problems.

Author Contributions: Conceptualization, S.S., F.S., X.G. and Z.F.; methodology, Z.F.; validation, Z.F., X.G., F.S. and S.S.; formal analysis, X.G., S.S., Z.F. and F.S.; investigation, X.G., S.S., Z.F., and F.S.; resources, F.S., X.G., S.S., and Z.F.; data curation, X.G., S.S., Z.F. and F.S.; writing—original draft preparation, Z.F.; writing—review and editing, S.S., F.S., Z.F. and X.G.; visualization, Z.F. and X.G.; supervision, X.G., S.S., Z.F. and F.S.; project administration, Z.F. All authors have read and agreed to the published version of the manuscript.

Funding: This study was conducted without external funding or financial support.

Data Availability Statement: The data analyzed during this study are available via public bibliographic databases and can be found on Github at: https://github.com/Zongwen-Fan/SamplingML (accessed on 7 March 2024).

Conflicts of Interest: The authors declare no conflicts of interest.

Abbreviations

The following abbreviations are used in this manuscript:

IoT	Internet of Things
SMOTE	Synthetic minority oversampling technique
SVM	Support vector machine
ADASYN	Adaptive synthetic sampling
DT	Decision tree
MLP	Multilayer perceptron
RF	Random forest
XGBoost	Extreme gradient boosting
CatBoost	Category boosting

References

1. Perwej, Y.; Haq, K.; Parwej, F.; Mumdouh, M.; Hassan, M. The internet of things (IoT) and its application domains. *Int. J. Comput. Appl.* **2019**, *975*, 182. [CrossRef]
2. Hafeez, I.; Antikainen, M.; Ding, A.Y.; Tarkoma, S. IoT-KEEPER: Detecting malicious IoT network activity using online traffic analysis at the edge. *IEEE Trans. Netw. Serv. Manag.* **2020**, *17*, 45–59. [CrossRef]
3. Farooq, U.; Tariq, N.; Asim, M.; Baker, T.; Al-Shamma'a, A. Machine learning and the Internet of Things security: Solutions and open challenges. *J. Parallel Distrib. Comput.* **2022**, *162*, 89–104. [CrossRef]
4. Shafiq, M.; Tian, Z.; Sun, Y.; Du, X.; Guizani, M. Selection of effective machine learning algorithm and Bot-IoT attacks traffic identification for internet of things in smart city. *Future Gener. Comput. Syst.* **2020**, *107*, 433–442. [CrossRef]
5. Rani, M. Effective network intrusion detection by addressing class imbalance with deep neural networks multimedia tools and applications. *Multimed. Tools Appl.* **2022**, *81*, 8499–8518. [CrossRef]
6. Pirizadeh, M.; Alemohammad, N.; Manthouri, M.; Pirizadeh, M. A new machine learning ensemble model for class imbalance problem of screening enhanced oil recovery methods. *J. Pet. Sci. Eng.* **2021**, *198*, 108214. [CrossRef]
7. Chawla, N.V.; Bowyer, K.W.; Hall, L.O.; Kegelmeyer, W.P. SMOTE: Synthetic minority over-sampling technique. *J. Artif. Intell. Res.* **2002**, *16*, 321–357. [CrossRef]
8. Moreo, A.; Esuli, A.; Sebastiani, F. Distributional random oversampling for imbalanced text classification. In Proceedings of the 39th International ACM SIGIR Conference on Research and Development in Information Retrieval, Pisa, Italy, 17–21 July 2016; pp. 805–808.

9. Han, H.; Wang, W.Y.; Mao, B.H. Borderline-SMOTE: A new over-sampling method in imbalanced datasets learning. In *Advances in Intelligent Computing, Proceedings of the International Conference on Intelligent Computing, ICIC 2005, Hefei, China, 23–26 August 2005*; Springer: Berlin/Heidelberg, Germany, 2005; Part I, pp. 878–887.
10. Nguyen, H.M.; Cooper, E.W.; Kamei, K. Borderline over-sampling for imbalanced data classification. *Int. J. Knowl. Eng. Soft Data Paradig.* **2011**, *3*, 4–21. [CrossRef]
11. He, H.; Bai, Y.; Garcia, E.A.; Li, S. ADASYN: Adaptive synthetic sampling approach for imbalanced learning. In Proceedings of the 2008 IEEE International Joint Conference on Neural Networks (IEEE World Congress on Computational Intelligence), Hong Kong, China, 1–8 June 2008; pp. 1322–1328.
12. Siddappa, N.G.; Kampalappa, T. Adaptive condensed nearest neighbor for imbalance data classification. *Int. J. Intell. Eng. Syst.* **2019**, *12*, 104–113. [CrossRef]
13. Elhassan, T.; Aljurf, M. Classification of imbalance data using tomek link (T-Link) combined with random under-sampling (RUS) as a data reduction method. *Glob. J. Technol. Optim S* **2016**, *1*, 1–11
14. Putrada, A.G.; Abdurohman, M.; Perdana, D.; Nuha, H.H. Shuffle Split-Edited Nearest Neighbor: A Novel Intelligent Control Model Compression for Smart Lighting in Edge Computing Environment. In *Information Systems for Intelligent Systems, Proceedings of the ISBM 2022*; Springer: Singapore, 2023; pp. 219–227.
15. Kubat, M.; Matwin, S. Addressing the curse of imbalanced training sets: One-sided selection. In Proceedings of the 14th International Conference on Machine Learning, San Francisco, CA, USA, 8–12 July 1997; Volume 97, p. 179.
16. Smith, M.R.; Martinez, T.; Giraud-Carrier, C. An instance level analysis of data complexity. *Mach. Learn.* **2014**, *95*, 225–256. [CrossRef]
17. Shelke, M.S.; Deshmukh, P.R.; Shandilya, V.K. A review on imbalanced data handling using undersampling and oversampling technique. *Int. J. Recent Trends Eng. Res* **2017**, *3*, 444–449.
18. Wongvorachan, T.; He, S.; Bulut, O. A Comparison of Undersampling, Oversampling, and SMOTE Methods for Dealing with Imbalanced Classification in Educational Data Mining. *Information* **2023**, *14*, 54. [CrossRef]
19. Liu, A.Y.c. The Effect of Oversampling and Undersampling on Classifying Imbalanced Text Datasets. Ph.D. Thesis, The University of Texas at Austin, Austin, TX, USA, 16 August 2004.
20. Negandhi, P.; Trivedi, Y.; Mangrulkar, R. Intrusion detection system using random forest on the NSL-KDD dataset. In *Emerging Research in Computing, Information, Communication and Applications, Proceedings of the ERCICA 2018*; Springer: Singapore, 2019; Volume 2, pp. 519–531.
21. Panigrahi, R.; Borah, S.; Bhoi, A.K.; Ijaz, M.F.; Pramanik, M.; Kumar, Y.; Jhaveri, R.H. A consolidated decision tree-based intrusion detection system for binary and multiclass imbalanced datasets. *Mathematics* **2021**, *9*, 751. [CrossRef]
22. Yin, Y.; Jang-Jaccard, J.; Xu, W.; Singh, A.; Zhu, J.; Sabrina, F.; Kwak, J. IGRF-RFE: A hybrid feature selection method for MLP-based network intrusion detection on UNSW-NB15 Dataset. *J. Big Data* **2023**, *10*, 15. [CrossRef]
23. Chaithanya, P.; Gauthama Raman, M.; Nivethitha, S.; Seshan, K.; Sriram, V.S. An efficient intrusion detection approach using enhanced random forest and moth-flame optimization technique. In *Computational Intelligence in Pattern Recognition, Proceedings of the CIPR 2019*; Springer: Singapore, 2020; pp. 877–884.
24. Choraś, M.; Pawlicki, M. Intrusion detection approach based on optimised artificial neural network. *Neurocomputing* **2021**, *452*, 705–715. [CrossRef]
25. Hodo, E.; Bellekens, X.; Hamilton, A.; Dubouilh, P.L.; Iorkyase, E.; Tachtatzis, C.; Atkinson, R. Threat analysis of IoT networks using artificial neural network intrusion detection system. In Proceedings of the 2016 International Symposium on Networks, Computers and Communications (ISNCC), Yasmine Hammamet, Tunisia, 11–13 May 2016; pp. 1–6. [CrossRef]
26. Sohail, S.; Fan, Z.; Gu, X.; Sabrina, F. Multi-tiered Artificial Neural Networks model for intrusion detection in smart homes. *Intell. Syst. Appl.* **2022**, *16*, 200152. [CrossRef]
27. Jabbar, M.; Aluvalu, R.; Reddy, S.S.S. RFAODE: A novel ensemble intrusion detection system. *Proc. Comput. Sci.* **2017**, *115*, 226–234. [CrossRef]
28. Jabbar, M.A.; Aluvalu, R.; Reddy, S.S.S. Cluster based ensemble classification for intrusion detection system. In Proceedings of the 9th International Conference on Machine Learning and Computing, Singapore, 24–26 February 2017; pp. 253–257.
29. Gaikwad, D.; Thool, R.C. Intrusion detection system using bagging ensemble method of machine learning. In Proceedings of the 2015 International Conference on Computing Communication Control and Automation, Pune, India, 26–27 February 2015; pp. 291–295.
30. Karthik, M.G.; Krishnan, M.M. Hybrid random forest and synthetic minority over sampling technique for detecting internet of things attacks. *J. Ambient. Intell. Humaniz. Comput.* **2021**, 1–11. [CrossRef]
31. Bej, S.; Davtyan, N.; Wolfien, M.; Nassar, M.; Wolkenhauer, O. LoRAS: An oversampling approach for imbalanced datasets. *Mach. Learn.* **2021**, *110*, 279–301. [CrossRef]
32. Qaddoura, R.; Al-Zoubi, A.M.; Almomani, I.; Faris, H. A Multi-Stage Classification Approach for IoT Intrusion Detection Based on Clustering with Oversampling. *Appl. Sci.* **2021**, *11*, 3022. [CrossRef]
33. Farah, A. Cross Dataset Evaluation for IoT Network Intrusion Detection. Ph.D. Thesis, University of Wisconsin Milwaukee, Milwaukee, WI, USA, December 2020.
34. Krishnan, S.; Neyaz, A.; Liu, Q. IoT Network Attack Detection using Supervised Machine Learning. *Int. J. Artif. Intell. Expert Syst.* **2021**, *10*, 18–32.

35. Grina, F.; Elouedi, Z.; Lefevre, E. Evidential undersampling approach for imbalanced datasets with class-overlapping and noise. In *Modeling Decisions for Artificial Intelligence, Proceedings of the 18th International Conference, MDAI 2021, Umeå, Sweden, 27–30 September 2021*; Springer: Cham, Switzerland, 2021; pp. 181–192.
36. Onan, A. Consensus clustering-based undersampling approach to imbalanced learning. *Sci. Program.* **2019**, *2019*, 5901087. [CrossRef]
37. Tsai, C.F.; Lin, W.C.; Hu, Y.H.; Yao, G.T. Under-sampling class imbalanced datasets by combining clustering analysis and instance selection. *Inf. Sci.* **2019**, *477*, 47–54. [CrossRef]
38. Fan, Z.; Sohail, S.; Sabrina, F.; Gu, X. The Code of Sampling-Based Machine Learning Models for Intrusion Detecion. 2024. Available online: https://github.com/Zongwen-Fan/SamplingML (accessed on 8 April 2024).
39. Imbalanced-Learn Documentation. Available online: https://imbalanced-learn.org/stable/ (accessed on 20 December 2023).
40. Zhou, H.; Zhang, J.; Zhou, Y.; Guo, X.; Ma, Y. A feature selection algorithm of decision tree based on feature weight. *Expert Syst. Appl.* **2021**, *164*, 113842. [CrossRef]
41. Rosenblatt, F. The perceptron: A probabilistic model for information storage and organization in the brain. *Psychol. Rev.* **1958**, *65*, 386–408. [CrossRef] [PubMed]
42. Ho, T.K. Random decision forests. In Proceedings of the 3rd International Conference on Document Analysis and Recognition, Montreal, QC, Canada, 14–16 August 1995; Volume 1, pp. 278–282.
43. Chen, T.; Guestrin, C. Xgboost: A scalable tree boosting system. In Proceedings of the 22nd ACM SIGKDD International Conference on Knowledge Discovery and Data Mining, San Francisco, CA, USA, 13–17 August 2016; pp. 785–794.
44. Zhang, Y.; Zhao, Z.; Zheng, J. CatBoost: A new approach for estimating daily reference crop evapotranspiration in arid and semi-arid regions of Northern China. *J. Hydrol.* **2020**, *588*, 125087. [CrossRef]
45. Batista, G.E.; Prati, R.C.; Monard, M.C. A study of the behavior of several methods for balancing machine learning training data. *ACM SIGKDD Explor. Newsl.* **2004**, *6*, 20–29. [CrossRef]
46. Moustafa, N.; Slay, J. UNSW-NB15: A comprehensive dataset for network intrusion detection systems (UNSW-NB15 network dataset). In Proceedings of the 2015 Military Communications and Information Systems Conference (MilCIS), Canberra, ACT, Australia, 10–12 November 2015; pp. 1–6. [CrossRef]
47. Koroniotis, N.; Moustafa, N.; Sitnikova, E.; Turnbull, B. Towards the development of realistic botnet dataset in the Internet of Things for network forensic analytics: Bot-IoT dataset. *Future Gener. Comput. Syst.* **2019**, *100*, 779–796. [CrossRef]
48. Ullah, I.; Mahmoud, Q. A Scheme for Generating a Dataset for Anomalous Activity Detection in IoT Networks. In *Advances in Artificial Intelligence, Proceedings of the Canadian Conference on AI, Ottawa, ON, Canada, 13–15 May 2020*; Springer: Cham, Switzerland, 2020.
49. Fan, Z.; Gou, J. Predicting body fat using a novel fuzzy-weighted approach optimized by the whale optimization algorithm. *Expert Syst. Appl.* **2023**, *217*, 119558. [CrossRef]
50. Fan, Z.; Gou, J.; Weng, S. A Novel Fuzzy Feature Generation Approach for Happiness Prediction. *IEEE Trans. Emerg. Top. Comput. Intell.* **2024**, *8*, 1595–1608. [CrossRef]
51. McDonnell, K.; Murphy, F.; Sheehan, B.; Masello, L.; Castignani, G. Deep learning in insurance: Accuracy and model interpretability using TabNet. *Expert Syst. Appl.* **2023**, *217*, 119543. [CrossRef]
52. Lim, W.; Yong, K.S.C.; Lau, B.T.; Tan, C.C.L. Future of generative adversarial networks (GAN) for anomaly detection in network security: A review. *Comput. Secur.* **2024**, *139*, 103733. [CrossRef]
53. Liu, L.; Wang, P.; Lin, J.; Liu, L. Intrusion detection of imbalanced network traffic based on machine learning and deep learning. *IEEE Access* **2020**, *9*, 7550–7563. [CrossRef]
54. Pan, Z.; Niu, L.; Zhang, L. UniGAN: Reducing mode collapse in GANs using a uniform generator. *Adv. Neural Inf. Process. Syst.* **2022**, *35*, 37690–37703.
55. Kim, J.; Jeong, K.; Choi, H.; Seo, K. GAN-based anomaly detection in imbalance problems. In *Proceedings of the Computer Vision–ECCV 2020 Workshops: Glasgow, UK, 23–28 August 2020*; Springer: Cham, Switzerland, 2020; Part VI, pp. 128–145.

Disclaimer/Publisher's Note: The statements, opinions and data contained in all publications are solely those of the individual author(s) and contributor(s) and not of MDPI and/or the editor(s). MDPI and/or the editor(s) disclaim responsibility for any injury to people or property resulting from any ideas, methods, instructions or products referred to in the content.

Article

Learn-IDS: Bridging Gaps between Datasets and Learning-Based Network Intrusion Detection

Minxiao Wang [1], Ning Yang [2,*], Yanhui Guo [3] and Ning Weng [1,*]

1. The Computer Engineering Program in the School of Electrical, Computer, and Biomedical Engineering, Southern Illinois University, Carbondale, IL 62901, USA; minxiao.wang@siu.edu
2. The Information Technology Program in the School of Computing, Southern Illinois University, Carbondale, IL 62901, USA
3. Department of Computer Science, University of Illinois, Springfield, IL 62703, USA; yguo56@uis.edu
* Correspondence: nyang@siu.edu (N.Y.); nweng@siu.edu (N.W.)

Citation: Wang, M.; Yang, N.; Guo, Y.; Weng, N. Learn-IDS: Bridging Gaps between Datasets and Learning-Based Network Intrusion Detection. *Electronics* 2024, *13*, 1072. https://doi.org/10.3390/electronics13061072

Academic Editors: Abdussalam Elhanashi and Pierpaolo Dini

Received: 7 February 2024
Revised: 6 March 2024
Accepted: 10 March 2024
Published: 14 March 2024

Copyright: © 2024 by the authors. Licensee MDPI, Basel, Switzerland. This article is an open access article distributed under the terms and conditions of the Creative Commons Attribution (CC BY) license (https://creativecommons.org/licenses/by/4.0/).

Abstract: In an era marked by the escalating architectural complexity of the Internet, network intrusion detection stands as a pivotal element in cybersecurity. This paper introduces Learn-IDS, an innovative framework crafted to bridge existing gaps between datasets and the training process within deep learning (DL) models for Network Intrusion Detection Systems (NIDS). To elevate conventional DL-based NIDS methods, which are frequently challenged by the evolving cyber threat landscape and exhibit limited generalizability across various environments, Learn-IDS works as a potent and adaptable platform and effectively tackles the challenges associated with datasets used in deep learning model training. Learn-IDS takes advantage of the raw data to address three challenges of existing published datasets, which are (1) the provided tabular format is not suitable for the diversity of DL models; (2) the fixed traffic instances are not suitable for the dynamic network scenarios; (3) the isolated published datasets cannot meet the cross-dataset requirement of DL-based NIDS studies. The data processing results illustrate that the proposed framework can correctly process and label the raw data with an average of 90% accuracy across three published datasets. To demonstrate how to use Learn-IDS for a DL-based NIDS study, we present two simple case studies. The case study on cross-dataset sampling function reports an average of 30.3% OOD accuracy improvement. The case study on data formatting function shows that introducing temporal information can enhance the detection accuracy by 4.1%.The experimental results illustrate that the proposed framework, through the synergistic fusion of datasets and DL models, not only enhances detection precision but also dynamically adapts to emerging threats within complex scenarios.

Keywords: network intrusion detection; datasets engineering; deep learning; feature fusion

1. Introduction

Network Intrusion Detection Systems (NIDS) are crucial components in ensuring the security and integrity of computer networks. By continuously monitoring and analyzing network traffic for suspicious or malicious activities, NIDSs help detect and prevent unauthorized access and attacks. NIDSs play a vital role in safeguarding sensitive data [1], protecting against cyber threats, and maintaining the confidentiality, integrity, and availability of network resources. With the ever-increasing sophistication of cyber attacks and the proliferation of network-connected devices, the importance of NIDS in identifying and mitigating security risks cannot be overstated. NIDSs provide organizations with early detection capabilities, allowing them to respond promptly to potential threats, minimize damage, and maintain the trust and reliability of their network infrastructure.

Deep Learning (DL) has been increasingly applied to NIDS due to its ability to automatically learn complex patterns and representations from data [2,3]. In DL-based NIDS methods, neural networks are used for mining the inherent patterns and characteristics within network traffic flow to distinguish malicious traffic from benign traffic. Compared with traditional data mining and machine learning-based NIDS, DL-based methods hold the potential to revolutionize intrusion detection in aspects such as streamlining the feature engineering process, reducing the manual effort required for designing handcrafted features, automatically extracting hierarchical representations, and adapting dynamically to new and unseen threats.

As a data-driven method [4], DL-based NIDSs require large amounts of labeled data to learn meaningful patterns and representations. Standardized datasets also allow researchers and practitioners to compare the effectiveness of different algorithms, architectures, and techniques on a common ground. Although many large-scale public NIDS datasets exist, we have observed many gaps between existing NIDS datasets and the DL-based NIDS studies.

The first gap is the mismatch between the provided tabular format and the diversity of DL models. In order to apply different types of neural networks, researchers have to design and implement a pre-processing module, which converts the raw traffic flows into samples with a particular data format [5–7]. Even for the provided tabular format, researchers also need to use the pre-processing module if they want to use some extra features. Using various data formats can introduce the missing information of tabular data, for example, the dynamic trends of traffic flow or the raw malicious bytes. Therefore, the public datasets need to be present in different formats.

The second gap is the mismatch between the fixed data instances (traffic flow) and the dynamic network scenarios. Although DL-based NIDSs achieved many successes, employing DL in NIDSs still has many limitations, open questions, and the unique challenges of [8,9]. For example, the dynamic network environments challenge the capability of DL-based NIDS to detect intrusion behaviors early and in real-time [10]. However, the fixed data samples cannot be used to study a dynamic case. Meanwhile, adversarial attacks challenge the robustness of DL-based NIDS [11], and researchers need to perturb the behavior of network traffic. Therefore, each data sample needs an interface to be modified.

The third gap is the mismatch between the isolated datasets and the cross-dataset requirement of DL-based NIDS studies. Given the various NIDS application scenarios, DL-based NIDS should be functional for different environments. The generalization of DL-based NIDS against distribution shifts is one of the most urgent problems [12]. The isolated datasets, which include unique handcraft features and different attack types, should be usable for the same DL-based NIDS model to study the generalization problem. Therefore, different datasets need to be uniformly processed for cross-dataset usage.

In order to bridge all the gaps at the same time and solve other challenges of DL-based IDS in Figure 1, this paper presents a novel tool, called Learn-IDS. Learn-IDS provides a comprehensive, efficient, and extensible framework for preparing customized NIDS data for diverse DL-based study purposes. The structure of Learn-IDS is shown in Figure 2. This framework can efficiently process the raw packet capture (PCAP) traces with a C language-based parser. Both the raw PCAP and tabular data are utilized to construct and label traffic flow for multiple public datasets correctly. By using the raw data, Learn-IDS can uniformly process and store multiple datasets together in a network traffic data bank to solve the third isolated dataset gap. The data sampling module of Learn-IDS supports variable data sampling functions. Hence, the stored data can be further selected and processed based on users' requirements to solve the second gap. The data formatting module provides flexible network traffic representation formats, such as time series, raw bytes arrays, and graphs, for solving the first gap between tabular data and the diversity of DL models.

The main contribution of this paper is as follows:

- We developed a comprehensive, efficient, and extensible tool (Learn-IDS) to prepare customized data for DL-based NIDS studies.
- Learn-IDS can uniformly process the raw PCAP from different datasets, sampling data samples based on research purposes, and generating different traffic representation formats to meet the DL model input requirements.

- We show that Learn-IDS can correctly process the raw data from multiple datasets and present two case studies, in which Learn-IDS plays an important role. The source code is available at https://github.com/wangminxiao/Learn_IDS.git (accessed on 9 March 2024).

The rest of this paper is presented as follows, in Section 2, we introduce the backgrounds of the NIDS dataset and DL-based NIDSs. In Section 3, we focus on analyzing the related work of DL-based IDS to show the gaps and our motivations. In Section 4, we design the details of our Learn-IDS framework, which includes four components: Raw Data Pre-Processing, Data Customizing, DL-based IDS Models, and Training and Evaluation. In Section 5, we evaluate the effect of Learn-IDS and its impact on DL-based NIDS. Section 6 is the conclusion.

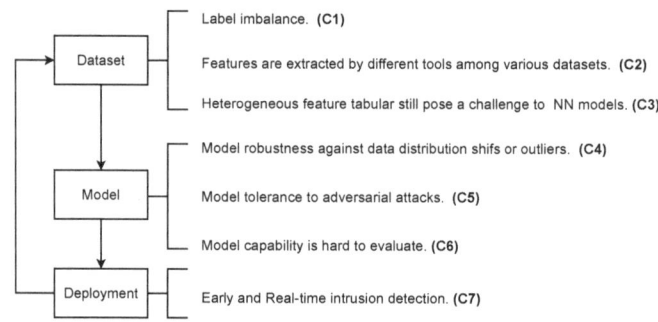

Figure 1. The challenges within the three aspects of DL-based NIDS.

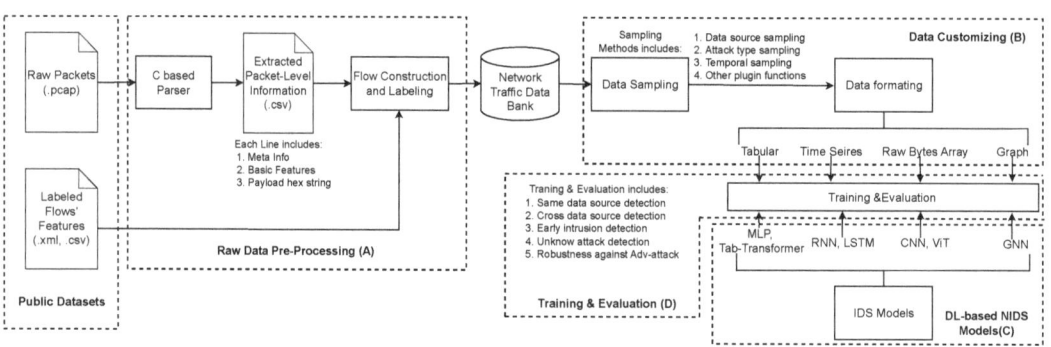

Figure 2. Overview of the framework of Learn-IDS.

2. Background

2.1. NIDS Datsets

The NIDS datasets are crucial for data-driven intrusion detection methods. These datasets serve as the foundation for training, validating, and testing NIDS models, allowing researchers and practitioners to develop and assess the effectiveness of various techniques. The examples of the most popular datasets are introduced as follows, the UNSW-NB15 intrusion detection system dataset [13] encompasses source files in various formats, including PCAP, BRO, and CSV. The dataset was generated from an IXIA traffic generator. Within this data-generating testbed, two servers were linked to a router equipped with a TCP dump, interfacing with three clients and producing PCAP files. Simultaneously, the third server was connected to a distinct router, interacting with its own set of three clients. CIC-IDS2017 and CSE-CIC-IDS2018 datasets are presented by the Canadian Institute of Cybersecurity (CIC) and Communications Security Establishment (CSE) [14]. The CIC17 dataset consists of PCAP and CSV data and the CIC18 dataset consists of PCAP, system log, and CSV data. The two datasets both use a B-profile system to generate network behaviors, such as HTTP, HTTPS, FTP, SSH, and email protocols. The attack traffic is generated by different attack tools within Kali and Windows systems. The difference is CIC17 dataset is collected from a small testbed environment including 2 Web servers, 16 Public, 6 Ubuntu servers, 5 Windows servers, and a MAC server but the CIC18 dataset is collected from an AWS environment, the network topology includes an attack network of 50 machines, five departments holding 100 machines each and a server with 30 machines.

2.2. Deep Learning-Based NIDS Models and Formats

In recent years, Deep Learning (DL) has demonstrated significant success and maintained its influence across diverse fields, including computer vision (CV) [15], natural language processing (NLP) [16], and medical science [17]. Using different types of deep learning models for different types of data is a common and effective strategy in deep learning. Diverse data formats may require specialized architectures to extract meaningful patterns and features. Unlike other areas, network traffic data are uniquely shaped by human design, exhibiting full deformability into various formats such as tabular, time series, raw bytes array, or graphs. This distinct characteristic introduces a noteworthy aspect of DL-based NIDS.

For using different formats and DL-based NIDS models, KitNET [5] employs a collection of neural networks known as autoencoders to collaboratively distinguish between regular and anomalous traffic patterns. He et al. [18] present MS-DHPN, an LSTM model with the feature embedding technique IDS. This model uses feature embedding to combine categorical features with sequential information from network traffic data captured by the LSTM. Wang et al. [19] design HAST-IDS, a CNN-based model for raw byte array format. Another one is LUCID which is presented in [20], the CNN model uses a time window-based feature matrix for observing network traffic behavior. For the GNN-based model, E-GraphSAGE NIDS [21], a GNN approach that allows capturing both edge features of a graph as well as the topological information for network intrusion detection in IoT networks. Hu et al. [22] propose a graph embedding method to model the packet interactions in network traffic for NIDS.

3. Related Work and Motivations

Three fundamental aspects for constructing a deep learning-based IDS are as follows: (1) Dataset: Involves IDS data preparation, encompassing data generation/collection, annotation, and preprocessing. (2) Model: Encompasses the design, training, and evaluation of a DL-based IDS model. (3) Deployment: involves the integration of the trained model with the entire system for monitoring streaming network data.

3.1. Dataset Challenges

Commonly, existing datasets exhibit imbalances not only between benign and malicious traffic but also notably among different attack types. For example, in the CIC-IDS-2017 dataset [14], the "BENIGN" class has 2,359,087 flow records (83.34%), the majority of malicious classes is "DoS Hulk", with 231,072 flow records (8.16%), and the minority malicious class "Heartbleed" only has eight flow records (0.00039%). Training an IDS model on an imbalanced dataset will directly impact its performance, particularly in multi-class classification tasks (**C1** in Figure 1).

Meanwhile, prevalent published datasets are typically presented in the form of tabular data, where network traffic flows serve as instances. Each instance is accompanied by a label and multiple features, providing the necessary information for training IDS models. For instance, the UNSW-NB15 dataset [13] has 47 features and the CIC-IDS-2017 dataset has 80 features, while only six features overlap between them. One reason for the difference is these two datasets use different network traffic analysis tools. UNSW-NB15 uses the matched features, which have been extracted by both the Argus tool and Bro-IDS tool. CIC-IDS-2017 used their own CICFlowMeter tool, whose lower version was known as the

ISCXFlowMeter which was used by the ISCX2012 dataset. The isolation of feature sets makes it challenging to train a model simultaneously on multiple datasets, hindering the potential to address the data distribution problem associated with a single dataset (**C2** in Figure 1).

Additionally, the most commonly used tabular data form of features poses challenges to convolutional neural networks (CNN) [23], even being called the last "unconquered castle" [24]. Because tabular data features for IDS are heterogeneous data with different attribute types, value scales, and a mix of discrete and continuous variables, there is often a lack of strong correlation among features (different from language data) and no spatial dependencies (different from image data). Many proven powerful methods based on CNN cannot be adopted for modeling this data form (**C3** in Figure 1).

3.2. Model Challenges

Enhancing the robustness of IDS models to accommodate data distribution shifts is crucial. In the realm of intrusion detection, models trained on curated datasets may prove unreliable and inefficient when deployed in real-network scenarios. Certain classical machine learning-based methods, such as Decision Trees, exhibit considerable strength in handling heterogeneous tabular data. However, they often face a significant limitation in terms of robustness when confronted with shifts in data distribution [25]. Given the inevitability of data distribution shifts in real-world deployments spanning both cross-time [10,26] and cross-domains [12], it becomes imperative to include robustness verification in the model assessment process (**C4** in Figure 1).

Adversarial attacks pose a significant threat to DL-based IDS models [27]. A network adversarial attack involves the intentional modification of a network attack to induce mispredictions in the IDS model while preserving the original impact of the network attack. This underscores the importance of fortifying IDS models against such adversarial manipulations to ensure their reliability and effectiveness in detecting network threats (**C5** in Figure 1).

The primary goal of DL-based IDS models, as well as other applications was to outperform the state-of-the-art on a single large-scale dataset [28]. However, outperforming on an IDS dataset has become challenging in the current landscape, given the multitude of works showcasing flawless performance across various datasets. This widespread success makes it increasingly difficult to assess the true potential capabilities of IDS models, crucial for fostering realistic deployments [29], as their evaluation becomes intricate on the majority of existing datasets (**C6** in Figure 1).

3.3. Deployment Challenges

The implementation of DL-based IDS in real-network deployments is still in its nascent stages [28]. The prompt detection of intrusions is paramount for the effective deployment of DL-based IDS. This capability allows for the timely execution of accurate counter-attack responses, preventing potential damage before it occurs [30]. Prompt detection entails discerning the nature of network traffic as benign or malicious in the initial stages of traffic flows. However, achieving early intrusion detection presents challenges, primarily stemming from the limited information available in the form of raw packets [31]. Many existing Deep Learning-based NIDS methods depend on handcrafted features as input, and the restricted nature of raw packets can potentially impact the quality of these input features (**C7** in Figure 1).

4. Framework

In this section, we will introduce the structure and modules of Learn-IDS. Learn-IDS is a general framework for DL-based IDS study. Its whole framework is shown in Figure 2, the whole pipeline consists of four components: (A) Raw Data Pre-Processing; (B) Data Customizing; (C) DL-based IDS Models; (D) Training and Evaluation.

4.1. Raw Data Pre-Processing

The raw data pre-processing component serves three main objectives: (1) the efficient extraction of raw information from PCAP files; (2) accurate labeling of traffic flow and (3) the storage of processed data for swift retrieval during both training and testing phases. To achieve these goals, we have developed modules such as parser and flow construction and labeling.

4.1.1. Parser Module

The parser module constructs traffic flow based on traffic termination and a customized timeout threshold; it labels traffic flow using public tabular data and independently stores all the extracted information for each traffic flow. This approach facilitates quick loading of data by the PyTorch Dataloader (in PyTorch, a DataLoader is a utility that helps efficiently load and iterate over a dataset during the training or evaluation of a machine learning model). The specific details of each module are elaborated below: This module is a specialized C program designed for parsing and processing PCAP files, enabling the effective utilization of extensive raw PCAP data from public datasets. Leveraging the PCAP library, the program meticulously reads and analyzes packets, extracting vital information such as IP addresses, ports, protocols, timestamps, and payload data. The parsed data are thoughtfully organized and exported to a CSV file, laying the groundwork for further analysis or visualization.

Key features of this program include the inclusion of header files for networking and PCAP functionalities, a robust packetHandler function for detailed packet processing, and thorough data extraction covering attributes like IP details, ports, timestamps, and payload. Notably, the code implements payload truncation for TCP and UDP packets, limiting it to a maximum length of 8000 bytes in hexadecimal format within the CSV output.

In addition to facilitating detailed packet analysis, the code maintains statistics by tracking the occurrences of TCP, UDP, ICMP, and other packet types during parsing. Furthermore, the program's offline processing capability makes it a valuable tool for historical network traffic data analysis, contributing to comprehensive retrospective examinations.

4.1.2. Flow Construction and Labeling Module

The flow construction module further processes the raw packet information in the CSV output of the parser module with the pandas library to construct and label network traffic flow. As shown in the (A) part of Figure 2, the construction module also takes the tabular data from datasets as inputs to group the packets into the unit of flow. The flow construction module uses seven fields (source IP, destination IP, source port, destination port, protocol, start timestamp, duration) as flow IDs to match packets to labeled flows in tabular data. An alternative method for determining the start timestamp and duration involves following public tabular data. However, this approach proves to be inefficient as it requires filtering the output CSV files N times, with N being the line number of the public tabular data. Moreover, it is worth noting that some errors have been identified in certain public tabular datasets [32,33] recently, further complicating the efficiency and reliability of this method.

Therefore, the process of flow construction plays a crucial role in determining the termination of a network flow, relying on both connection closure and session timeout considerations. In the context of connection-oriented protocols like TCP, the conclusion of a network flow often occurs through mutual agreement or upon the completion of the communication task. This can transpire when one entity sends a TCP FIN (finish) packet, and the other entity acknowledges it with an ACK packet, signaling the conclusion of the connection. Alternatively, network flows may terminate due to session timeout, which takes place when there is no activity or communication between entities for a specified duration, prompting the network infrastructure or application layer to close the connection.

Once all packets of a network flow are appropriately grouped, the flow is labeled, and the information in the output CSV file is extracted and saved into a TXT file named after the unique flow ID. All flow file paths and their respective labels are recorded in the network traffic data bank, and stored as a JSON file that preserves a Python dictionary structure for subsequent use. The data bank dictionary structure comprises three levels of keys: "dataset name", "attack type name", and "flow data path".

4.2. Data Customizing

In the network traffic data bank, all raw information of network flows from different public datasets is well organized and stored for further NIDS studies. The data customizing component includes a data sampling module and a data formatting module to prepare training and testing sets for different NIDS studies.

4.2.1. Data Sampling

The preparation of datasets for DL-based NIDS can vary based on the specific focus and requirements of the study. Therefore, we provide data sampling functions to select the needed data samples to ensure that the model is trained, validated, and tested on relevant and representative data.

Data Source and Attack Types Sampling: The data source and attack types sampling module provides an interface for users to obtain specific types of network flow from single or multiple datasets. Sampling instances from the Data Bank requires two elements: dataset name and class label. For a given class, sampling from different datasets can bring more information about how a type of traffic behaves in different environments. The information can help DL-based IDS models learn more general patterns to distinguish this type of instance from others. Meanwhile, in the original datasets, some classes of instances only occupy a small percentage. The imbalance class distribution harms IDS models' training effect. Instead of using an over-sampling method [34], a cross-dataset sample mitigates the imbalance problem by combining the same class from different datasets.

Temporal sampling for early detection: To simulate and study the early intrusion detection situation, we use the temporal sampling function to re-generate the labeled flow. We defined a concept of sampling window size, which is the number of the first of several packets. The generated flows are cropped out by the sampling window, and the parts in the sampling window will be regarded as the early state of flows, based on which tabulars or other formats are built for early intrusion detection.

4.2.2. Data Formatting

The network flow is the actual instance for a flow-based IDS model to "observe". However, there is no particular ML model that can directly process raw packets. Hence, the raw information in packets needs further processing to transform into a format that can be fed into the ML model, but what kind of format to use is still open. As shown in the (B) part of Figure 2, the data formatting module includes four typical data formats: (1) Tabular; (2) Time Series; (3) Raw Array and (4) Graph.

Tabular: The tabular format is the most common data format, which is normally provided by existing datasets. However, tabular contains a mix of discrete and continuous variables as well as various attribute types and value scales. Those characters challenge most neural network models [23]. Meanwhile, existing datasets normally use unique features that barely overlap. The impact of features for IDS receives a lot of attention, Binbusayyis et al. [35] presented an optimal features study to enhance IDS performance; Adhao et al. [36] argued that feature engineering is essential for DL-based IDS. Therefore, we provided API for users to define and extract customized feature sets for further IDS features to study.

Time Series: The time series format introduces temporal information that can show the behavior of malicious traffic better. The FTime Series is generated by dynamically extracting features when packets arrive. Hence, the sequence has the same length as the number of packets belonging to a flow. This type of data form is more suitable for recurrent neural networks (RNN) to learn the temporal pattern. For example, He et al. [18] proposed MS-DHPN, a combination of AE and LSTM-based models, to detect time seriess.

Raw Bytes Array: The raw bytes array format is proposed for taking advantage of powerful CNN and Vision Transformer (ViT). One benefit is this data format does not require extracting handcrafted features. Since the length of a packet is not fixed, its first N bytes are used to represent it as a whole. The headers of a packet can be fully included upon taking the appropriate value N, as well as some or all of the payload data. N is set to be a square number to make processing more convenient [7].

Graph: Many current graph-based NIDS transform network traffic flows into dynamic IP address graphs, serving as a representation of network topology. However, in the context of the data formatting module, the term "formats" pertains to the depiction of an individual traffic flow. The graph format provided in this context captures diverse correlations within dynamic feature series. Notably, graph formats and graph neural networks excel in explicitly modeling inter-temporal and inter-variable relationships, a task that proves challenging for traditional methods and other deep neural network-based approaches [37].

4.3. DL-Based NIDS Models

To accelerate the NIDS model building progress, Learn-IDS contains implementations of IDS methods with different basic neural network blocks: (1) MLP, (2) RNN, (3) CNN, (4) Transformer, and (5) GNN. We choose those models as representative models to perform DL-based NIDSs. In this phase, we only implement one or two methods for each block type to test the function of the toolbox, instead of duplicating more existing methods. Hence, we duplicated four existing IDS methods, which correspond to four different data formats mentioned in the last subsection. The following Table 1 lists the supported DL-based NIDS models.

Table 1. Supported DL-based NIDS models.

Formats	Basic Block	Models
Tabular	MLP	KitNET [5]
	Transformer	FT-Transformer [38]
Time Series	LSTM	MS-DHPN [18]
	Transformer	Informer [39]
Raw Byte Array	CNN	HAST-IDS [19]
	CNN	LUCID [20]
	Transformer	ViT [40]
Graph	GNN	E-GraphSAGE [21]

4.4. Training and Evaluation

In Learn-IDS, we design a unified Pytorch workflow pipeline for training and evaluating NIDS with different datasets, different data formats, and different IDS models so that studying different NIDS topics, for example, cross-data source detection generalization problems, early intrusion detection problems, unknown/unseen attack detection problem, and the robustness of DL-based NIDS model against adversarial attacks.

Most DL-based IDS methods share a similar workflow. Hence, we can summarize a basic training pipeline for most work in Figure 3. In addition, Learn-IDS also provides many customized training options for improving the training effect, for instance, learning rate schedule, optimizer, regularization, training logs, and checkpoints save/reload. We also introduce many existing methods on IDS models and IDS data structure, which are not our contribution.

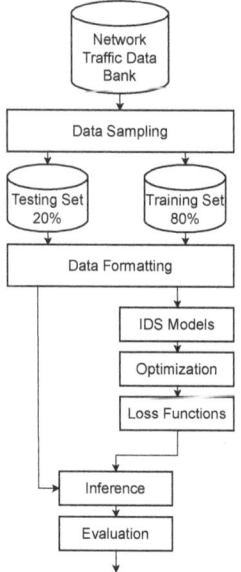

Figure 3. The training and evaluation workflow pipeline.

The metrics for evaluating models' detection performance in terms of

- **Precision:** The number of predictions of the positive class that actually belong to the positive class.

$$Precision = \frac{TP}{TP + FP} \quad (1)$$

- **Recall:** The number of predictions of the positive class out of the total number of positive class items in the dataset.

$$Recall = \frac{TP}{TP + FN} \quad (2)$$

- **F1-score:** A score that balances precision and recall.

$$F1 = \frac{2 \times Precision \times Recall}{Precision + Recall} = \frac{2 \times TP}{2 \times TP + FP + FN} \quad (3)$$

5. Experimental Results and Discussion

In this section, we begin by demonstrating the proficiency of Learn-IDS's raw data pre-processing component in accurately handling raw PCAP data. Subsequently, we delve into two illustrative examples that leverage Learn-IDS to investigate the cross-dataset generalization of DL-based IDS models and assess the impact of network flow representation formats.

5.1. Datasets Pre-Processing Result

Correct raw data pre-processing is the ground for the rest of the stages. The majority of IDS works utilized the labeled features tables to train their DL-based models. Taking into account the need for comparison with existing works, new methods with novel data formats should use the same sample instances as prior methods to isolate the improvement of performance from changing data instances. Almost all published datasets include Pcap files and flow-based feature tables, but only the tabular data are labeled. In order to benefit from utilizing the raw Pcap data, Learn-IDS needs to match those raw packets to the labeled flows recorded in the table, so that raw packets also have a label.

The flow can normally be identified by seven sets (source IP, destination IP, source port, destination port, protocol, start timestamp, duration), in which the element duration is chosen by receiving the FIN flag or flow time-out threshold. The time-out threshold is a configurable parameter [41] variable among different datasets. Learn-IDS will maintain the given flow duration threshold of each dataset instead of unifying it. Therefore, the Raw Data pre-processing module matches the raw packets to the flows defined by the original dataset.

We test Learn-IDS's raw data pre-processing module on three datasets: CIC-IDS-2017, ISCX-2012, and UNSW-15. After generating raw packet lists of each flow, we calculate the generated results with the original flow tables. To analyze the characteristics of datasets, we define several measures, which are "unique payload contents amount", "unique source IP number", "average duration per flow", and "average packets number per flow". Considering that different attacks have different behaviors, we count those measures independently among different classes.

The statistical information of three datasets is shown in Tables 2–4. For the unique source IP address, we found both UNSW-15 and CIC-IDS-17 have very few source IPs for most of their attack traffic flows. The UNSW-15's attack flows have four fixed source IPs and most of the attack traffic flows in CIC-IDS-17 only have one fixed source IP, which belongs to the gateway of the testbed. These fixed-source IPs could work as dataset bias, which can hinder the DL-based IDS model from learning the real features of attack behaviors.

Table 2. ISCX-2012 dataset.

Flow Class	Pub Num	Ext Num	Acc	Uniq Payload	Uniq SIP	Total Dur	Total Pkts	Ave Dur	Ave Pkts	Unique Ratio
DDoS_Botnet	37,460	22,595	60.32%	30,684	11	9,074,275.38	6,229,132	401.61	275.69	1.3579
Inside_Attack	20,358	6950	34.14%	20,695	160	62,459.89	53,343	8.99	7.68	2.9776
BruteForce_SSH	5219	5180	99.25%	30,029	1	14,602.88	94,266	2.82	18.20	5.7971
HTTP_DoS	3776	1412	37.39%	53,704	75	43,761.77	124,199	30.99	87.96	38.0339

Table 3. UNSW-15 dataset.

Flow Class	Pub Num	Ext Num	Acc	Uniq Payload	Uniq SIP	Total Dur	Total Pkts	Ave Dur	Ave Pkts	Unique Ratio
Generic	215,481	215,360	99.94%	135,321	4	15,903.51	745,353	0.073	3.46	18.15
Exploits	44,525	44,392	99.70%	947,990	4	100,209.53	2,587,983	2.25	58.29	36.63
Fuzzers	24,246	24,216	99.87%	150,088	4	65,600.74	488,487	2.70	20.17	30.72
DoS	16,353	16,292	99.62%	276,828	4	44,369.21	766,163	2.72	47.02	36.13
Reconnaissance	13,987	13,970	99.87%	63,920	4	14,854.26	174,148	1.06	12.46	36.70
Analysis	2677	2674	99.88%	6507	4	4643.56	26,136	1.73	9.77	24.89
Backdoor	1795	1791	99.77%	6510	4	5273.26	20,864	2.94	11.64	31.20
Shellcode	1511	1511	100.00%	6636	4	548.27	13,981	0.36	9.25	47.46
Backdoors	534	534	100%	735	4	691.42	3954	1.29	7.40	18.58
Worms	174	172	98.85%	6722	4	231.51	14,102	1.34	81.98	47.66

Table 4. CIC-IDS-2017 dataset.

Flow Class	Pub Num	Ext Num	Acc	Uniq Payload	Uniq SIP	Total Dur	Total Pkts	Ave Dur	Ave Pkts	Unique Ratio
Hulk	159,048	134,699	84.69%	81,294	1	9,854,295.53	2,247,117	73.15	16.68	3.61
PortScan	159,023	66,486	41.80%	2281	1	25,651.55	543,961	0.11	2.40	0.41
DDoS	95,123	93,425	98.21%	6151	2	2,833,760.22	1,280,601	30.33	13.70	0.48
GoldenEye	7647	7948	96.06%	11,771	1	226,891.78	111,174	28.54	13.98	10.58
FTP-Patator	3984	4007	99.42%	11,919	1	35,997.33	111,610	8.98	27.84	10.67
SSH-Patator	2988	3195	93.07%	85,424	1	36,978.08	169,583	11.57	53.06	50.37
slowloris	5707	7172	74.32%	983	1	415,723.94	58,612	58.08	8.18	1.67
Slowhttptest	5109	5742	87.61%	7867	1	324,997.07	51,534	56.59	8.97	15.26
Bot	738	735	99.59%	1580	5	72.47	9870	0.09	13.41	16.00
Brute Force	1365	1359	99.56%	9128	1	9773.93	30,121	7.18	22.14	30.30
XSS	679	678	99.85%	2780	1	4770.07	9637	7.02	14.19	28.84
Infiltration	48	33	68.75%	976	1	2818.63	59,753	82.90	1757.44	1.63
Sql Injection	12	12	100%	97	1	164.01	389	4.82	11.44	24.93
Heartbleed	11	11	100%	20,922	1	1217.80	49,295	110.70	4481.36	42.44

For the unique payload contents amount, we found both ISCX-2012 and UNSW-15 datasets always have large amounts of unique payload contents for all classes. However, CIC-IDS-17 has some classes that only own a very limited number of unique payload contents such as the DoS-slowloris attack in CIC-IDS-17 which only has 96 unique payloads for 46,327 packets, and the PortScan attack has 168 unique payloads for 321,208 packets. Meanwhile, we define "Unique ratio", which is the number of unique payload contents divided by the total flow amount. The unique ratio shows the diversity of packet content, and we believe that the training data with less diversity may lead the IDS model to be short of generalization.

Other measures "average duration per flow", and "average packets number per flow" are also shown as dataset bias because they are directly affected by the traffic-generating method or the bottleneck of testbed scenarios. We found that the traffic in UNSW-15 has a shorter average duration than the other two datasets. In the next subsection, we will show how the dataset bias "unique payload contents amount", and "unique source IP number" affect the DL-based model's performance.

5.2. Cross-Dataset Sampling Effect

To demonstrate Learn-IDS's cross-dataset sampling function, we adopt Learn-IDS to prepare training data across multiple datasets for studying the out-of-distribution (OOD) [42] generalization of DL-based NIDS.

To evaluate the generalization of trained IDS models, we design experiments to determine the robustness of trained models to distribution shifts. We generate distribution shifts by testing trained models on samples from a new dataset, which is different from the training data. The data samples from the same dataset that is used to train the model have the same distribution, known as in-distribution (ID). In contrast, data samples from another dataset have a different distribution, known as out-of-distribution. We call the data samples set using cross-dataset sampling "mixed" set, and the data samples that are all from the same dataset "single" set. Our purpose is to observe the OOD test performance of models that are trained on mixed ID and single ID.

Experiment Setup. The training data were derived from three large-scale IDS datasets: ISCX-12, UNSW-15, and CIC-IDS-17. The training dataset included 8000 normal and 8000 DoS samples. Specifically, the ID train set was composed of data samples solely from dataset A, while the mixed ID train set included samples from both datasets A and B (4000 samples from each dataset). We pre-processed the raw data using the Learn-IDS, extracting unified tabulars for each instance. For the tabular format, we use an auto-encoder (AE) based model to conduct the DoS detection. The auto-encoder model has two components: the auto-encoder and a linear prediction head. Similar to prior methodology [43,44] on OOD performance evaluation. We adopted the Adam optimizer and employed L1 regularization during training. The use of L1 regularization was motivated by its ability to induce sparsity in model parameters, aiding feature selection. The training procedure involved 30 epochs, and we ensured that the model was exposed to a diverse range of normal and DoS flow patterns for robust learning.

To evaluate the generalization of our trained IDS models, we performed both OOD and ID on a DoS detection task.

- **ID train set:** Single ID set contains 8000 data samples from dataset A; Mixed ID set contains 8000 data samples from dataset A and B.
- **ID test set:** Same as an ID train set, different samples set. (The model trained on dataset A will be tested on 2000 testing samples from dataset A. The model trained on datasets A and B will be tested on 2000 testing samples from datasets A and B)
- **OOD test set:** 2000 testing samples from dataset C for both single and mixed OOD.

We repeat the OOD evaluation experiments three times on different dataset combinations: (a) dataset C: ISCX-2012, A: UNSW-15, B: CIC-IDS-17; (b) dataset C: UNSW-15, A: CIC-IDS-17, B: ISCX-2012; (c) dataset C: CIC-IDS-17, A: ISCX-2012, B: UNSW-15; the performance results are shown in Figure 4. For each dataset combination, the left part shows the OOD test performance and the right part is the ID test performance. We can observe that models can achieve very good results (over 90% accuracy) on the ID test set, but poor performance (less than 70%) on the OOD test set. Although distribution shift will reduce the trained models' performance, we also can see that the model trained on a mixed ID set performs much better than the single one among all three experiments. The OOD test improvement from the mixed to the single shows that the model trained on the mixed dataset has better robustness against the distribution shift. Furthermore, it is noteworthy that the magnitudes of improvement vary among the three combinations. These discrepancies are likely due to the influence of dataset cross-correlation, a factor that diverges across the different dataset combinations. However, quantifying this cross-dataset correlation proves challenging at present. These more robust pre-trained models can work as a better starting point for further fine-tuning the target environment, which has a different distribution from the training data.

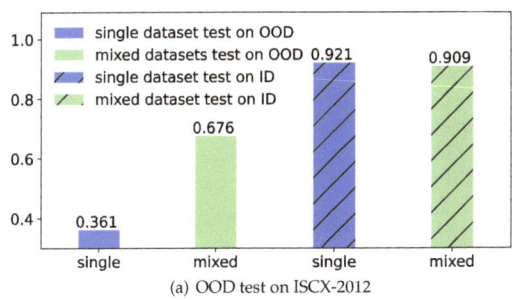

(a) OOD test on ISCX-2012

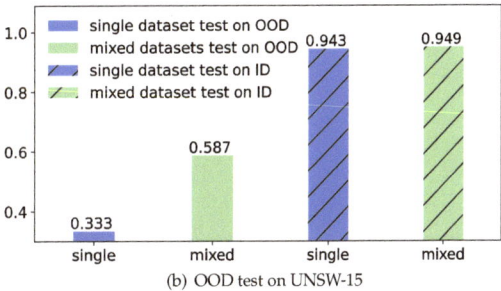

(b) OOD test on UNSW-15

Figure 4. *Cont.*

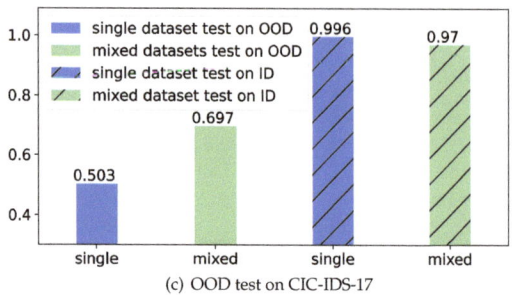

(c) OOD test on CIC-IDS-17

Figure 4. OOD performance of models trained on single dataset or mixed dataset for DoS detection task. For (**a**), we conduct OOD test on ISCX-2012 with models trained on UNSW-15 (single) and UNSW-15 + CIC-IDS-17 (mixed). For (**b**), we conduct OOD test on UNSW-15 with models trained on CIC-IDS-17 (single) and CIC-IDS-17 + ISCX-2012 (mixed). For (**c**), we conduct OOD test on CIC-IDS-17 with models trained on ISCX-2012 (single) and ISCX-2012 + UNSW-15 (mixed).

5.3. Influence of Data Format

To demonstrate Learn-IDS's data formatting function, we adopt Learn-IDS to prepare training data in both the tabular format and time series format for studying the impact of data format on DL-based NIDS.

To analyze the influence of data format for DL-based IDS, we design an experiment to show that a novel data format can provide more valuable information for the IDS model, such as time series which can provide additional temporal information. The novel data format is denoted relative to the traditional tabular format, which is statistical values extracted from network traffic. We can notice that the tabular works on projecting the attack's anomaly behaviors to the manual feature dimensions, yet cannot show the behaviors directly. It also should be noted that those statistical values can only be calculated after the traffic finishes its transmission. In the case of attack traffic, it means the attack has been executed. However, a novel data format, such as time series, can present the behaviors by showing features varying with time. Meanwhile, time series also can be partly obtained by the first few packets. Hence, we design experiments to compare the IDS models' performance with the tabular format and features sequence format.

To compare the performance of two different data formats, we noticed two important problems: First, the different data formats may lead the IDS models to have different structures and numbers of weights. If comparing the performance of different models, we cannot claim that additional temporal information can improve the performance; second, there is a conflict between the random packet number of each flow and the requirement of fixed input size of the IDS model. In order to exclude the influence of other factors, we use the same input data structure to represent both tabular and sequence as in Figure 5. As the figure shows, each flow can be represented as a 2D tensor with the shape $N \times L$, where N is the number of features and L is the length of the sequence. The features sequence is combined by L tabulars, the first vector is calculated from the first n packets of this flow, the second one is calculated based on the first $2n$ packets, and so on. The n should meet the requirement that $L \times n$ is less or equal to the total number of packets that belong to this flow. For the flow that only has less than L packets, $n = 1$ and the insufficient parts will be padded with the last tabular. Similarly, the tabular is calculated with n equals the number of packets and the insufficient parts will be padded with the tabular itself.

Experiment Setup. To show the effect of additional temporal information, we ran the comparison experiments with tabular format and features sequence format on four datasets: ISCX-12, UNSW-15, and CIC-IDS-17. On each dataset, each targeted model is trained on 20,000 benign data samples and 20,000 malicious samples and is tested on 5000 benign data samples and 5000 malicious samples. We adopted the Adam optimizer and cross-entropy loss function. Since we have already constructed the tabular and sequence into the 2D tensor with the same shape, we train the 2D CNN model to conduct the multi-class classification IDS task.

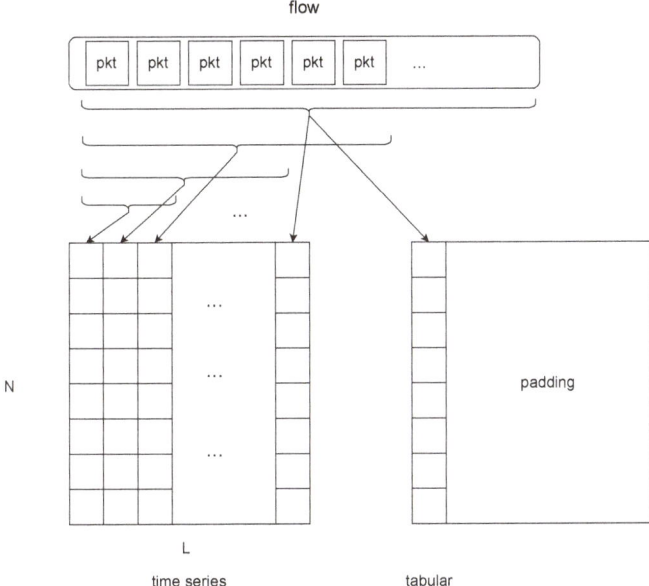

Figure 5. Represent flow in the form of tabular and features sequence in the same shape 2D tensor structure.

The main results are shown in Figure 6, the same CNN model can achieve better performance when using the features sequence format than tabular. We believe the reason is that the time series can provide additional temporal information for the IDS model.

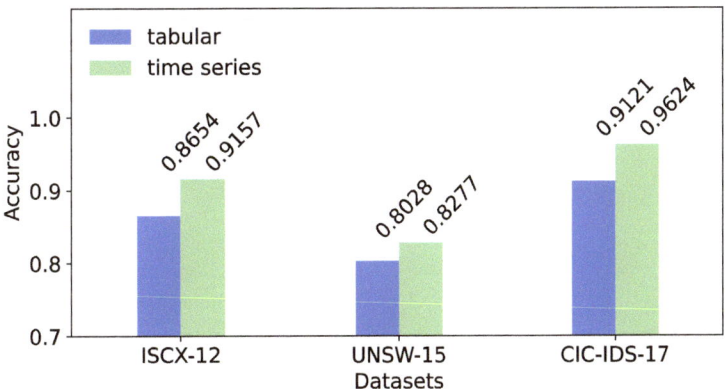

Figure 6. The performance comparison on three datasets with two formats of tabular and time series. The results show that using a features sequence format can achieve better performance than a tabular.

5.4. Results Summary and Discussion

The Learn-IDS framework presents comprehensive capabilities for raw data preprocessing, cross-dataset sampling, and flexible data format representation. The experimental results highlight the potential benefits of each component in enhancing the effectiveness and generalization of DL-based IDS models.

The cross-dataset sampling functionality addresses the challenge of model generalization across different datasets. The experiments in Tables 5 and 6 demonstrate that exposing models to mixed datasets during training improves their ability to handle OOD scenarios. This is particularly valuable in real-world situations where diverse and evolving threats may differ significantly from training data.

The influence of data format on IDS model performance emphasizes the importance of considering the temporal aspects of network traffic. The features sequence format, capturing temporal patterns in network flows, outperformed tabular representations. The results in Table 7 suggest that incorporating time series information can enhance the model's ability to detect complex and dynamic intrusion patterns.

Table 5. OOD performance of models trained on single dataset or mixed dataset for DoS detection task.

	Single				Mixed			
Dataset	Acc	Prec	Rec	F1	Acc	Prec	Rec	F1
ISCX-12	0.361	0.358	0.335	0.346	0.676	0.691	0.682	0.686
UNSW-15	0.333	0.347	0.358	0.353	0.587	0.578	0.597	0.576
CIC-IDS-2017	0.503	0.496	0.509	0.502	0.697	0.711	0.685	0.698

Table 6. ID performance of models trained on single dataset or mixed dataset for DoS detection task.

	Single				Mixed			
Dataset	Acc	Prec	Rec	F1	Acc	Prec	Rec	F1
ISCX-12	0.921	0.898	0.944	0.921	0.909	0.916	0.899	0.908
UNSW-15	0.943	0.931	0.949	0.94	0.949	0.960	0.923	0.941
CIC-IDS-2017	0.996	1.00	0.991	0.995	0.97	0.967	0.974	0.971

Table 7. Multi-class detection performance on three datasets with two formats of tabular and time series.

Format	Tabular				Time Series			
Dataset	Acc	Prec	Rec	F1	Acc	Prec	Rec	F1
ISCX-12	0.865	0.803	0.819	0.778	0.915	0.924	0.926	0.921
UNSW-15	0.808	0.799	0.796	0.912	0.827	0.841	0.827	0.829
CIC-IDS-2017	0.912	0.918	0.924	0.912	0.962	0.958	0.945	0.945

In summary, Learn-IDS effectively handles raw network data, demonstrating its potential to address challenges in intrusion detection. The framework's modular components work synergistically to provide a comprehensive solution. The experimental results affirm its capacity to enhance the robustness and performance of DL-based IDS models in diverse and evolving network environments.

6. Conclusions

In this paper, we introduced Learn-IDS, a framework designed to address critical challenges in the realm of deep learning-based Network Intrusion Detection Systems (NIDS). Recognizing the escalating architectural complexity of the Internet and the evolving cyber threat landscape, we identified three key gaps in existing datasets used for training deep learning models. These gaps encompassed issues related to data format diversity, static data instances, and the lack of cross-dataset compatibility. Learn-IDS emerges as a comprehensive, efficient, and extensible tool tailored to bridge these gaps simultaneously. Through a structured framework encompassing Raw Data Pre-Processing, Data Customizing, DL-based IDS Models, and Training and Evaluation, Learn-IDS facilitates the preparation of customized data for diverse DL-based NIDS studies. It efficiently processes raw PCAP traces, supports variable data sampling functions, and provides flexible network traffic representation formats, such as time series, raw byte arrays, and graphs, aligning with the diverse requirements of deep learning models. Our contribution lies in the development of Learn-IDS, a tool that not only overcomes the identified gaps but also showcases its efficacy through the correct processing of raw data from multiple datasets. Two case studies presented in this paper demonstrate Learn-IDS's pivotal role in enhancing the OOD accuracy by around 30% through the cross-dataset sampling function and improving multi-class detection accuracy by an average of 4.1% across three datasets.

Author Contributions: Methodology, software, validation, formal analysis, investigation, resources, data processing, writing—original draft preparation, visualization, M.W.; methodology, supervision, writing—original draft preparation, review and editing, project administration, funding acquisition, N.Y.; writing—review, supervision, Y.G.; writing-review, supervision, N.W. All authors have read and agreed to the published version of the manuscript.

Funding: This work is supported in part by the US National Science Foundation under Grant CC-2018919.

Data Availability Statement: The datasets used in this work are available as the following links: https://www.unb.ca/cic/datasets/ids.html, https://research.unsw.edu.au/projects/unsw-nb15-dataset, https://www.unb.ca/cic/datasets/ids-2017.html, https://intrusion-detection.distrinet-research.be/CNS2022/Dataset_Download.html (accessed on 9 March 2024).

Conflicts of Interest: The authors declare no conflict of interest.

References

1. Shu, X.; Yao, D.; Bertino, E. Privacy-Preserving Detection of Sensitive Data Exposure. *IEEE Trans. Inf. Forensics Secur.* **2015**, *10*, 1092–1103. [CrossRef]
2. Goodfellow, I.; Bengio, Y.; Courville, A. *Deep Learning*; MIT Press: Cambridge, MA, USA, 2016.
3. Yi, T.; Chen, X.; Zhu, Y.; Ge, W.; Han, Z. Review on the application of deep learning in network attack detection. *J. Netw. Comput. Appl.* **2023**, *212*, 103580. [CrossRef]
4. Chou, D.; Jiang, M. A survey on data-driven network intrusion detection. *ACM Comput. Surv. (CSUR)* **2021**, *54*, 1–36. [CrossRef]
5. Mirsky, Y.; Doitshman, T.; Elovici, Y.; Shabtai, A. Kitsune: An ensemble of autoencoders for online network intrusion detection. *arXiv* **2018**, arXiv:1802.09089.
6. Zhang, Y.; Chen, X.; Jin, L.; Wang, X.; Guo, D. Network Intrusion Detection: Based on Deep Hierarchical Network and Original Flow Data. *IEEE Access* **2019**, *7*, 37004–37016. [CrossRef]
7. Xu, C.; Shen, J.; Du, X. A method of few-shot network intrusion detection based on meta-learning framework. *IEEE Trans. Inf. Forensics Secur.* **2020**, *15*, 3540–3552. [CrossRef]
8. Sommer, R.; Paxson, V. Outside the Closed World: On Using Machine Learning for Network Intrusion Detection. In Proceedings of the 2010 IEEE Symposium on Security and Privacy, Oakland, CA, USA, 16–19 May 2010; pp. 305–316. [CrossRef]
9. Quiring, E.; Pendlebury, F.; Warnecke, A.; Pierazzi, F.; Wressnegger, C.; Cavallaro, L.; Rieck, K. Dos and don'ts of machine learning in computer security. In Proceedings of the 31st USENIX Security Symposium (USENIX Security 22), USENIX Association, Boston, MA, USA, 10–12 August 2022.
10. Guarino, I.; Bovenzi, G.; Di Monda, D.; Aceto, G.; Ciuonzo, D.; Pescapé, A. On the use of machine learning approaches for the early classification in network intrusion detection. In Proceedings of the 2022 IEEE International Symposium on Measurements & Networking (M&N), Padua, Italy, 18–20 July 2022; IEEE: Piscataway, NJ, USA, 2022; pp. 1–6.
11. Sharon, Y.; Berend, D.; Liu, Y.; Shabtai, A.; Elovici, Y. Tantra: Timing-based adversarial network traffic reshaping attack. *IEEE Trans. Inf. Forensics Secur.* **2022**, *17*, 3225–3237. [CrossRef]
12. Layeghy, S.; Baktashmotlagh, M.; Portmann, M. DI-NIDS: Domain invariant network intrusion detection system. *Knowl.-Based Syst.* **2023**, *273*, 110626. [CrossRef]
13. Moustafa, N.; Slay, J. UNSW-NB15: A comprehensive data set for network intrusion detection systems (UNSW-NB15 network data set). In Proceedings of the 2015 Military Communications and Information Systems Conference (MilCIS), Canberra, Australia, 10–12 November 2015; IEEE: Piscataway, NJ, USA, 2015; pp. 1–6.
14. Sharafaldin, I.; Lashkari, A.H.; Ghorbani, A.A. Toward generating a new intrusion detection dataset and intrusion traffic characterization. *ICISSp* **2018**, *1*, 108–116.

15. Krizhevsky, A.; Sutskever, I.; Hinton, G.E. ImageNet classification with deep convolutional neural networks. In Proceedings of the Advances in Neural Information Processing Systems 25 (NIPS 2012), Lake Tahoe, NV, USA, 3–6 December 2012.
16. Hannun, A.; Case, C.; Casper, J.; Catanzaro, B.; Diamos, G.; Elsen, E.; Prenger, R.; Satheesh, S.; Sengupta, S.; Coates, A.; et al. Deep speech: Scaling up end-to-end speech recognition. *arXiv* **2014**, arXiv:1412.5567.
17. Shen, D.; Wu, G.; Suk, H.I. Deep learning in medical image analysis. *Annu. Rev. Biomed. Eng.* **2017**, *19*, 221–248. [CrossRef]
18. He, H.; Sun, X.; He, H.; Zhao, G.; He, L.; Ren, J. A novel multimodal-sequential approach based on multi-view features for network intrusion detection. *IEEE Access* **2019**, *7*, 183207–183221. [CrossRef]
19. Wang, W.; Sheng, Y.; Wang, J.; Zeng, X.; Ye, X.; Huang, Y.; Zhu, M. HAST-IDS: Learning hierarchical spatial-temporal features using deep neural networks to improve intrusion detection. *IEEE Access* **2017**, *6*, 1792–1806. [CrossRef]
20. Doriguzzi-Corin, R.; Millar, S.; Scott-Hayward, S.; Martinez-del Rincon, J.; Siracusa, D. LUCID: A practical, lightweight deep learning solution for DDoS attack detection. *IEEE Trans. Netw. Serv. Manag.* **2020**, *17*, 876–889. [CrossRef]
21. Lo, W.W.; Layeghy, S.; Sarhan, M.; Gallagher, M.; Portmann, M. E-graphsage: A graph neural network based intrusion detection system for IoT. In Proceedings of the NOMS 2022-2022 IEEE/IFIP Network Operations and Management Symposium, Budapest, Hungary, 25–29 April 2022; IEEE: Piscataway, NJ, USA, 2022; pp. 1–9.
22. Hu, X.; Gao, W.; Cheng, G.; Li, R.; Zhou, Y.; Wu, H. Towards Early and Accurate Network Intrusion Detection Using Graph Embedding. *IEEE Trans. Inf. Forensics Secur.* **2023**, *18*, 5817–5831. [CrossRef]
23. Borisov, V.; Leemann, T.; Sessler, K.; Haug, J.; Pawelczyk, M.; Kasneci, G. Deep Neural Networks and Tabular Data: A Survey. *IEEE Trans. Neural Networks Learn. Syst.* **2022**, early access.
24. Kadra, A.; Lindauer, M.; Hutter, F.; Grabocka, J. Regularization is all you need: Simple neural nets can excel on tabular data. *arXiv* **2021**, arXiv:2106.11189.
25. Lu, J.; Liu, A.; Dong, F.; Gu, F.; Gama, J.; Zhang, G. Learning under concept drift: A review. *IEEE Trans. Knowl. Data Eng.* **2018**, *31*, 2346–2363. [CrossRef]
26. Schwengber, B.H.; Vergütz, A.; Prates, N.G.; Nogueira, M. A method aware of concept drift for online botnet detection. In Proceedings of the GLOBECOM 2020-2020 IEEE Global Communications Conference, Taipei, Taiwan, 7–11 December 2020; IEEE: Piscataway, NJ, USA, 2020; pp. 1–6.
27. Apruzzese, G.; Laskov, P.; de Oca, E.M.; Mallouli, W.; Rapa, L.B.; Grammatopoulos, A.V.; Franco, F.D. The Role of Machine Learning in Cybersecurity. *Digit. Threat. Res. Pract.* **2023**, *4*, 1–38. [CrossRef]
28. Apruzzese, G.; Pajola, L.; Conti, M. The cross-evaluation of machine learning-based network intrusion detection systems. *IEEE Trans. Netw. Serv. Manag.* **2022**, *19*, 5152–5169. [CrossRef]
29. Heine, F.; Laue, T.; Kleiner, C. On the evaluation and deployment of machine learning approaches for intrusion detection. In Proceedings of the 2020 IEEE International Conference on Big Data (Big Data), Atlanta, GA, USA, 10–13 December 2020; IEEE: Piscataway, NJ, USA, 2020; pp. 4594–4603.
30. Pivarníková, M.; Sokol, P.; Bajtoš, T. Early-stage detection of cyber attacks. *Information* **2020**, *11*, 560. [CrossRef]
31. Ahmad, T.; Truscan, D.; Vain, J.; Porres, I. Early Detection of Network Attacks Using Deep Learning. In Proceedings of the IEEE International Conference on Software Testing, Verification and Validation Workshops (ICSTW), Valencia, Spain, 4–13 April 2022; pp. 30–39. [CrossRef]
32. Engelen, G.; Rimmer, V.; Joosen, W. Troubleshooting an Intrusion Detection Dataset: The CICIDS2017 Case Study. In Proceedings of the 2021 IEEE Security and Privacy Workshops (SPW), San Francisco, CA, USA, 27 May 2021; pp. 7–12. [CrossRef]
33. Liu, L.; Engelen, G.; Lynar, T.; Essam, D.; Joosen, W. Error Prevalence in NIDS datasets: A Case Study on CIC-IDS-2017 and CSE-CIC-IDS-2018. In Proceedings of the 2022 IEEE Conference on Communications and Network Security (CNS), Austin, TX, USA, 3–5 October 2022; pp. 254–262. [CrossRef]
34. Yulianto, A.; Sukarno, P.; Suwastika, N.A. Improving adaboost-based intrusion detection system (IDS) performance on CIC IDS 2017 dataset. *J. Phys. Conf. Ser.* **2019**, *1192*, 012018. [CrossRef]
35. Binbusayyis, A.; Vaiyapuri, T. Identifying and benchmarking key features for cyber intrusion detection: An ensemble approach. *IEEE Access* **2019**, *7*, 106495–106513. [CrossRef]
36. Adhao, R.; Pachghare, V. Feature selection based on hall of fame strategy of genetic algorithm for flow-based ids. In *Data Science and Security: Proceedings of IDSCS 2021*; Springer: Berlin/Heidelberg, Germany, 2021; pp. 310–316.
37. Jin, M.; Koh, H.Y.; Wen, Q.; Zambon, D.; Alippi, C.; Webb, G.I.; King, I.; Pan, S. A survey on graph neural networks for time series: Forecasting, classification, imputation, and anomaly detection. *arXiv* **2023**, arXiv:2307.03759.
38. Gorishniy, Y.; Rubachev, I.; Khrulkov, V.; Babenko, A. Revisiting deep learning models for tabular data. *Adv. Neural Inf. Process. Syst.* **2021**, *34*, 18932–18943.
39. Zhou, H.; Zhang, S.; Peng, J.; Zhang, S.; Li, J.; Xiong, H.; Zhang, W. Informer: Beyond efficient transformer for long sequence time-series forecasting. In Proceedings of the AAAI Conference on Artificial Intelligence, Vancouver, BC, Canada, 2–9 February 2021; Volume 35, pp. 11106–11115.
40. Dosovitskiy, A.; Beyer, L.; Kolesnikov, A.; Weissenborn, D.; Zhai, X.; Unterthiner, T.; Dehghani, M.; Minderer, M.; Heigold, G.; Gelly, S.; et al. An image is worth 16x16 words: Transformers for image recognition at scale. *arXiv* **2020**, arXiv:2010.11929.
41. Vormayr, G.; Fabini, J.; Zseby, T. Why are my flows different? a tutorial on flow exporters. *IEEE Commun. Surv. Tutorials* **2020**, *22*, 2064–2103. [CrossRef]

42. Hendrycks, D.; Gimpel, K. A baseline for detecting misclassified and out-of-distribution examples in neural networks. *arXiv* **2016**, arXiv:1610.02136.
43. Shah, H.; Tamuly, K.; Raghunathan, A.; Jain, P.; Netrapalli, P. The pitfalls of simplicity bias in neural networks. *Adv. Neural Inf. Process. Syst.* **2020**, *33*, 9573–9585.
44. Shi, Y.; Daunhawer, I.; Vogt, J.E.; Torr, P.; Sanyal, A. How robust are pre-trained models to distribution shift? In Proceedings of the ICML 2022: Workshop on Spurious Correlations, Invariance and Stability, Baltimore, MD, USA, 22 July 2022.

Disclaimer/Publisher's Note: The statements, opinions and data contained in all publications are solely those of the individual author(s) and contributor(s) and not of MDPI and/or the editor(s). MDPI and/or the editor(s) disclaim responsibility for any injury to people or property resulting from any ideas, methods, instructions or products referred to in the content.

Review

Detection of DoS Attacks for IoT in Information-Centric Networks Using Machine Learning: Opportunities, Challenges, and Future Research Directions

Rawan Bukhowah *, Ahmed Aljughaiman and M. M. Hafizur Rahman

Department of Computer Networks and Communications, College of Computer Sciences and Information Technology, King Faisal University, Al-Ahsa 31982, Saudi Arabia; aaaljughaiman@kfu.edu.sa (A.A.); mhrahman@kfu.edu.sa (M.M.H.R.)
* Correspondence: 222402836@student.kfu.edu.sa

Citation: Bukhowah, R.;
Aljughaiman, A.; Rahman, M.M.H.
Detection of DoS Attacks for IoT in
Information-Centric Networks Using
Machine Learning: Opportunities,
Challenges, and Future Research
Directions. Electronics 2024, 13, 1031.
https://doi.org/10.3390/
electronics13061031

Academic Editors: Abdussalam
Elhanashi and Pierpaolo Dini

Received: 29 January 2024
Revised: 28 February 2024
Accepted: 28 February 2024
Published: 9 March 2024

Copyright: © 2024 by the authors. Licensee MDPI, Basel, Switzerland. This article is an open access article distributed under the terms and conditions of the Creative Commons Attribution (CC BY) license (https:// creativecommons.org/licenses/by/ 4.0/).

Abstract: The Internet of Things (IoT) is a rapidly growing network that shares information over the Internet via interconnected devices. In addition, this network has led to new security challenges in recent years. One of the biggest challenges is the impact of denial-of-service (DoS) attacks on the IoT. The Information-Centric Network (ICN) infrastructure is a critical component of the IoT. The ICN has gained recognition as a promising networking solution for the IoT by supporting IoT devices to be able to communicate and exchange data with each other over the Internet. Moreover, the ICN provides easy access and straightforward security to IoT content. However, the integration of IoT devices into the ICN introduces new security challenges, particularly in the form of DoS attacks. These attacks aim to disrupt or disable the normal operation of the ICN, potentially leading to severe consequences for IoT applications. Machine learning (ML) is a powerful technology. This paper proposes a new approach for developing a robust and efficient solution for detecting DoS attacks in ICN-IoT networks using ML technology. ML is a subset of artificial intelligence (AI) that focuses on the development of algorithms. While several ML algorithms have been explored in the literature, including neural networks, decision trees (DTs), clustering algorithms, XGBoost, J48, multilayer perceptron (MLP) with backpropagation (BP), deep neural networks (DNNs), MLP-BP, RBF-PSO, RBF-JAYA, and RBF-TLBO, researchers compare these detection approaches using classification metrics such as accuracy. This classification metric indicates that SVM, RF, and KNN demonstrate superior performance compared to other alternatives. The proposed approach was carried out on the NDN architecture because, based on our findings, it is the most used one and has a high percentage of various types of cyberattacks. The proposed approach can be evaluated using an ndnSIM simulation and a synthetic dataset for detecting DoS attacks in ICN-IoT networks using ML algorithms.

Keywords: Internet of Things; denial-of-service attack; Information-Centric Network; machine learning

1. Introduction

With the rapid development of new technologies, the Internet environment is changing accordingly. The Internet of Things (IoT) is one of those technologies. It creates new challenges including content, service, and device-naming challenges that aim to connect a massive number of heterogeneous devices. Consequently, there is a need for an infrastructure in which the content is the main element [1]. The IoT is an ecosystem, and most of its applications follow content-oriented usage patterns. The IoT is a vast network of interconnected devices that collect and exchange data. Most IoT applications follow content-oriented usage patterns, meaning that users are primarily interested in obtaining specific data items, rather than communicating with specific devices. The Information-Centric Network (ICN) is a new networking paradigm that can improve the performance and security of IoT applications. The ICN replaces the traditional host-centric networking approach with a content-centric approach, meaning that users request specific data items, rather than communicating with specific devices. The ICN offers a number of benefits for IoT applications, including content-based security, reduced network traffic, improved performance, retrieval of the content fast, multi-casting, and in-network caching [2,3].

IoT technology plays a key role in our daily lives as the IoT has opened new doors for better communication and better ways to deal with the massive data obtained through these heterogeneous devices [1]. Based on the application domain, IoT sensors can be utilized for monitoring medical services, smart homes, smart environments, smart cities, and smart enterprises, as shown in Figure 1. Consequently, these applications have led to an increase in interconnected devices, including smart and sensing devices. With technology development, IoT applications surround us from every side. For example, in healthcare, there is a huge interest in using IoT as it reflects good services for patients and hospitals in general. Wireless sensor networks (WSNs) and wireless body area network (WBANs) are essential components used in healthcare environments [4]. The implementation of the IoT has led to many security and privacy challenges [5]. Therefore, nowadays the IoT is one of the main focuses of research across the world.

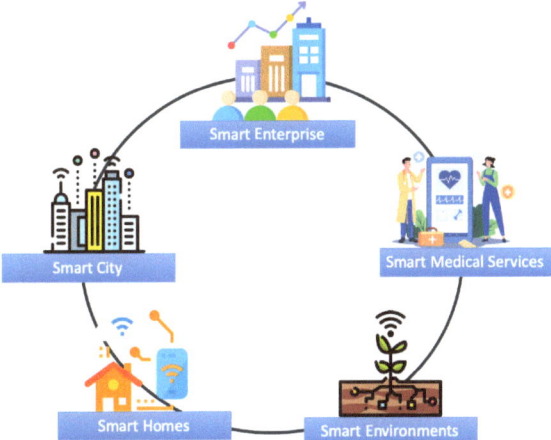

Figure 1. IoT applications.

The goal is to make IoT resources accessible anytime, anywhere, and by anything in a secure manner. The old Transmission Control Protocol-/Internet Protocol-based networks (TCP/IP) appear to have serious problems in the IoT field, so to overcome these problems we need a network with a big capacity to address these issues. In light of this, the ICN architecture is an alternative solution to the oldest TCP/IP networks. The ICN secures content itself instead of the communication channel, so it replaces the host-centric model with the content-centric model [6]. In the traditional TCP/IP protocol, it is not possible to connect a massive number of devices to the system. In contrast, the ICN can accommodate a vast number of devices. Therefore, network routing, stability, scalability, and mobility management have been enhanced for various IoT applications through the ICN [3].

Moreover, ICN architectures consist of Named-Data Networking (NDN), Content-Centric Networking (CCN), the Publish–Subscribe Internet Routing Paradigm (PSIRP), Data-Oriented Network Architecture (DONA), and Network of Information (NetInf) [7]. The ICN enhances the user experience with security, privacy, and access controls. However, it also introduces new challenges. Therefore, the challenge of this paper is to define and present all possible issues, attacks, and cybersecurity problems that take place in the ICN architectures of the IoT. Another challenge is to give potential solutions, using techniques to defend against defined cybersecurity problems.

There are several techniques to detect and mitigate DoS attacks. Some of the mitigation techniques have been suggested by [8], including artificial intelligence technology, probabilistic and statistical methods, and miscellaneous approaches. These methods are significant methods to detect and mitigate DoS attacks such as the interest flooding attack (IFA). Other methods are recommended by [9], including the statistical approach, heuristic approach, machine learning (ML), and deep learning (DL). Consequently, there are various anomaly-based attack detection techniques to use. Table 1 shows a comparison between ML, DL, and statistical approaches based on their features and limitations.

Table 1. Comparison between mitigation techniques.

Techniques	Features	Limitations
Statistical approach	• These statistical rests are used to prove that the observed pattern and expected pattern is different based on historical data • Real-time detection • Low false positive rate	• Misclassification • Very complex • Accuracy depends on the mathematical model
ML	• Can learn complex traffic patterns quickly • Can automatically extract relevant features from network traffic data • High detection accuracy • Real-time detection	• Long training time • Need large dataset for better results
DL	• Can extract features of supervised and unsupervised learning • Can directly process raw data	• Data dependency • May leads to overfitting • Very complex • Computational overhead • Need large dataset for better results

Detecting cyberattacks is challenging, especially in ICN networks, as it is a new concept, and based on recent research there are various types of attacks, including DoS attacks, that successfully breach the network. Therefore, it is critical to detect these attacks and develop better methods to minimize the likelihood of a system being compromised. According to the papers that we have reviewed, ML technology is considered an efficient way to address ICN-IoT challenges. However, none of these papers applied ML to detecting DoS attacks in the ICN-IoT network. Therefore, our proposed paper combines these technologies to enhance security.

1.1. Problem Statement

In the contemporary IP-based network infrastructure, DDoS attacks represent a pervasive threat, inflicting substantial harm upon the network's functionality and integrity. However, the ICN exhibits promising capabilities in thwarting numerous prevalent DoS attack vectors. Common DoS attacks afflicting IP-based networks include reflection attacks, bandwidth depletion, and black-holing via prefix hijacking [10]. Initially, there was a prevailing belief that the IoT realm could withstand DDoS attacks due to the inherent design of NDN. This belief was challenged upon the discovery of specific DoS attack vectors, such as IFAs and state-based attacks. Additionally, the proliferation of content requests and user-generated names has led to the overloading of forwarding information base (FIB) and pending interest table (PIT) structures [11]. Consequently, the integration of the IoT with the ICN has emerged as a focal point for researchers, necessitating a deeper exploration of the associated challenges and vulnerabilities.

DoS represents a significant security concern within the ICN-IoT paradigm, where attackers exploit vulnerabilities to deplete network resources by flooding malicious content over the infrastructure. For example, IFAs involve a large volume of diverse or identical interest packets, often containing non-existent data. The aim of those packets is to overwhelm the resources of the data producer or the entire network infrastructure [10], thus causing significant disruption and damage within the ICN-IoT environment.

To mitigate the impact of such attacks, various methods have been proposed. The token bucket with per-interface-fairness approach, for instance, restricts the number of outgoing interest packets by limiting tokens associated with the outgoing interface. Similarly, the satisfaction-based interest acceptance method allocates tokens to incoming interfaces based on their interest–satisfaction ratio. Another strategy involves disabling PIT exhaustion. However, these mitigation methods exhibit certain limitations and may not be sufficiently effective in practice [10]. Consequently, there is a pressing need for more robust and efficient solutions to bolster the resilience of IoT-ICN integration against DoS attacks. The proposed detection system, leveraging ML technology, aims to provide a more effective solution. By harnessing ML algorithms and advanced analytics, our proposed approach endeavors to enhance the security posture of ICN-integrated IoT environments, offering proactive threat detection and mitigation capabilities.

1.2. Why ICN for IoT?

The IoT deals with numerous and diverse networked devices and complex traffic patterns that issue unique challenges. Applying ICN principles in this context creates new opportunities [12]. Recently, the ICN has gained recognition as a promising networking solution for the IoT. In the context of the IoT, the data generated by smart devices are treated as content. This means that network users can request IoT data without needing to know their specific sources, like sensors or actuators. The ICN approach focuses on content names rather than network addresses, eliminating the need for end-to-end session requirements.

The ICN offers several advantages to the IoT, including improved data retrieval, reduced latency, an efficient receiver-driven design, request aggregation, and support for mobility and multicast at the network layer. One additional feature is in-network caching, which enhances data availability by caching the requested contents near users. This technique reduces traffic near the IoT data publisher and leads to lower energy requirements [2]. Consequently, the ICN is often considered more suitable for IoT applications than traditional IP-based networking. Moreover, the ICN provides robust content-based security and encryption, ensuring data integrity and authenticity [13]. The ICN also addresses issues in the IoT by using content-based naming and a name-resolution system (NRS). In addition, within the application layer, there are various trust models implemented to serve multiple purposes, such as content encryption and security [13].

The remainder of this paper is organized as follows. In Section 2, we discuss IoT three-layer architectures, ICN architectures, related attacks in the ICN and the IoT, and the importance of integrating the ICN with the IoT. Section 3 discusses machine learning techniques in the ICN-IoT environments. Section 4 conducts a comprehensive analysis of related studies. Section 5 presents results and discussions based on the ICN and the IoT. Section 6 includes the recommendation and future research directions. Section 7 concludes the paper.

2. Background

2.1. Internet of Things (IoT)

The IoT refers to the "Internet of things". The IoT is a technological paradigm that revolves around the seamless integration of computational devices and physical objects. These objects are equipped with unique identifiers, sensors, and the capability to collect, process, and share data. Regarded as one of the most groundbreaking technological advancements in recent years, the IoT has far-reaching implications in various domains. IoT systems encompass a wide range of applications. They find their place in automated vehicles, where they enhance safety and efficiency, in the energy sector, optimizing the supply transmission, distribution and consumption of energy and in surveillance systems, and exemplified by the deployment of drones for enhanced monitoring and security. However, the influence of the IoT extends far beyond these sectors. It has infiltrated academic institutions, industries, factories, agricultural settings, healthcare organizations, and more [14]. Nowadays, IoT security is a big concern and the security vulnerabilities of IoT are unique because they are complex and heterogeneous in technology and data.

2.1.1. IoT Applications

The IoT comprises a wide range of applications that have increased very rapidly. These applications require additional security support from various new technologies. Below, some of the security applications in the IoT are discussed:

- Smart medical services: This includes monitoring respiration, drug delivery systems, body position, quality of sleep, and early detection of illness [4].
- Smart homes: For example, a smart camera captures an image around the door and then sends it to the owner when someone acts in unusual behavior near the door [4].
- Smart environments: This includes anything that is connected to and affected by humans, animals, and plants. These include air quality and water quality monitoring, natural disaster monitoring, and smart farming [4]. Moreover, smart environments include monitoring the level of snow in high-altitude regions, pollution monitoring, early detection of earthquakes, fire detection, and landslides prevention [15].
- Smart cities: This involves emerging computation and communication resources. Although, the goal is to increase the quality of life of people, it comes with a lot of challenges. For example, Air Tag from Apple could be used to enable parents to track their children; this feature will become risky if it is hacked because it breaches the privacy of citizens [15].
- Smart enterprises: These organizations can maintain real-time shipment tracking in transportation and logistics areas and smart farming under energy and production and resource management [4].

2.1.2. IoT Layers

Various architectures have been proposed by researchers, the most fundamental architecture developed is the three-layer model (Figure 2). Every layer has many security issues and attacks that can target each layer.

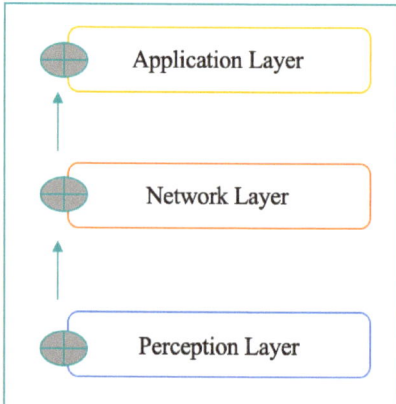

Figure 2. The three layers of the IoT.

Denial-of service attacks in the IoT layer: The IoT is a rapidly growing network of physical devices that are connected to the Internet and can collect and exchange data. While the IoT has the potential to revolutionize many industries and aspects of our lives, it also introduces new security challenges. One of the most significant challenges is DoS attacks. DoS attacks aim to overwhelm the target system with too much traffic, rendering it unavailable to legitimate users. These attacks can be particularly effective against IoT devices due to their often poor security and limited resources. IoT systems are structured into three layers: the perception layer, the network layer, and the application layer. According to a recent report by Cloudflare, in 2023, the number of DoS attacks targeting the network layer increased by 85% compared to 2022 [16]. The most targeted industries were the telecommunications industry, gaming and gambling companies, and then the information technology and services industry. Consequently, DoS attacks against the IoT are a serious concern and may have a significant impact on both businesses and individuals, potentially resulting in financial losses and reputational damage [17]. DoS attacks target IoT layers, including:

- Perception layer DoS attacks: In this layer, DoS attacks target sensors and other devices, aiming to send invalid data or overwhelm traffic within IoT devices [18].
- Network layer DoS attacks: In this layer, the routers, switches, and other networking devices are targeted by DoS attacks in order to disrupt the communication between the perception layer and the application layer.
- Application layer DoS attacks: In this layer, applications are targeted. DoS attacks aim to overwhelm the applications with traffic or to send them invalid data [18].

2.2. Information-Centric Network (ICN)

The ICN has been proposed as a future internet architecture. Unlike IP-based networks, which rely on host addresses for communication, the ICN adopts a content-centric communication model. In this model, the focus is on the content itself, instead of host addresses. The ICN's objective is to separate content from its hosting locations, allowing ICN forwarders to cache content within the network and fulfill subsequent client requests without the need to forward them to the original producer [7].

In the ICN, content discovery and delivery operate in a receiver-driven manner, as shown in Figure 3. It all begins when a consumer initiates a content request, referred to as an interest packet, which contains the name of the desired content. This interest packet is routed hop-by-hop, based on the content name, until it reaches the content producer. The producer can be either the original data source or an intermediate node that stores the requested content. Moreover, the data packet is sent back as a response along the reverse path of the interest to fulfill the request. Unlike host-based networks, on which securing the communication channel between hosts is paramount, the ICN employs a content-centric security model. This model directly secures the content itself instead of the communication channel [6].

Implementing the ICN in wireless environments can cause various security concerns caused by the inherent characteristics of wireless communication and the content-centric communication model of the ICN. These issues include content name attacks, which can impact user privacy, the management of access control rules tied to content names, the establishment of trust in wireless settings, and the authentication of data originating from content stores [6].

2.2.1. ICN Architectures

As shown in Figure 4, which illustrates the timeline of ICN projects, throughout the years, different countries have supported different ICN projects. The United States supported the following projects: Content-Centric Networking (CCN), Named-Data Networking (NDN), MobilityFirst, and ICE-AR. In contrast, the European Union funded different ICN projects, such as the Publish–Subscribe Internet Technology (PURSUIT), Scalable and Adaptive Internet Solutions (SAIL), Content Mediator Architecture for Content-Aware Networks (COMET), CONVERGENCE, and ANR Connect. Multiple countries collaborated to fund the following ICN projects: GreenICN and ICN-2020. The architectures of some of the above projects are discussed below.

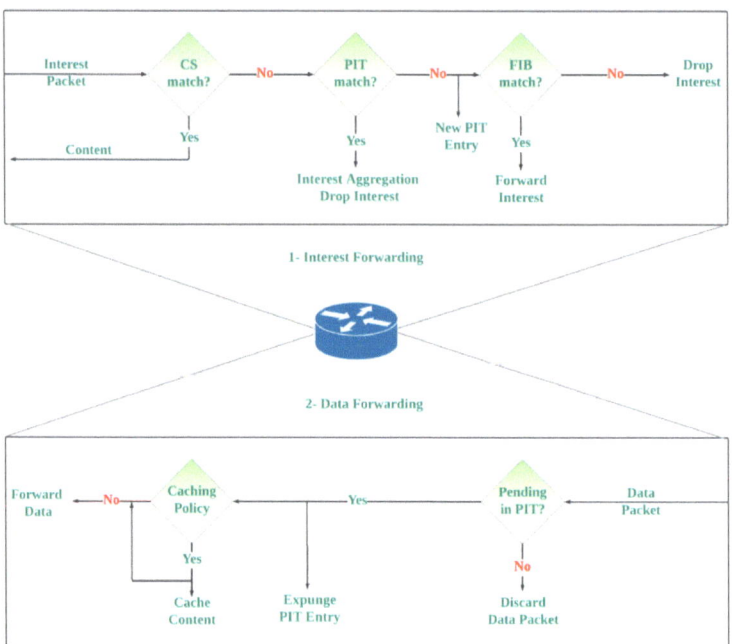

Figure 3. ICN process.

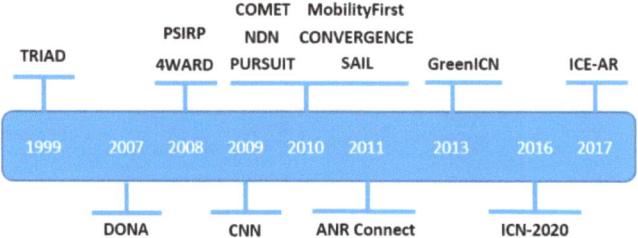

Figure 4. Timeline of ICN projects [6].

Data-Oriented Network Architecture (DONA): DONA was marked as the first project in the development of the ICN, as it became the first ICN architecture to reach completion. DONA was an evolution of TRIAD's foundational design principles, particularly transforming from hierarchical URLs to a flat naming scheme, which makes DONA no longer reliant on a single host to serve content [19]. In the DONA scheme, content names are comprised of a cryptographic hash of the publisher's key and an object ID assigned by the publisher. This naming scheme ensures uniqueness within the publisher's domain [18].

Publish–Subscribe Internet Technology (PURSUIT): PURSUIT is an initiative within the European Union. This project follows the Publish–Subscribe Internet Routing architecture like its predecessor. This architecture features a comprehensive publish–subscribe protocol stack, instead of the conventional IP protocol stack. In the context of PURSUIT, its core functions encompass three key aspects: rendezvous, topology management, and forwarding. In 2008, PSIRP was introduced, and around the same time, another promising project known as Architecture and Design for the Future Internet (4WARD) also emerged [18].

The Network of Information (NetInf): This network was started in the 4WARD architecture. NetInf uses a publish–subscribe scheme as well as flat naming, which maps names to locators. Content is retrieved by having the publisher first publish its data objects to the network, and NRS stores the name and locator mapping. Data can be stored in cache in many places, and hence, can be available to many locators. When a subscriber requests the content, the routing forwarders deliver the data [18].

Content-Centric Networking (CCN): CCN and NDN exhibit similarities in their architectural design, encompassing elements such as a hierarchical naming scheme, content caching, and named content routing.

Named-Data Networking (NDN): CCN was introduced in 2009 by researchers at the Palo Alto Research Center. In CCN, known as a data-named communication architecture, the packet is addressed with content names instead of locations. This procedure was also applied to other objects, with NDN being the most famous one. They are similar, a client requests content by sending an interest packet into the network with the content's name. In the NDN network, the subscribers send the interest packet to request data objects. This interest packet goes to the content router (CR) then it will be forwarded hop-by-hop across CRs. A CR contains three data structures: content store (CS), FIB, and PIT. The CR stores the content that travels around the interface. FIB storage also stores pairs of names and then forwards them to the target address. Consistently, interest packets are forwarded to the destination based on the requested content. Moreover, the PIT, as well as FIB, stores pairs of names in the interface that request the content. It will then be used to propagate data objects to the subscriber. NDN is the chosen architecture for the project's implementation within the field of the ICN. Content storage has consistently created challenges, including managing large-scale network caches, determining what content to discard when new content arrives, and efficiently storing content in the cache [18]. Figure 5 illustrates the components of NDN's interest packet and data packet.

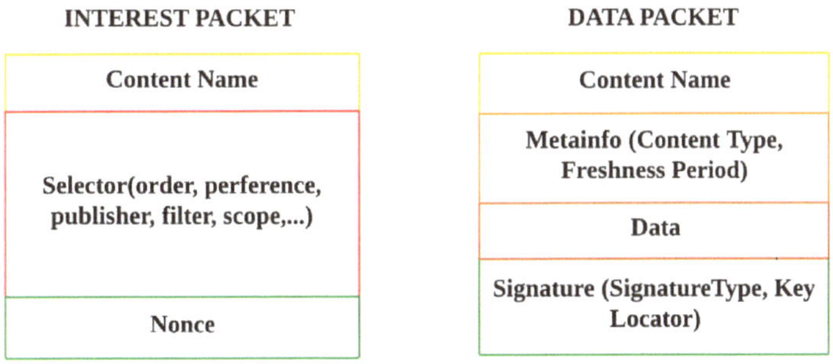

INTEREST PACKET	DATA PACKET
Content Name	Content Name
Selector(order, perference, publisher, filter, scope,...)	Metainfo (Content Type, Freshness Period)
	Data
Nonce	Signature (SignatureType, Key Locator)

Figure 5. NDN's interest packets and data packets.

2.2.2. Taxonomy of DoS Attacks in the ICN

The ICN poses numerous security challenges that require attention. It introduces new types of attacks that did not occur previously. Moreover, ICN environments may also be susceptible to attacks commonly observed in other networking environments. The ICN is considered a new architecture in terms of naming, routing, and caching. DoS attacks aim to disrupt or disable the normal operation of a network or service. DoS attacks can be launched against a variety of targets, including servers, websites, and routers. In ICNs, DoS attacks can be particularly disruptive due to the reliance of the ICN on secure and reliable communication between network entities. Figure 6 illustrates the list of attacks that can target the ICN.

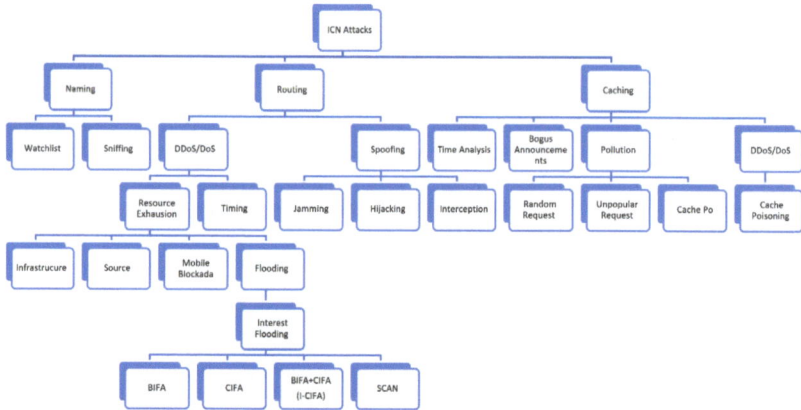

Figure 6. Taxonomy of ICN attacks in the ICN.

Naming-Related Attacks

In the naming domain, the biggest concern is that the ICN will face a threat with respect to privacy, such as a sniffing attack, in which attackers try to monitor Internet usage. In naming-related attacks, the watchlist attack occurs when attackers attempt to prevent the distribution of particular content. This is achieved by blocking the delivery of the content and potentially identifying who is requesting it [20].

Routing-Related Attacks

Attacks that belong to routing include DoS attacks, spoofing, timing, and jamming attacks. The attackers' goal is to cause unwanted traffic flows and DoS. Other attacks, including infrastructure and flooding attacks, try to exhaust the resources such as memory and processing power, that are used to support, maintain, and exchange content states.

Flooding is a special type of attack that only happens in the ICN environment. There are four types of IFAs: basic IFA (BIFA), collusive IFA (CIFA), smart collaborative attack in NDN (SCAN) and BIFA+CIFA [20]. These types of attacks aim to disrupt the network functionality of the ICN by flooding it with a large number of illegitimate interest messages to overwhelm the network's infrastructure. These attacks may cause network congestion, resource depletion, DoS, and increased energy consumption in IoT devices [21].

Caching-Related Attacks

Caching plays a crucial role in the ICN by improving infrastructure performance through receiver-driven caching. The goal of this approach is to offer users the nearest available copy. Consequently, the ICN is vulnerable to any actions that pollute or corrupt the caching system. Notably, cache pollution attacks and interest poisoning attacks pose substantial threats to the effectiveness of ICN caching [20].

3. Machine Learning Techniques in the ICN-IoT

The ICN is a new networking paradigm that only focuses on the information instead of the location of the information, so it is well-suited to the IoT. However, ICN-IoT networks are vulnerable to various attacks, including DoS, malware, and data breaches. ML is an optimal solution for detecting anomalies in ICN-IoT networks in order to mitigate these attacks. ML algorithms can be trained on historical data to learn normal network behavior. Then, they can monitor network traffic against anomalies. ML-based anomaly detection systems are very helpful in the detection of a variety of attacks, such as DoS attacks, malware attacks, and data breaches [22]. There are various ML techniques; each of them has its advantages and limitations, so we should choose between them based on the specific requirements of the ICN-IoT network. The most common techniques include the following:

- Unsupervised learning: In unsupervised learning algorithms, labeled data for training is not required. Instead, this type focuses on analyzing unlabeled data to identify patterns and anomalies. This approach is particularly useful in ICN-IoT networks.
- Supervised learning: In supervised learning algorithms, labeled data for training is required. Therefore, categorizing labeled data as either normal or abnormal, and identifying real-time network traffic as either normal or abnormal. This approach can be beneficial in detecting anomalies in ICN-IoT networks.
- Semi-supervised learning: In semi-supervised learning techniques, ML algorithms can be trained using a small amount of labeled data along with a large amount of unlabeled data. Since labeled data may be scarce, this approach is useful in ICN-IoT networks [22].

3.1. Supervised Learning Algorithms

There are two fundamental types of machine learning algorithms: supervised and unsupervised. Supervised algorithms employ pre-labeled (classified) objects to predict the class of objects. In contrast, unsupervised algorithms aim to discover the natural grouping or patterns within unlabeled, new, and unseen objects. Supervised learning algorithms can be adapted to the unique characteristics of NDN-IoT environments. By tailoring the features and labels to capture relevant aspects of the network traffic, these algorithms can be customized for specific IoT scenarios. There are three types of classifier algorithms.

3.1.1. K-Nearest Neighbor (KNN)

KNN is valuable for classification as it does not assume underlying data distributions, making it a non-parametric and lazy learning algorithm. Non-parametric statistics operate under the assumption that there is no predefined data distribution [23]. In addition, the KNN algorithm takes as input the k-nearest training samples in the feature space. The accuracy of this algorithm can be significantly enhanced by normalizing the training data [24]. Furthermore, in KNN, the critical factor is the number of nearest neighbors [23].

3.1.2. Support Vector Machine (SVM)

The SVM framework serves as an intersection between machine learning and embedded systems. SVM functions as both a linear and nonlinear classifier, acting as a mathematical function capable of distinguishing between two types of objects, categorized as classes [23]. The reason for utilizing the SVM algorithm is its efficient capability to differentiate between the characteristics of traffic flow in normal and abnormal scenarios [24].

3.1.3. Random Forest (RF)

RF is a supervised classification algorithm that constructs a random forest. The accuracy of the results improves with an increase in the number of trees in the forest [23]. In addition, during the training phase, RF builds numerous decision trees and provides an output based on the mode of classes (for classification) or the mean prediction (for regression) of the individual trees [24].

Several ML algorithms are used for attack detection in ICN-IoT. Related studies indicate that KNN, SVM, and RF yield optimal results and are popularly used. Kumar et al. [25] evaluated various shallow ML algorithms including SVM and KNN, showing their effectiveness in attack detection. In [8], the authors highlighted SVM as the primary method, using Jensen–Shannon (JS) divergence, for detecting IFA attacks. Consistently, the parameters utilized include the number of incoming and outgoing data packets and interest packets, alongside the size of the PIT entries. Furthermore, classification metrics such as accuracy = $\frac{TP+TN}{TP+TN+FP+FN}$, precision = $\frac{TP}{TP+FP}$, recall (sensitivity) = $\frac{TP}{TP+FN}$, and F-measure are commonly employed for evaluation purposes [8,25].

ML is a subset of AI that focuses on the development of algorithms. In the context of IFA over the NDN platform, three popular ML algorithms, which are SVM, RF, and KNN, have been highlighted in the papers that either reviewed or tested IFA. The literature has explored several ML algorithms, including neural networks, decision trees (DTs), clustering algorithms, XGBoost, J48, multilayer perceptron (MLP) with backpropagation (BP), deep neural networks (DNNs), MLP-BP, RBF-PSO, RBF-JAYA, and RBF-TLBO. Researchers compare these detection approaches using classification metrics such as accuracy.

The reviewed studies indicate that SVM, RF, and KNN demonstrated superior performance in IFA detection. In the KNN method, attacks were localized with 98.35% accuracy. As for the SVM, it achieved a detection accuracy of 99.4%. Moreover, RF had the best performance in terms of accuracy, precision, and recall. Overall, the results highlight the effectiveness of the KNN, SVM, and RF models compared to other alternatives [25]. Consistently, the reviewed studies considered several parameters, including the number of incoming and outgoing data packets and interest packets, alongside the size of the PIT entries [8].

In addition, to deploy a proposed project using ML several key stages are essential, as shown in Figure 7. These start with the processing phase, which includes ndnSIM to train the system and dataset preprocessing to clean and transform data. Also, there is a need to check and remove any unnecessary or null values because ML algorithms have a difficult time handling them, which can lead to incorrect results. The next stage is feature selection, which also includes feature preprocessing. In this stage, the initial dataset should be built. After the features are selected, the dataset is divided into two subsets: the training subset and the testing subset. By selecting the right testing and training data, classification accuracy can be improved [26]. The training data are the set of instances trained on the model, while the test data are used to determine the model's ability or execution. Classification determines whether the information belongs to a normal class or a DoS attack. To achieve the best classification approach, a variety of algorithms, such as KNN, SVM, and RF, should be compared.

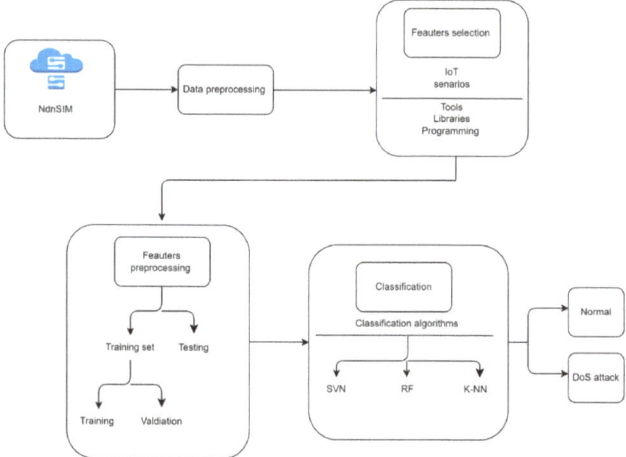

Figure 7. System architecture.

4. Related Work

In this section, we review several research papers related to security threats associated with IoT and ICN technologies. The research addresses its aims by undertaking a systematic review of the literature, following the PRISMA 2020 protocol. The procedures involve collecting, reading, selecting, and analyzing various papers' results using Google Scholar and the Saudi Digital Library databases.

4.1. Threats on the IoT

Samundra et al. [27] presented security and privacy challenges related to the IoT in general and then described each layer of the IoT architecture, with its related security issues. These IoT layers are the perception layer, network layer, middleware layer, and application layer. The paper suggested several solutions to address the challenges in the IoT.

Azzam et al. [28] reviewed threats and attacks that are related to IoT and classified them according to the three layers of the IoT: perception or the physical, network, and application. In addition, the paper explained mitigation techniques that could be applied to IoT threats and attacks. Some of these countermeasure techniques are encryption, access control, AI, and blockchain technology.

Waseem et al. [29] presented the threats to the IoT, as well as the challenges, risks, and security requirements for the IoT in general and each of its layers. They also highlighted most of the well-known and generic attacks on all types of IoT layers, such as DoS attacks in the network layer, SQL injection in the application layer, and battery drainage attacks in the physical layer. In addition, the paper reviewed and analyzed SDN as a network that can deploy an IoT architecture. The paper suggested solving security issues through ML.

Pintu et al. [5] reviewed the challenges, risks, and issues of the IoT from different perspectives. The authors also discussed the four layers of the IoT that have been described in detail, along with the main attacks against each of them. Moreover, the paper presented relevant countermeasures and solutions. The proposed solutions include PUF-based, blockchain, ML, and others. Finally, the paper discussed the pros and cons of these security solutions to facilitate the process of presenting future recommendations.

Thalawattha et al. [30] reviewed and analyzed security issues and requirements in the IoT to defend against IoT attacks and identified current vulnerabilities. Moreover, the paper specified attacks for each layer of the IoT and discussed the challenges in the healthcare and education fields and more. The current security techniques and their challenges were also discussed.

Dhuha et al. [31] provided a literature review to clarify related works and discuss IoT security and privacy issues. The paper then presented security attacks as layer-wise attacks and attack taxonomy perspectives and analyzed them based on the IoT layers. In addition, the paper suggested solutions and stated strategies against these attacks.

Muhammed et al. [32] proposed a comprehensive survey of security issues in the IoT. This paper presented an overview of different layers in the IoT and classified the attacks on each layer. Furthermore, mitigation techniques for these attacks were reviewed. The paper suggested new IoT architecture layers to overcome these issues, the security mechanisms used are authentication, encryption, decryption, and encrypted hash.

Garima et al. [33] discussed a comprehensive study and undertook a literature review to determine threats and security aspects for all IoT layer types. They also explained the various parameters of IoT protocols, including common characteristics, limitations, advice, and countermeasures. In addition, this paper focused on exploring all related IoT security issues concerning various layers.

Muath et al. [15] reviewed IoT applications, limitations, and security issues related to each domain. In addition, it provided a survey on IoT security, its threats and vulnerabilities, its privacy challenges, and corresponding countermeasures. The paper presented the Internet of Vulnerabilities (IoV) as the growth of the IoT in our daily lives has increased. Furthermore, it discussed various cybersecurity attacks, threats, and vulnerabilities based on confidentiality, integrity, and availability. Finally, mitigation and countermeasures to these security concerns were also discussed in the paper.

Shantanu et al. [34] analyzed security issues of different layers in the Industrial Internet of Things (IIoT) architecture. In addition, they presented a list of IIoT security requirements; then, the paper looked at future research directions and countermeasures against IIoT issues. Table 2, below, presents a summary of the addressed threats for every layer of the IoT.

Table 2. Summary of addressed threats for every layer of IoT.

Author	Publication Year	Application Layer Threats	Network Layer Threats	Perception Layer Threats
Samundra et al. [27]	2022	Data access control issues, authentications issues, data protection and recovery issues, phishing attacks problems, vulnerabilities, clone attack	Cluster security issues, DoS spoofed and replayed routing information or altered them	Fake or node capture, mass node authentication, key management mechanism, cryptographic algorithm
Azzam et al. [28]	2022	Jamming physical attack, injection attacks, cloning attacks, eavesdropping and tampering	DoS attacks, DDoS attacks, MITM attacks, sinkhole attack, traffic analysis	Phishing attacks, malware attacks, DoS/DDoS, buffer overflow attacks, spyware attacks
Waseem et al. [29]	2020	Malicious code, weak application security cross-site scripting, eavesdropping and MITM	DoS attacks (collision attack and channel congestion), escalating frame counter value, spoofing like battery exhaustion attack, message alteration attacks, replication of nodes and storage attack	Eavesdropping, battery drainage attack, hardware malfunctioning, malign data injection, node cloning, gaining unauthorized access to the device
Pintu et al. [5]	2022	Spyware and adware, Trojan horse, Botnet attack, DoS, brute-force password attack, firmware hijacking, malicious scripts, phishing attacks, worms, virus	Traffic analysis attacks, RFID spoofing, RFID cloning, RFID unauthorized access, MITM, DoS, sinkhole attack, routing information attack, Sybil attack, replay attack, hello flood attack, blackmail attack, blackhole attack, wormhole attack, grayhole attack	Node jamming, physical damage, node tampering, social engineering, malicious node injection, sleep deprivation attack, RF interference, tag cloning, eavesdropping, tag tampering outage attack, object replication, hardware Trojan
Thalawattha et al. [30]	2021	DoS attacks	DoS attacks	Leakage of privacy, sniffing service information manipulation, lack of encryption, weak default password, the rise of Botnets
Dhuha et al. [31]	2020	Malicious code injection, DoS attack, spear-phishing attack, sniffing attack, overwhelm, reprogram	Sybil attack, sinkhole attack, sleep deprivation attack, DoS attack, malicious code injection, MITM attack, traffic analysis, passive monitoring, eavesdropping	Unauthorized access to tags, tag cloning, eavesdropping, spoofing, RF jamming, timing attack, replay attack, node capture attack, malicious node injection attack, brute-force attack, radio interference, tampering
Muhammed et al. [32]	2018	Cross-site scripting attack, malicious code injection attacks, the ability to deal with mass data spear-phishing, social engineering	DoS attacks, MITM attacks, storage attacks, exploit attacks	Eavesdropping, node capture, fake node injection, replay attack, timing attack
Garima et al. [33]	2021	DoS attack, data leakage, data transit attacks	Routing attacks, data transit attacks	Node tempering, impersonation
Muath et al. [15]	2020	DoS attacks, software vulnerabilities, malicious virus/worm/Trojan horse, code injection, buffer overflow, phishing attack, spyware, malicious scripts, sensitive data manipulation, data leakage	Spoofing attacks, selective forwarding, DoS attacks, MITM attacks, sinkhole attacks, packet replication attacks, wormhole attacks, Sybil attacks, traffic analysis, sniffing attacks, routing information attacks	False data injection attacks, eavesdropping, interference, social engineering, node jamming attack, malicious node injection, sleep deprivation, tag cloning
Shantanu et al. [34]	2021	Injecting malicious codes, MITM attacks, DoS/DDoS attacks	Session hijacking, blackhole attack, wormhole attack, DNS spoofing, DoS/DDoS attacks, MITM attacks	Jamming attack, eavesdropping

4.2. Threats to the ICN

Bengt et al. [35] observed and compared ICN architectures, including CCN, PSIRP, NetInf, and DONA. Under these approaches, some challenges and issues for the ICN were discussed. One of the main issues is security; mobility, privacy, and scalability are also concerns. The implantation challenges for implementing the generic building blocks of these four approaches were introduced in the paper, and deployment issues were also discussed. Finally, the remaining challenges were discussed.

Boubakr et al. [6] presented all related security issues of deployed ICNs in wireless environments, especially for NDN architecture. The paper discussed attacks that may take place on the ICN network and its applications. A variety of issues and attacks were focused on in the paper, including security risks with attacks on content names, a list of attacks in the ICN, issues that relate to the design of NDN, such as unclear interest aggregation and illegal content caching, as well as some other attacks, such as DoS, content poisoning, and more. In addition, the paper provided solutions and countermeasures to present issues and attacks.

Reza et al. [7] provided an overview survey of related works about security and privacy issues in the ICN. This paper focused on three areas of security threats that review DoS, content poisoning, and cache pollution attacks. Then, it discussed privacy risks that concern user privacy and anonymity, content privacy, and name and signature privacy. At the end, the paper reviewed all available access control mechanisms.

Muhammed et al. [36] presented a new scheme called the secure distribution of protected content (SDPC) to ensure that only authenticated consumers can access the content. Furthermore, the paper presented all relevant security issues in the ICN by performing a formal analysis using BAN logic Scyther implementation.

Rao et al. [37] observed that the CCN is the extraction of the ICN so it analyzes the basic CCN principles and proprieties of three areas: caching, mobility, and security. This paper presented a quantitative comparison and discussion of related studies. In addition, the analysis of CCN security schemes and relevant risks was included in the paper. Last, DoS attacks were discussed in detail.

Yong et al. [38] presented an architecture, and then, examined content security to classify the security concerns of NDN content, including naming, caching, and routing-related attacks, as the paper gives all possible solutions to these problems. Moreover, the paper suggested some digital signature schemes, such as network coding signatures. Table 3, below, presents a summary of ICN-related work.

Table 3. Summary of classification of ICN architectures based on related attacks.

Author Publication	Year	Classification of ICN Architectures	Related Attacks for Each Architecture
Bengt et al. [35]	2012	NetInf	Human friendly issues, data integrity issues
		CCN	security issues
		PSIRP	Human readable names issues (DoS attacks)
		DONA	Multiple names issues
Boubakr et al. [6]	2021	NDN	Passive attack, cache pollution attack, DoS attack, content poisoning attack

Table 3. *Cont.*

Author Publication	Year	Classification of ICN Architectures	Related Attacks for Each Architecture
Reza et al. [7]	2018	Applicable to all ICN architectures	DoS attack
		Applicable to all ICN architectures	Content poisoning attack
		Applicable to all ICN architectures	Cache pollution attack
		Feasible for all ICN architectures that employ caching, so the PSIRP and PURSUIT architectures are excepted	Timing attack
		Only applicable to CCN and NDN architectures	Discovery and protocol attacks
Muhammed et al. [36]	2020	NDN	Named based attacks (watchlist, sniffing attack), DDoS, time analysis attack, unauthorized access, traffic monitoring attack
Rao et al. [37]	2020	CCN	DoS attacks
Yong et al. [38]	2018	NDN	Watchlist, sniffing attacks, DoS, time analysis, caching pollution, bogus announcements attacks, jamming attack, time attack, flooding attack

4.3. Related Work on ICN-Based IoT with Machine Learning Technologies

Various research studies are reviewed concerning the ICN and the IoT with the classification of ML techniques and others. Next, the Results and Discussion are presented.

Gang et al. [39] proposed an enhanced distributed low-rate attack mitigating (eDLAM) mechanism that works on resisting new DoS attack types existing in the IoT. The paper established a game model as the first stage. This model analyzes the attack and helps to explain the benefit behind this attack between the attacker and the defender. Furthermore, the evaluation criteria of eDLAM's performance are in terms of the false negative rate and false positive rate. The paper mentioned that SDN is useful for separating control and forward functions to be applied to various IoT scenarios. In addition, this mechanism is implemented in the NDN architecture by NDN-Cxx and ndnSIM simulators.

Michael et al. [11] described a robust, efficient, and secure IoT network by analyzing the potential of the ICN. Then, they compared IP-based approaches with the ICN. They suggested using NDN to deploy the content-centric security model to resist DoS and enhance IoT industrial security. Furthermore, they suggested using the RIOT operating system for content security to deploy scenarios.

Tehseen et al. [40] reviewed IoT security issues, threats, and cyberattacks and presented an innovative method to protect all IoT devices by using ML and DL techniques. Moreover, this paper examined how ML and DL can be used to enhance attack detection and mitigation.

Kazeem et al. [9] reviewed methods used to detect and mitigate DDoS attacks on many Internet-enabled networks, including NDN, SDN, and IoT. In addition, it presented attack scenarios in some domains, such as the Internet of Drones, routing protocol-based IoT, and named-data networking. Furthermore, the paper observed that DDoS attacks are the most prevalent attack in the IoT, SDN, RPL-IoT, and NDN.

Haoyue et al. [10] presented a new mechanism to detect and mitigate one of the DoS attacks, called an IFA, by using ndnSIM simulation that realizes NDN and runs on an SDN-based platform.

A new app design called NDN4IoT was presented by [41], where the authors focused their design specifically on enhancing security and efficient analysis for the effectiveness of NDN-IoT devices. The FIWARE IoT platform is integrated with this app for some purposes, such as storing information about NDN-IoT devices and then making an analysis to decide whether the system or device fails. FIWARE is an open-source platform that provides APIs, a set of standards, and tools to develop IoT applications and enhance their security.

Azana et al. [42] presented the ICN as a solution for the IoT environment to enhance mobility and security. The network performance was evaluated based on general metrics, including service recovery time and packet loss cost.

Ravindran et al. [43] discussed ICN for the IoT in detail, including what the IoT requirements are, how ICN features will enhance the IoT environment, and what the challenges of an ICN-based IoT are.

Akhila et al. [44] proposed a strategy called UTS-LRU cache replacement that helps in evaluating the performance implications of one of the IoT network scenarios in the ICN. Then, the paper discussed the result that was defined by using the CCNx 1.0 protocol.

Sarika et al. [45] presented a network intrusion detection system (IDS) to be applied to the IoT. This paper focused on some attacks such as DoS and DDoS attacks using ML algorithms to detect all possible intrusions, such as deep neural network (DNN) and SVM. Finally, this paper only focused on the IoT.

Abdelhak et al. [8] introduced detection and mitigation mechanisms to identify malicious attacks in the NDN architecture by using an IDS. The paper discussed several attacks, such as DoS/DDoS, cache pollution attacks (CPAs), cache privacy attacks, cache poisoning attacks, and IFA. These attacks affect NDN operations. Furthermore, this paper presented the challenges and issues of NDN IDS in detail. Moreover, in one of the suggested models, the two main phases are the pre-attack phase and the main attack phase. In the pre-attack phase, the minimum re-transmission wait time, the minimum interest frequency, the topology characteristics, and the minimum number of pieces of content stored by malicious producers are identified. Accordingly, in the main attack phase, collecting the prefixes that are stored by the malicious producers is an important step, and then these malicious consumers set the interest frequency. The paper describes each attack mentioned in depth and how to mitigate them using IDS, but how to integrate NDN with the IoT and what potential issues and challenges there are were not mentioned.

Raneem et al. [46] proposed an approach that contains a multi-stage process for distinguishing intrusion activities from normal activities by using some techniques based on ML, such as k-means clustering that used the SLFN-SVM-SMOTE algorithm, and many others. The paper mentioned that the datasets CICIDS2017, UNSWNB15, and ISCX2012 are the most beneficial and well-known datasets. The NSL-KDD dataset is utilized to detect DoS attacks in the IoT environment. Furthermore, the most recent databases are LITNET-2020 and IoTD20; also, the BoT-IoT dataset is used in IoT environments. The paper discussed all these techniques and technologies for its proposed approach to mitigate attacks that pose a threat to the IoT system.

Sobia et al. [47] introduced the famous ICN architectures such as NDN that fit the requirements of IoT architectures with respect to their suitability in terms of caching, security, naming, and mobility handling schemes. In addition, they provided a classification of the ICN-based security scheme for IoT and mentioned that DoS may rise when the producer or intermediate routers set the freshness value for the required content. The paper reviewed ICN-IoT simulators and OSs, then discussed them; it identified the ndnSIM simulator as the most explored tool that goes well with implementing the IoT over the ICN. Then, issues, challenges, and future directions for the ICN in the IoT were discussed.

Ridha et al. [48] simply provided a review of ML algorithms and caching methods to integrate them with future network architectures such as SDN, ICN, NDN, 5G-ICN cellular networks, and edge computing.

Sedat et al. [21] provided a practical implementation of NDN-based IoT using the IfNoT mechanism to mitigate IFA. However, the paper mentions that in addition to their study, only one other study presents how to mitigate IFA in ICN-IoT. This proposed project was conducted using the ndnSIM simulator and the Contiki NG OS, which comes with the Cooja Network Simulator. The paper also reviewed other mitigation techniques and categorized them as statistical-based, collusive statistical-based, ML-based, and cryptography-based solutions. Regarding ML, the authors recommended several algorithms as the most effective ones in detecting and preventing IFA, including graph neural networks, RF, isolation forest, long short-term memory, and SVM. This is because ML algorithms differ from other traditional statistical approaches as they are adaptive and have a large capacity to detect attack patterns and prevent IFA from exposing the NDN-IoT networks. Table 4, presents a summary of ICN-based IoT threats and mitigation techniques.

Table 4. Summary of ICN-based IoT threats and mitigation techniques.

Author	Publication Year	Methodology	Technology	Threats and Challenges	Simulators and Datasets	Methods and Techniques	Limitations
Gang et al. [39]	2019	Quantitative	NDN-IoT integrated by SDN	DoS attacks	NDN-Cxx and ndnSIM simulations	eDLAM mechanism	ML technology is not mentioned
Michael et al. [11]	2018	Qualitative	NDN-IoT	DoS attack	CCN-Lite simulator	RIOT OS	ML technology is not mentioned
Tehseen et al. [40]	2023	Qualitative	IoT	Spoofing attacks, DoS, malicious code usage, data injection, MITM	Datasets such as NSL-KDD, UNSW-NB15, DARPA, CAIDA	DL, ML includes SVM, logistic regression, intrude tree, and behave DT	ICN including NDN is not mentioned
Kazeem et al. [9]	2023	Mixed	IoT NDN SDN RPL-IoT	DDoS attacks	Datasets such as NSL-KDD, DARPA'09 and KKD, drone data simulations	ML includes RF, DT, KNN, XGBoost, DT, J48, MLP + BP, SVM, and RF algorithms	No discussion of integrating IoT with NDN
Haoyue et al. [10]	2017	Mixed	NDN integrated with SDN	IFA	ndnSIM simulation	-	IoT technology and ML technology are not mentioned
Mohamed [41]	2023	Qualitative	NDN-IoT	-	NDN4IoT app	FIWARE platform	ML technology is not mentioned and no attack type is explored
Azana et al. [42]	2019	Qualitative	ICN-IoT	DoS	Simulators (NS-3, OMNeT, OPNET, GNS3, QualNet)	Wireshark	ML technology is not mentioned
Ravindran et al. [43]	2019	Qualitative	ICN-IoT	DoS, name spoofing attack, message manipulation attack, information modification attack	-	-	ML technology is not mentioned and no suggested simulator or dataset
Akhila et al. [44]	2016	Qualitative	ICN(CCN)-IoT	-	CCN-lite simulator	-	ML technology is not mentioned and no attack type is explored

Table 4. *Cont.*

Author	Publication Year	Methodology	Technology	Threats and Challenges	Simulators and Datasets	Methods and Techniques	Limitations
Sarika et al. [45]	2021	Qualitative	IoT	DoS, DDoS, replay attack, routing attacks	IDS KDD-Cup dataset	DNN, SVM, KNN, neural network, and random forest ML algorithms	ICN including NDN is not mentioned
Abdelhak et al. [8]	2022	Qualitative	NDN SDN	DoS, DDoS, cache pollution attacks, cache privacy attacks, cache poisoning attacks, and IFA	IDS	DNN, SVM, KNN, neural network, and random forest ML algorithms	IoT technology is not mentioned
Raneem et al. [46]	2021	Mixed	IoT	DoS, DDoS, and MITM	IDS datasets (CICID2017, UNSWNB15, ISCX2012, BoT-IoT, LITNET-2020, IoTD20)	SVM, SLFN-SVM-SMOTE, LR	ICN including NDN is not mentioned
Sobia et al. [47]	2018	Qualitative	ICN (NDN)-IoT	DoS	Simulations (CCN-lite, ndnSIM, NS-3, and Cooja) IoT OS (RRIOT OS, Contiki OS, FreeRTOS, OpenWSN, TinyOS)	-	ML technology is not mentioned
Ridha et al. [48]	2020	Qualitative	ICN-IoT SDN-NDN NDN ICN	-	-	ML type DQL (deep Q-learning), ILP (integer linear programming) ANN Deep RNN, RL, SNN	No attack type is explored and no suggested simulator or dataset
Sedat et al. [21]	2024	Mixed	NDN-IoT	IFA	ndnSIM simulation Contiki NG OS, Cooja	IfNoT mechanism	ML technology is mentioned as one of the solutions, but the proposed solution is built by the IfNoT mechanism

5. Results and Discussion

In the previous section, we reviewed comprehensive research papers about common cyberattacks in each ICN architecture: NetInf, NDN, DONA, CCN, PSIRP, and PURSUIT. The results showed that DoS attacks were the most executed attacks in all these architectures. On the NDN platform, while DoS attacks are prevalent, other attacks are also frequent, including watchlist sniffing, time analysis, caching pollution, bogus announcements, jamming, flooding, traffic monitoring, and passive and content poisoning. Some attacks do not happen traditionally; they are only applicable to ICN architectures. The other main architecture is CNN, which has features similar to NDN since they use the same hierarchical naming scheme. On the other hand, DONA, PSIRP, and PURSUIT use a flat naming scheme. DONA is the least prevalent and PSIRP faces attacks directly and indirectly because of hash content and hash from the publisher key. Moreover, NetInf is vulnerable to DoS/DDoS attacks. As a result, CCN and NDN are more frequently exposed and serve as rich platforms for DoS/DDoS attacks.

The most common cyberattacks are DoS, content poisoning, and cache pollution, which can be carried out on almost all architectures. According to our findings, as shown in Figure 8, first, NDN, the most common type, suffers 35% of the total attacks, such as DoS/DDoS attacks, watchlist sniffing attacks, and time analysis attacks. Second, CCN has 25% of the total attacks and DoS attacks are dominant. Third, DONA is the oldest platform, so many researchers avoid using it because more effective and secure architectures have been developed. Its percentage is only 15%. Finally, the results of PSIRP and PURSUIT, both posed attacks with only 10%. However, NetInf is in last place, with only 5%. According to recommendations, the architecture that should be used is NDN, which fits very well with the IoT because both are concerned with content.

Figure 8. Cyberattacks on ICN architectures.

IoT three-layer architectures are a common way to structure IoT systems. The results of these IoT-related studies showed that there are common threats at each layer of IoT architecture. In the application layer, cloning attacks, eavesdropping, phishing attacks, malicious code injection, DoS attack, malicious viruses/worms/Trojan horses, and MITM attacks are considered the greatest threats because they pose a threat to the whole network.

Moreover, the main threats in the network layer include DoS/DDoS attacks, spoofing attacks, malicious code injection attacks, eavesdropping, sniffing attacks, and routing information attacks. Most of the analyzed studies focused on DoS/DDoS attacks and presented them as a major threat in this layer. DoS/DDoS attacks are easy to target and the first attack type that comes to the mind of attackers. They are attacks that aim to prevent authentic users from accessing devices or other network resources.

In the perception layer, some common threats are phishing attacks, malware attacks, DoS/DDoS, malicious script attacks, eavesdropping, node cloning, malicious code injection attacks, sniffing, spoofing, timing attacks, malicious node injection attacks, jamming attacks, and tag cloning. DoS/DDoS, malicious node injection attacks, and sniffing attacks are the major threats to the perception layer based on the analyses performed in these studies. Usually, attackers choose the easiest ways to gain access through devices and networks and attack the desired component. Figure 9 shows the most common threats based on application, network, and perception layers.

As a result, we deduce that the most dangerous cyberattacks are DoS/DDoS attacks, spoofing attacks, routing information attacks, and malicious code injection posing on the network layer. This is the reason why we worked on the network layer. Additionally, all layers share the following attacks: DoS/DDoS, malicious code injection, eavesdropping, and cloning attacks. As a result, these types of IoT-layer attacks were compared with the ICN architecture threats and the statistical results are shown in Figure 10.

Figure 9. The most common IoT threats based on three layers.

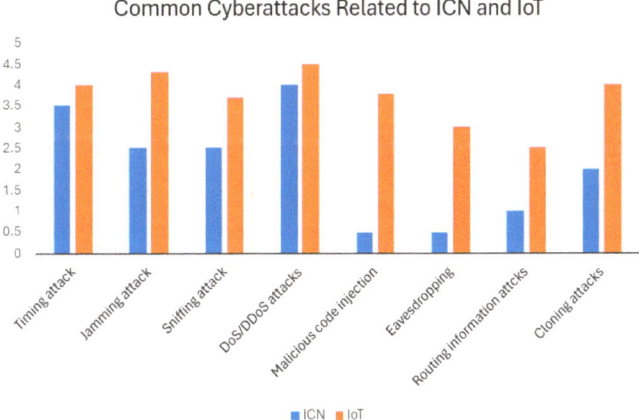

Figure 10. Common cyberattacks related to the ICN and the IoT.

According to the analyzed studies, the most common cyberattacks that have targeted ICN and IoT architectures are DoS and DDoS attacks. It was observed that DoS attacks pose significant challenges in both domains, emphasizing the necessity for finding effective solutions. Figure 10 categorizes these attacks based on the frequency of their mentions in the literature. For instance, the jamming attack is mentioned less frequently in ICN-related papers compared to IoT-related papers. However, DoS and DDoS attacks emerge as the most prevalent attack types in both the ICN and IoT domains.

As a result, the most common threats are DoS/DDoS attacks, and to detect and mitigate these types of attacks on ICN-based IoT, we must use one of the technologies such as artificial intelligence (AI), including DL and ML, edge computing, fog computing, and blockchain. There are also conventional solutions, but these are not recommended by researchers as technologies become very innovative. There are many algorithms that can be used with ML, such as RF, DT, KNN, XGBoost, DT, J48, MLP + BP, SVM, and RF. Using ML algorithms can be useful for anomaly detection in ICN-IoT networks. We select the SVM, RF, and KNN methods, as they go well with the ndnSIM simulator that we work with. Table 5 clarifies the purpose, advantages, and disadvantages of the selected ML methods, along with providing accuracy percentages to offer insights into the effectiveness of these methods. According to our findings, we work on the NDN architecture to enhance IoT networks by detecting and mitigating DoS attacks using ML technology. In addition, we recommend that researchers in this area focus on ML technologies.

Overall, ICN-IoT could be beneficial not only to the technical domain but also in various fields such as healthcare and smart homes. For safeguarding sensitive information in IoT applications, the ICN can be an effective solution due to its inherent data-centric approach, which provides better privacy protection. ICN networks can enhance security in various other domains such as supply chain management, finance, government, transportation, and utilities services fields. Furthermore, network administrators and policymakers should learn more about the benefits and challenges of adopting ICN-IoT technology to create appropriate strategies. For network administrators, this involves adapting to the specific characteristics of ICN-IoT networks, including understanding routing and caching mechanisms and how to analyze this type of network. However, policymakers may need to establish new guidelines for data privacy and security when using ICN-IoT.

Table 5. Comparison between SVM, KNN, and RF ML algorithms.

ML Technique	Purpose	Advantages	Disadvantages	Accuracy Percentage
SVM	• To classify various attack types • Features selection and intrusion detection	• Robust in high-dimensional spaces • Powerful algorithm	• May require significant memory resources, particularly in scenarios with a high number of features • Ineffective for datasets with a large number of rows	• 99.4%
KNN	• Network IDS • To reduce the false alarm rate	• Simple implementation	• Long prediction time	• 98.35%
RF	• To build network IDS	• More robust and accurate predictions	• Complexity may affect the training time of the model	• 99.98% better than SVM and KNN

6. Recommendations and Future Research Directions

The previous papers contained most of the information we needed to present our paper. The most used ICN architecture, known as NDN, is suitable for integrating with the IoT. NDN can easily operate above the network layer as many simulators allow us to experiment with an NDN-based IoT immediately. In addition, there are many DoS attack types that were included and discussed in the previous papers and some of them take place only when the two architectures are integrated, such as in the IFA.

The reviewed literature showed that high numbers of ML algorithms have been experimented with on either IoT, one of the ICN architectures, or in SDN. However, in the case of integrated NDN with the IoT, there are few studies and no implementation of ML algorithms. On the other hand, no article explains or applies how can we mitigate and prevent DoS attacks by using ML technology for ICN-IoT networks. Accordingly, no dataset is right for us, and most of the papers that worked with the ICN deal with simulators. We summarize the following future directions of ICN-IoT security:

- We recommend more future research investigations to enhance ICN-IoT security.
- We recommend more future research investigations concerning ICN-IoT implementation in the security field.
- We recommend more future research investigations into the use of ML technology to improve the security of ICN-IoT.
- We recommend more future research investigations to test and train a variety of ML classifiers to identify the most effective option used to improve the security of the ICN-IoT network.
- We recommend more future research investigations to explore other attack types and threats on the ICN-IoT.
- We recommend more future research investigations to expand the implementation process by using DL technology after the initial results of ML classifiers by training various DL classifier types to increase the security level of ICN-based IoT.
- We recommend more future research investigations using other technologies such as DL, fog computing, and blockchain to compare the results between them and identify which one provides the most effective results to improve the security of ICN-IoT against cybersecurity attacks.

7. Conclusions

In conclusion, this paper discusses ICN architectures and how the ICN integrates with the IoT. The paper provides a comprehensive literature review, including common cybersecurity threats, with particular emphasis on DoS attacks. The analysis highlights that DoS attacks pose a significant challenge to both the ICN and IoT, so we present some detecting and mitigating techniques through integrated ML technologies, such as SVM, RF, and KNN to enhance security measures against these attacks. Further investigations are discussed and emphasize ML-based approaches for ICN-IoT security. The new proposed approach is based on the NDN architecture that relies on the ndnSIM simulator, which is considered an IDS, to prevent DoS attacks from exposing IoT devices. Overall, future studies should expand their research, explore additional attack types and threats on the ICN-IoT, and enhance security using ML technology by advancing innovative solutions for robust and resilient network architectures.

Author Contributions: Conceptualization, R.B. and A.A.; methodology, R.B., A.A. and M.M.H.R.; software, R.B. and A.A.; validation, R.B., A.A. and M.M.H.R.; formal analysis, R.B., A.A. and M.M.H.R.; investigation, R.B. and A.A.; resources, R.B. and A.A.; writing—original draft preparation, R.B.; writing—review and editing, R.B. and A.A.; supervision, A.A. and M.M.H.R.; project administration, A.A. and M.M.H.R.; funding acquisition, A.A. and M.M.H.R. All authors have read and agreed to the published version of the manuscript.

Funding: This paper was funded by the Deanship of Scientific Research, Vice Presidency for Graduate Studies and Scientific Research, King Faisal University, Saudi Arabia [GRANT 5933].

Data Availability Statement: Data available in a publicly accessible repository.

Acknowledgments: The authors would like to thank the anonymous reviewers for their insightful scholastic comments, directions and suggestions, which improved the quality of the paper.

Conflicts of Interest: The authors declare no conflicts of interest.

Abbreviations

The following abbreviations are used in this manuscript:

IoT	Internet of Things
ICN	Information center network
IFA	Interest flooding attack
ML	Machine learning
DoS	Denial-of-service attack
DDoS	Distributed denial-of-service
IP	Internet Protocol
TCP	Transmission Control Protocol
AI	Artificial intelligence
NDN	Named-Data Networking
CCN	Content-Centric Networking
PSIRP	Publish–Subscribe Internet Routing Paradigm
DONA	Data-Oriented Network Architecture
NetInf	Network of information
SVM	Support vector machine
KNN	K-nearest neighbor
RF	Random forest
MITM	Man-in-the-middle
PURSUIT	Publish–Subscribe Internet Technology

SAIL	Scalable & Adaptive Internet Solutions
COMET	Content Mediator Architecture for Content-Aware Networks
4WARD	Architecture and Design for the Future Internet
CR	Content router
CS	Content store
FIB	Forwarding information base
PIT	Pending interest table
IFA	Interest flooding attack
BIFA	Basic IFA
CIFA	Collusive IFA
SCAN	Smart collaborative attack in NDN
NRS	Name-resolution system
IoV	Internet of Vulnerabilities
IIoT	Industrial Internet of Things
SDPC	Secure distribution of protected content
eDLAM	Enhanced distributed low-rate attack mitigating
DL	Deep learning
IDS	Intrusion detection system
DNN	Deep neural network
CPA	Cache pollution attacks
WSN	Wireless sensor network
WBAN	Wireless body area network
JS	Jensen–Shannon
DT	Decision tree
MLP	Multilayer perceptron
BP	Backpropagation
SDPC	Secure distribution of protected content

References

1. Mohanta, B.K.; Jena, D.; Satapathy, U.; Patnaik, S. Survey on IoT security: Challenges and solution using machine learning, artificial intelligence and blockchain technology. *Internet Things* **2020**, *11*, 100227. [CrossRef]
2. Mishra, S.; Jain, V.K.; Gyoda, K.; Jain, S. An efficient content replacement policy to retain essential content in information-centric networking based internet of things network. *Ad Hoc Netw.* **2024**, *155*, 103389. [CrossRef]
3. Rahman, A.; Hasan, K.; Kundu, D.; Islam, M.J.; Debnath, T.; Band, S.S.; Kumar, N. On the ICN-IoT with federated learning integration of communication: Concepts, security-privacy issues, applications, and future perspectives. *Future Gener. Comput. Syst.* **2023**, *138*, 61–88. [CrossRef]
4. Krishna, B.V.S.; Gnanasekaran, T. A systematic study of security issues in Internet-of-Things (IoT). In Proceedings of the 2017 International Conference on I-SMAC (IoT in Social, Mobile, Analytics and Cloud) (I-SMAC), Palladam, India, 10–11 February 2017; pp. 107–111. [CrossRef]
5. Sadhu, P.K.; Yanambaka, V.P.; Abdelgawad, A. Internet of Things: Security and Solutions Survey. *Sensors* **2022**, *22*, 7433. [CrossRef] [PubMed]
6. Nour, B.; Mastorakis, S.; Ullah, R.; Stergiou, N. Information-Centric Networking in Wireless Environments: Security Risks and Challenges. *IEEE Wirel. Commun.* **2021**, *28*, 121–127. [CrossRef]
7. Tourani, R.; Misra, S.; Mick, T.; Panwar, G. Security, Privacy, and Access Control in Information-Centric Networking: A Survey. *IEEE Commun. Surv. Tutor.* **2018**, *20*, 566–600. [CrossRef]
8. Hidouri, A.; Hajlaoui, N.; Touati, H.; Hadded, M.; Muhlethaler, P. A Survey on Security Attacks and Intrusion Detection Mechanisms in Named Data Networking. *Computers* **2022**, *11*, 186. [CrossRef]
9. Adedeji, K.B.; Abu-Mahfouz, A.M.; Kurien, A.M. DDoS Attack and Detection Methods in Internet-Enabled Networks: Concept, Research Perspectives, and Challenges. *J. Sens. Actuator Netw.* **2023**, *12*, 51. [CrossRef]
10. Xue, H.; Li, Y.; Rahmani, R.; Kanter, T.; Que, X. A mechanism for mitigating DoS attack in ICN-based internet of things. In Proceedings of the Proceedings of the 1st International Conference on Internet of Things and Machine Learning, New York, NY, USA, 17–18 October 2017. [CrossRef]
11. Frey, M.; Gündoğan, C.; Kietzmann, P.; Lenders, M.; Petersen, H.; Schmidt, T.C.; Juraschek, F.; Wählisch, M. Security for the Industrial IoT: The Case for Information-Centric Networking. In Proceedings of the 2019 IEEE 5th World Forum on Internet of Things (WF-IoT), Limerick, Ireland, 15–18 April 2019; pp. 424–429. [CrossRef]
12. Amadeo, M.; Campolo, C.; Quevedo, J.; Corujo, D.; Molinaro, A.; Iera, A.; Aguiar, R.L.; Vasilakos, A.V. Information-centric networking for the internet of things: Challenges and opportunities. *IEEE Netw.* **2016**, *30*, 92–100. [CrossRef]
13. Nour, B.; Sharif, K.; Li, F.; Biswas, S.; Moungla, H.; Guizani, M.; Wang, Y. A survey of Internet of Things communication using ICN: A use case perspective. *Comput. Commun.* **2019**, *142–143*, 95–123. [CrossRef]

14. Ahmed, S.F.; Shuravi, S.; Bhuyian, A.; Afrin, S.; Mehjabin, A.; Kuldeep, S.A.; Alam, M.S.B.; Gandomi, A.H. Navigating the IoT landscape: Unraveling forensics, security issues, applications, research challenges, and future. *arXiv* **2023**, arXiv:cs.NI/2309.02707.
15. Obaidat, M.A.; Obeidat, S.; Holst, J.; Al Hayajneh, A.; Brown, J. A Comprehensive and Systematic Survey on the Internet of Things: Security and Privacy Challenges, Security Frameworks, Enabling Technologies, Threats, Vulnerabilities and Countermeasures. *Computers* **2020**, *9*, 44. [CrossRef]
16. Yoachimik, O.; Pacheco, J. *DDoS Threat Report for 2023 Q4*; Cloudflare: San Francisco, CA, USA, 2024. Available online: https://blog.cloudflare.com/ddos-threat-report-2023-q4 (accessed on 28 January 2024).
17. Wang, J.; Jiang, C.; Zhang, H.; Ren, Y.; Chen, K.C.; Hanzo, L. Thirty Years of Machine Learning: The Road to Pareto-Optimal Wireless Networks. *IEEE Commun. Surv. Tutor.* **2020**, *22*, 1472–1514. [CrossRef]
18. Dalmazo, B.L.; Marques, J.A.; Costa, L.R.; Bonfim, M.S.; Carvalho, R.N.; da Silva, A.S.; Fernandes, S.; Bordim, J.L.; Alchieri, E.; Schaeffer-Filho, A.; et al. A systematic review on distributed denial of service attack defense mechanisms in programmable networks. *Int. J. Netw. Manag.* **2021**, *31*, e2163. [CrossRef]
19. Srinivasan, K.; Mubarakali, A.; Alqahtani, A.S.; Dinesh Kumar, A. A Survey on the Impact of DDoS Attacks in Cloud Computing: Prevention, Detection and Mitigation Techniques. In *Intelligent Communication Technologies and Virtual Mobile Networks*; Balaji, S., Rocha, Á., Chung, Y.N., Eds.; Springer: Cham, Switzerland, 2020; pp. 252–270.
20. AbdAllah, E.G.; Hassanein, H.S.; Zulkernine, M. A Survey of Security Attacks in Information-Centric Networking. *IEEE Commun. Surv. Tutor.* **2015**, *17*, 1441–1454. [CrossRef]
21. Bilgili, S.; Demir, A.K.; Alam, S. IfNot: An approach towards mitigating interest flooding attacks in Named Data Networking of Things. *Internet Things* **2024**, *25*, 101076. [CrossRef]
22. Alashhab, A.A.; Zahid, M.S.M.; Azim, M.A.; Daha, M.Y.; Isyaku, B.; Ali, S. A Survey of Low Rate DDoS Detection Techniques Based on Machine Learning in Software-Defined Networks. *Symmetry* **2022**, *14*, 1563. [CrossRef]
23. Altulaihan, E.; Almaiah, M.A.; Aljughaiman, A. Anomaly Detection IDS for Detecting DoS Attacks in IoT Networks Based on Machine Learning Algorithms. *Sensors* **2024**, *24*, 713. [CrossRef]
24. Aslam, M.; Ye, D.; Tariq, A.; Asad, M.; Hanif, M.; Ndzi, D.; Chelloug, S.A.; Elaziz, M.A.; Al-Qaness, M.A.A.; et al. Adaptive Machine Learning Based Distributed Denial-of-Services Attacks Detection and Mitigation System for SDN-Enabled IoT. *Sensors* **2022**, *22*, 2697. [CrossRef] [PubMed]
25. Kumar, N.; Singh, A.K.; Srivastava, S. Evaluating machine learning algorithms for detection of interest flooding attack in named data networking. In Proceedings of the 10th International Conference on Security of Information and Networks, New York, NY, USA, 13–15 October 2017; pp. 299–302. [CrossRef]
26. Alabsi, B.A.; Anbar, M.; Rihan, S.D.A. Conditional Tabular Generative Adversarial Based Intrusion Detection System for Detecting Ddos and Dos Attacks on the Internet of Things Networks. *Sensors* **2023**, *23*, 5644. [CrossRef] [PubMed]
27. Deep, S.; Zheng, X.; Jolfaei, A.; Yu, D.; Ostovari, P.; Kashif Bashir, A. A survey of security and privacy issues in the Internet of Things from the layered context. *Trans. Emerg. Telecommun. Technol.* **2022**, *33*, e3935. [CrossRef]
28. Albalawi, A.M.; Almaiah, M. Assessing and reviewing of cyber-security threats, attacks, mitigation techniques in IoT environment. *J. Theor. Appl. Inf. Technol.* **2022**, *100*, 2988–3011.
29. Iqbal, W.; Abbas, H.; Daneshmand, M.; Rauf, B.; Bangash, Y.A. An In-Depth Analysis of IoT Security Requirements, Challenges, and Their Countermeasures via Software-Defined Security. *IEEE Internet Things J.* **2020**, *7*, 10250–10276. [CrossRef]
30. Jayasinghe, K.; Thalawattha, S.; Rodrigo, R.; Dissanayaka, D.; Kathriarachchi, R. A Defence Against an Internet of Things (IoT) Attacks Based on Current Vulnerabilities. In Proceedings of the International Conference on Advancement of Development Administration, Bangkok, Thailand, 28–30 May 2020.
31. Alferidah, D.K.; Jhanjhi, N.Z. A Review on Security and Privacy Issues and Challenges in Internet of Things. *Int. J. Comput. Sci. Netw. Secur.* **2020**, *20*, 263–286.
32. Burhan, M.; Rehman, R.A.; Khan, B.; Kim, B.S. IoT Elements, Layered Architectures and Security Issues: A Comprehensive Survey. *Sensors* **2018**, *18*, 2796. [CrossRef] [PubMed]
33. Verma, G.; Prakash, S. Emerging Security Threats, Countermeasures, Issues, and Future Aspects on the Internet of Things (IoT): A Systematic Literature Review. In *Advances in Interdisciplinary Engineering*; Kumar, N., Tibor, S., Sindhwani, R., Lee, J., Srivastava, P., Eds.; Springer: Singapore, 2021; pp. 59–66.
34. Pal, S.; Jadidi, Z. Analysis of Security Issues and Countermeasures for the Industrial Internet of Things. *Appl. Sci.* **2021**, *11*, 9393. [CrossRef]
35. Ahlgren, B.; Dannewitz, C.; Imbrenda, C.; Kutscher, D.; Ohlman, B. A survey of information-centric networking. *IEEE Commun. Mag.* **2012**, *50*, 26–36. [CrossRef]
36. Bilal, M.; Pack, S. Secure Distribution of Protected Content in Information-Centric Networking. *IEEE Syst. J.* **2020**, *14*, 1921–1932. [CrossRef]
37. Rais, R.N.B.; Khalid, O. Study and analysis of mobility, security, and caching issues in CCN. *Int. J. Electr. Comput. Eng. (IJECE)* **2020**, *10*, 1438–1453. [CrossRef]
38. Yu, Y.; Li, Y.; Du, X.; Chen, R.; Yang, B. Content Protection in Named Data Networking: Challenges and Potential Solutions. *IEEE Commun. Mag.* **2018**, *56*, 82–87. [CrossRef]
39. Liu, G.; Quan, W.; Cheng, N.; Zhang, H.; Yu, S. Efficient DDoS attacks mitigation for stateful forwarding in Internet of Things. *J. Netw. Comput. Appl.* **2019**, *130*, 1–13. [CrossRef]

40. Mazhar, T.; Talpur, D.B.; Shloul, T.A.; Ghadi, Y.Y.; Haq, I.; Ullah, I.; Ouahada, K.; Hamam, H. Analysis of IoT Security Challenges and Its Solutions Using Artificial Intelligence. *Brain Sci.* **2023**, *13*, 683. [CrossRef] [PubMed]
41. Hail, M.A.M. Efficient Management, Control and Analysis of IoT-NDN Devices through "NDN4IoT" App Integrated with FIWARE. In Proceedings of the 2023 12th Mediterranean Conference on Embedded Computing (MECO), Budva, Montenegro, 6–10 June 2023; pp. 1–4. [CrossRef]
42. Aman1, A.H.M.; Hassan, R. Internet Protocol Function Enhancement using Information Centric Approach to Solve Mobility and Security Problems for Internets of Things. In Proceedings of the 1st International Conference on Informatics, Engineering, Science and Technology, INCITEST 2019, Bandung, Indonesia, 18 July 2019. [CrossRef]
43. Ravindran, R.; Zhang, Y.; Grieco, L.A.; Lindgren, A.; Burke, J.; Ahlgren, B.; Azgin, A. Design Considerations for Applying ICN to IoT. In *Internet-Draft Draft-Irtf-Icnrg-Icniot-03, Internet Engineering Task Force*; Work in Progress; 2019. Available online: https://datatracker.ietf.org/doc/draft-irtf-icnrg-icniot/ (accessed on 28 January 2024).
44. Rao, A.; Schelén, O.; Lindgren, A. Performance implications for IoT over information centric networks. In Proceedings of the Eleventh ACM Workshop on Challenged Networks, New York, NY, USA, 3–7 October 2016; pp. 57–62. [CrossRef]
45. Choudhary, S.; Kesswani, N.; Majhi, S. An Ensemble Intrusion Detection Model For Internet of Things Network. *Res. Sq.* 2021, preprint. [CrossRef]
46. Qaddoura, R.; Al-Zoubi, A.M.; Almomani, I.; Faris, H. A Multi-Stage Classification Approach for IoT Intrusion Detection Based on Clustering with Oversampling. *Appl. Sci.* **2021**, *11*, 3022. [CrossRef]
47. Arshad, S.; Azam, M.A.; Rehmani, M.H.; Loo, J. Recent Advances in Information-Centric Networking-Based Internet of Things (ICN-IoT). *IEEE Internet Things J.* **2019**, *6*, 2128–2158. [CrossRef]
48. Negara, R.M.; Rachmana Syambas, N. Caching and Machine Learning Integration Methods on Named Data Network: A Survey. In Proceedings of the 2020 14th International Conference on Telecommunication Systems, Services, and Applications (TSSA), Bandung, Indonesia, 4–5 November 2020; pp. 1–6. [CrossRef]

Disclaimer/Publisher's Note: The statements, opinions and data contained in all publications are solely those of the individual author(s) and contributor(s) and not of MDPI and/or the editor(s). MDPI and/or the editor(s) disclaim responsibility for any injury to people or property resulting from any ideas, methods, instructions or products referred to in the content.

Article

EPA-GAN: Electric Power Anonymization via Generative Adversarial Network Model

Yixin Yang [1,2], Wen Shen [3], Qian Guo [3], Qiuhong Shan [1,2], Yihan Cai [1,2] and Yubo Song [1,2,*]

1. School of Cyber Science and Engineering, Southeast University, Nanjing 210003, China; 220224777@seu.edu.cn (Y.Y.); qiuhshan@seu.edu.cn (Q.S.); 220224837@seu.edu.cn (Y.C.)
2. Purple Mountain Laboratories, Nanjing 211189, China
3. State Grid Smart Grid Research Institute Co., Ltd., Nanjing 210003, China; shenwen@geiri.sgcc.com.cn (W.S.); guoqian@geiri.sgcc.com.cn (Q.G.)
* Correspondence: songyubo@seu.edu.cn

Abstract: The contemporary landscape of electricity marketing data utilization is characterized by increased openness, heightened data circulation, and more intricate interaction contexts. Throughout the entire lifecycle of data, the persistent threat of leakage is ever-present. In this study, we introduce a novel electricity data anonymization model, termed EPA-GAN, which relies on table generation. In comparison to existing methodologies, our model extends the foundation of generative adversarial networks by incorporating feature encoders and feedback mechanisms. This adaptation enables the generation of anonymized data with heightened practicality and similarity to the original data, specifically tailored for mixed data types, thereby achieving a deliberate decoupling from the source data. Our proposed approach initiates by parsing the original JSON file, encoding it based on variable types and features using distinct feature encoders. Subsequently, a generative adversarial network, enhanced with information, downstream, generator losses, and the Was + GP modification, is employed to generate anonymized data. The introduction of random noise fortifies privacy protection during the data generation process. Experimental validation attests to a conspicuous reduction in both machine learning utility and statistical dissimilarity between the data synthesized by our proposed anonymization model and the original dataset. This substantiates the model's efficacy in replacing the original data for mining analysis and data sharing, thereby effectively safeguarding the privacy of the source data.

Keywords: generative adversarial networks; data anonymization; privacy preservation; loss feedback; feature coding; power data protection

Citation: Yang, Y.; Shen, W.; Guo, Q.; Shan, Q.; Cai, Y.; Song, Y. EPA-GAN: Electric Power Anonymization via Generative Adversarial Network Model. *Electronics* **2024**, *13*, 808. https://doi.org/10.3390/electronics13050808

Academic Editors: Abdussalam Elhanashi and Pierpaolo Dini

Received: 9 January 2024
Revised: 11 February 2024
Accepted: 15 February 2024
Published: 20 February 2024

Copyright: © 2024 by the authors. Licensee MDPI, Basel, Switzerland. This article is an open access article distributed under the terms and conditions of the Creative Commons Attribution (CC BY) license (https://creativecommons.org/licenses/by/4.0/).

1. Introduction

Ensuring that the power system is secured and stable especially after it has been exposed to varieties of strains and different contingencies is a major challenge that energy stakeholders are facing in today's world [1]. With the evolution of online collaboration channels in marketing and the innovative transformation of power company business models, the Power Marketing 2.0 business system aims to establish seamless data connectivity and business interaction across all operational channels to meet differentiated user needs on the basis of undertaking traditional customer service business [2]. Throughout this progression, the increasing prominence of data assets' value is evident, accompanied by a substantial rise in the demand for data sharing. Concurrently, the risks associated with data security are becoming more pronounced.

To address potential privacy leakage issues stemming from the interaction of power data, scholars have proposed various solutions. Guan et al. [3] introduced an efficient communication scheme for safeguarding the privacy of meter data in smart grids. This scheme, requiring no trusted authority, ensures accurate power billing calculation by simply transmitting a set of data. Given its utilization of straightforward and computationally efficient operations, such as addition and hashing, it proves suitable for devices with limited computational resources. Jawureket et al. [4] presented a secure billing method employing homomorphic commitment technology. In this approach, measurement data are initially submitted and aggregated, with only the aggregated power data disclosed to the power company. Ultimately, the data's accuracy is verified through zero-knowledge proofs. Kong et al. [5] proposed a group blind signature scheme to achieve conditional anonymity in smart grids, ensuring the integrity of power consumption data through homomorphic encryption. Li Yuyuan et al. [6] suggested a privacy protection request scheme for smart grids based on group blind signatures. In this scheme, smart meters can send power requests to substations, and the control center, unable to identify the real user identity through blind signatures, thereby ensures the privacy and security of power users. Nevertheless, these schemes fall short in effectively addressing the security concern of power data being reverse analyzed, leading to potential privacy leakage.

Generative adversarial networks (GANs), recognized as an emerging unsupervised learning methodology, find extensive application in image generation and are progressively garnering attention and utility within the domain of data science. As the practical applications and theoretical advancements in the field of generative adversarial networks unfold, scholars are increasingly redirecting their focus toward exploring data science. However, the majority of research related to GANs is predominantly concentrated on continuous datasets. Challenges arise when dealing with discrete data, primarily due to GANs encountering difficulties in utilizing gradient backpropagation for model training, given the non-differentiability of sampling from discrete distribution layers. To address this, Jang et al. [7] and Kingma et al. [8] proposed the Gumbel-SoftMax and concrete distribution methods, respectively, to tackle the generation of discrete data within the framework of variational autoencoders (VAEs). Kusner et al. [9] extended these methodologies to GAN models, thereby generating discrete sequence data based on the introduced VAE approach. Yu et al. [10] put forth a random policy approach based on reinforcement learning to circumvent the backpropagation challenges associated with discrete sequences. Another approach, proposed by Zhao et al. [11], involves adversarially regularized autoencoders (AR-AE), which transforms discrete terms used for training into a continuous latent feature space. Subsequently, GAN is employed to generate the latent feature distribution. Edward Choi et al. [12] introduced a model, medical generative adversarial network (med-GAN), akin to ARAE, specifically designed for generating synthetic binary or numerical data. Vincent et al. [13] presented an encoder–decoder method wherein the encoder maps samples to a low-dimensional continuous space, and the decoder reverts to the original data space. Leveraging the advantages of GAN in generating continuous data, the model enhances the accuracy of synthetic data generation for label variables by decoding the low-dimensional continuous space corresponding to the high-dimensional continuous or discrete space. Mottini et al. [14] proposed a GAN-based data generation method aimed at generating personal name information composed of missing/NaN values in categorical and numerical features.

Existing GAN-based data generation models primarily focus on distinguishing between discrete and continuous data [15]. However, in the context of electricity marketing business systems, the prevalent format for electricity data is JSON, which encompasses nested, mixed-type, and missing values. Consequently, a need arises for a data generator proficient in generating privacy-secured JSON data for electricity data.

In summary, this paper addresses the potential privacy breaches associated with electricity data in exploration, analysis, and business interactions. Through an analysis of electricity data characteristics and types, we propose the electric power anonymization via generative adversarial network model (EPA-GAN). The key contributions are as follows:

1. To tackle the diverse characteristics of electricity data, a method employing different feature encoders for data preprocessing is introduced, enhancing EPA-GAN's resilience to imbalanced discrete variables and skewed continuous variables.
2. By integrating conditional generative adversarial networks, we achieve data synthesis through a zero-sum game. We enhance GAN training stability and effectiveness

through the utilization of loss feedback. Additionally, Renyi differential privacy is employed to control privacy budgets by introducing random noise to the generator during the data generation process.

3. Experimental results validate the practicality and similarity of the data synthesized by EPA-GAN proposed in this paper. It proves to be a viable substitute for original data in exploration, analysis, and data sharing, ensuring the privacy protection of the original data.

The subsequent sections of this paper are organized as follows: Section 1 provides an introduction to the fundamental concepts of conditional generative adversarial networks; Section 2 establishes the generative adversarial network-based electricity data anonymization model, elaborating on electricity data preprocessing and the data synthesis method based on generative adversarial networks; Section 3 conducts experiments and presents comparative analyses of the proposed EPA-GAN; finally, Section 4 offers a succinct summary.

2. Materials and Methods

2.1. Conditional Generative Adversarial Network

Conditional generative adversarial network (CGAN) [16] represents an expanded iteration of the generative adversarial network, strategically incorporating additional conditional information during sample generation to augment the control prowess of the generator. In conventional generative adversarial networks, the generator's input comprises random noise, endeavoring to derive realistic data samples from this stochastic input, while the discriminator undertakes the task of assessing distinctions between the generated samples and authentic data.

However, CGAN introduces an extra conditional vector, conventionally denoted as c, as one of the inputs to the generator. This conditional vector encompasses diverse forms of information, such as categorical labels, textual descriptions, or images. By collaboratively inputting the conditional vector and random noise into the generator, it enables the generation of samples imbued with specific attributes dictated by the provided conditions. This refinement imparts heightened controllability to the generation process, facilitating the production of samples tailored to meet predefined criteria.

Through the incorporation of conditional information, CGAN bestows upon the generated samples distinctive attributes, fostering increased controllability and personalization in the resultant outcomes. This methodological approach presents a formidable instrument applicable across various tasks, empowering users to finesse outputs with nuanced precision during the generation process.

2.2. Differential Privacy

Differential privacy (DP) is a widely applied concept in the fields of data analysis and privacy protection. Its core objective is to minimize, to the greatest extent possible, the potential risk of individual privacy leakage while participating in data analysis. DP achieves this by introducing moderate noise or randomness in a mathematically rigorous and controllable manner, ensuring that the analysis process does not reveal sensitive information about individual entities.

The fundamental idea of DP lies in the addition of noise during data analysis or queries to obfuscate the true characteristics of the data, thereby preventing precise inferences about specific individuals. The careful design of this noise is crucial to balance the trade-off between protecting privacy and maintaining the accuracy of data analysis results.

Key steps in introducing differential privacy in GAN include modifying the training of the generator, parameter clipping to limit the model's overfitting to individual data, injecting noise to reduce the model's dependency on individual data, and setting a privacy budget to balance privacy and practicality. Such introductions help ensure that the generated data maintain a level of sensitivity to individual privacy, thereby enhancing the level of privacy protection.

2.3. Wasserstein Distance with Gradient Penalty

Wasserstein distance is a metric used to quantify the difference between two probability distributions. In the context of GAN, Wasserstein distance is widely employed to assess the distance between the data distribution generated by the generator and the real data distribution, typically serving as a measure of the quality of generated data.

Gradient penalty (GP) is a technique used to enhance the stability of GAN training. It ensures that the discriminator satisfies Lipschitz continuity conditions, meaning the gradient of the discriminator is bounded across the entire input space. By introducing a gradient penalty term, the discriminator's gradient is constrained within a reasonable range, preventing pattern collapse and training instability issues.

Wasserstein distance with gradient penalty (Was + GP) [17] refers to the simultaneous use of Wasserstein distance as a loss function and gradient penalty in GAN training for both the generator and discriminator.

Was + GP combines the advantages of Wasserstein distance and the stability offered by gradient penalty, contributing to improved GAN training effectiveness and the quality of generated results. And this further stabilizes the training of the network and requires less hyper-parameter tuning.

3. Model Design

The configuration of EPA-GAN is illustrated in Figure 1. It consists of two main modules: a data preprocessing module and a data generation module. During the process of data anonymization, the original JSON file undergoes parsing using the JSON data parser within the data preprocessing module. Following this, encoding is carried out based on variable types and features, employing distinct feature encoders. Subsequently, the data generation module, comprising the generator G, discriminator D, and auxiliary components (classifier or regressor) C, is employed to generate anonymized data. Notably, given the reliance on conditional generative adversarial networks in the model algorithm, a noise vector and a conditional vector are introduced into the generator's input for enhanced generative capabilities.

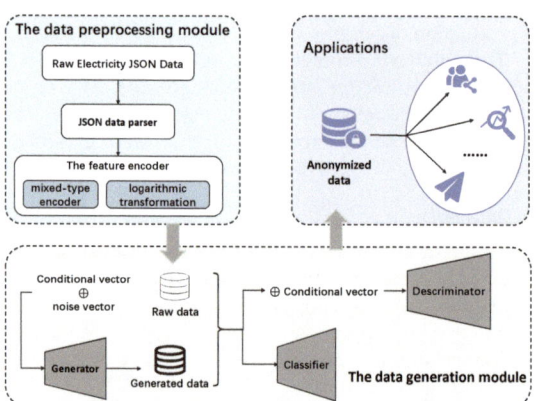

Figure 1. The general structure of EPA-GAN.

3.1. Data Preprocessing Module

Before transmitting the data to D and C, it is imperative to parse the data within the original JSON file and encode the variables using distinct feature encoders based on their types and characteristics.

3.1.1. JSON Data Parser

JSON (JavaScript Object Notation) format, as a lightweight data interchange format, has limitations such as supporting only a limited set of data types, lacking built-in metadata support, lack of standardized representations for dates and binary data, increased overhead for handling complex structures, and inability to handle circular references. However, despite these limitations, JSON is widely used in scenarios like power systems due to its simplicity, ease of use, and cross-platform compatibility.

The elements within the JSON file manifest in three forms: JSON objects, which hierarchically store data; JSON arrays, which concurrently store elements; and JSON primitives, directly encapsulating data.

This study initiates by parsing the original power data JSON file, denoted as J, into key-value pairs (denoted as P). Initializing P as an empty set, a tree structure T is then established for J, elucidating hierarchical relationships. Commencing from T, a comprehensive traversal of all nodes is undertaken. Upon encountering a JSON object during traversal, its key is appended to the existing prefix, initialized as an empty string. Subsequently, an underscore "_" is affixed to denote the hierarchical relationship, followed by continued exploration of the child nodes. In instances where a JSON array is encountered during parsing, parallel access to these elements suffices. Alternatively, when confronted with a JSON primitive, its key is appended to the prefix, facilitating backtracking and continued traversal until all nodes are visited. Figure 2 illustrates an exemplary scenario of JSON file parsing.

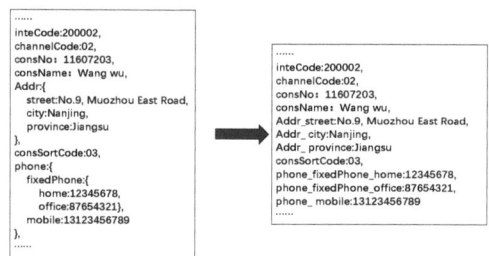

Figure 2. JSON file parsing example.

3.1.2. Feature Encoder

The feature encoder comprises three components: a hybrid encoder, a MinMax normalization encoder, and a logarithmic transformation encoder. During data preprocessing, the hybrid encoder is initially employed to individually encode variables. Subsequently, the MinMax normalization encoder is applied to discrete variables to address the issue of expanding dimensions due to a large number of discrete variable categories. The logarithmic transformation encoder is used to process continuous values, handling multi-modal data distribution and resolving the long-tail distribution—specifically, the challenge of rare points being distant from a significant amount of data.

- Hybrid Encoder

After parsing, data are encoded variable by variable. This paper categorizes variables into three types: discrete variables, continuous variables, and hybrid variables. A hybrid variable is defined if it contains both discrete and continuous values or if there are missing values within continuous values. To address such variables, this paper proposes a hybrid encoder. In this encoder, the values of hybrid variables are represented as sequences of value-pattern pairs. Here, "value" signifies specific values, and "pattern" denotes the nature of the variable, providing a more flexible representation of hybrid variables.

This manuscript elucidates the encoding methodology through the example of the mixed-variable distribution illustrated in Figure 3a in red. Notably, variable values can precisely be μ_0 or μ_3, representing discrete variables, or exhibit a distribution around the peaks of μ_1 and μ_2, indicative of continuous variables. When employing the variational

Gaussian mixture model (VGM) [18] to estimate the number of modes, denoted as k (e.g., in this instance, k = 2), and fitting the Gaussian mixture model, the study incorporates the modal-specific normalization (MSN) concept from [19] to address the continuous component. The Gaussian mixture acquired consists of $P = \sum_{k=1}^{2} \omega_k N(\mu_k, \sigma_k)$ components, with N representing the normal distribution, and ω_k, μ_k, and σ_k corresponding to the weights, means, and standard deviations of each mode.

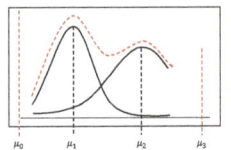

(a) Mixed variable distribution

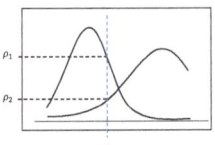

(b) Mode selection for discontinuous variables

Figure 3. Encoding of mixed data type variables.

To encode values of continuously distributed variables, each value is associated with the mode exhibiting the highest probability and subsequently normalized (refer to Figure 3b). For a given variable value τ, the probability densities of the two modes, labeled ρ_1 and ρ_2, are considered, and the mode with the highest probability is selected. In this specific example, given the higher probability of ρ_1, Mode 1 is utilized to normalize τ, resulting in the normalized value: $\alpha = \frac{\tau - \mu_1}{4\sigma_1}$. Furthermore, one-hot encoding is employed for encoding the mode β associated with τ. The final encoding is a union of α and β:$\alpha \oplus \beta$, where $\oplus$ denotes the vector union operator. The methodology for handling discrete components (e.g., μ_0 or μ_3 in Figure 3a) parallels that of the continuous range of mixed variables, with α set directly to 0. For instance, for a value $\beta = [0, 0, 0, 1]$ in μ_3, the ultimate encoding is $0 \oplus [0, 0, 0, 1]$. Missing values are treated as a distinct and unique category. A row encompassing $[1, \ldots, N]$ variables is constituted by concatenating the encodings of all variable values, where the encoding of continuous variables and mixed variables is denoted as ($\alpha \oplus \beta$), and the encoding of discrete variables is γ. In cases where there are n continuous/mixed variables and m discrete variables ($n + m = N$), the conclusive encoding is represented as: (1)

$$\bigoplus_{i=1}^{n} \alpha_i \oplus \beta_i \bigoplus_{j=n+1}^{N} \gamma_i \qquad (1)$$

- MinMax Normalization Encoder

To address the challenges associated with handling discrete variables with a high number of categories, especially in the context of simple distributions like single Gaussian, we introduce MinMax normalization (MMN) in this paper. MMN efficiently mitigates algorithmic complexity by mapping variable ranges to $(-1, 1)$ through a combination of shifting and scaling.

The fundamental principle behind MMN is to ensure compatibility between the encoding and the output range of a generator utilizing the tanh activation function. Mathematically, this normalization operation is expressed as $x_i^t = 2 \cdot \frac{x_i - min(x)}{max(x) - min(x)} - 1$, x_i represents the original variable's value, x_i^t denotes the normalized value, and $min(x)$ and $max(x)$ denote the minimum and maximum values of continuous variables.

For the reverse transformation of the normalized value X_i^t, it can be mapped back to the original range using the following formula: $X_i = (max(x) - min(x)) \cdot \frac{X_i^t + 1}{2} + min(x)$. Continuous variables can undergo direct normalization and reverse-normalization using this formula, while discrete variables, before normalization, are initially encoded as integers and rounded to integers after reverse-normalization.

Despite its utility, MMN comes with limitations. Specifically, its application leads to the loss of pattern indicators in the conditional vector, hindering the enhancement of correlation between specific category variables. As a result, MMN is best suited for simple distributions, such as single-mode Gaussian distributions, within continuous columns, and is less effective in handling more complex distributions.

By default, EPA-GAN prioritizes the use of MSN for handling variables, reserving the application of MMN exclusively for instances where the number of categories in discrete variables overwhelms existing models' capacity to train on encoded data.

- Logarithmic Transformation Encoder

In this study, we employ variational Gaussian mixture (VGM) for encoding continuous values, specifically designed for handling multimodal data distributions. It is noteworthy, however, that VGM encounters limitations in addressing long-tail distributions. To surmount this challenge, we integrate a preprocessing step involving a logarithmic transformation for variables exhibiting long-tail distributions. The logarithmic transformation serves the purpose of elongating the tail portion of the data, aligning it more closely with a normal distribution.

For variables constrained by lower bounds, we introduce a compressive alternative. The original value τ is replaced by the compressed logarithmic value τ^c:

$$\tau^c = \begin{cases} \log(\tau) & if\, l > 0 \\ \log(\tau - l + \varepsilon) & if\, l \leq 0, where\, \varepsilon > 0 \end{cases} \quad (2)$$

This compression strategy facilitates the uniformity of the entire data range, enhancing the capability of VGM to capture the intricate relationship between tail values and the overall data distribution. It is imperative to note that the term "VGM" refers to the variational Gaussian mixture model, a crucial component in our encoding methodology.

3.2. Data Generation Module

The data generation module employs a generative adversarial network (GAN) trained through a zero-sum game, where D (discriminator) aims to maximize a specified criterion, and G (generator) endeavors to minimize the same criterion. Concurrently, G is provided with additional feedback to enhance the quality and applicability of the generated data.

G and D utilize a CNN structure identical to that proposed by Park et al. in Table-GAN [20]. To adapt the CNN for handling row-record data stored in vector form, this paper transforms the row data into the nearest square matrix of dimension d × d, where d is the square root of the dimensionality of the row data, with missing values filled with zeros. C (classifier) employs a multilayer perceptron (MLP) with four hidden layers, each comprising 256 neurons. C is trained on the original data to capture semantic integrity effectively.

During the data generation process, a random noise vector combines with a conditional vector input to G, producing synthetic data. Synthetic and real data are encoded into matrix form and compared by D. The encoded data are then reverse transformed into vectors, serving as input for C to generate corresponding category label predictions.

3.2.1. Loss Feedback

This paper introduces additional feedback mechanisms for G by incorporating information loss, downstream loss, and generator loss. Information loss aims to enhance the quality of synthetic data by making it closer to the original data, emphasizing both semantic and statistical consistency. Downstream loss is introduced to align the generator with the requirements of downstream tasks, improving the practical applicability of generated data. Generator loss, implemented through adversarial training, encourages the generation of more realistic data, thereby enhancing the overall quality and robustness of the generative model, making it challenging to distinguish synthetic data from real data.

During training, these three loss terms are added to G's default loss terms (Was + GP). The calculation involves the use of f_x and $f_{G(x)}$ representing features of real and generated samples passed through the softmax layer of D.

- Information Loss

  Information loss L_{info}^G is computed based on first-order (mean) and second-order (standard deviation) statistical data of synthetic and original data. The objective is to optimize synthetic data to align statistically with the original data.

  $$L_{info}^G = ||E[f_x]_{x \sim p_{data}(x)} - E[f_{G(z)}]_{z \sim p(z)}||_2 + ||SD[f_x]_{x \sim p_{data}(x)} - SD[f_{G(z)}]_{z \sim p(z)}||_2$$

- Downstream Loss

  Downstream loss $L_{dstream}^G$ reflects the correlation between the target variable and other variable values. It checks the semantic integrity of synthetic data and penalizes semantically incorrect combinations of values in synthetic records.

  $$L_{dstream}^G = E[|l(G(z)) - C(fe(G(z)))|]_{z \sim p(z)}$$

- Generator Loss

  Generator loss $L_{generator}^G$ is the cross-entropy between the given condition vector and the generated output class. It ensures the generator produces output classes consistent with the given condition vector.

  $$L_{generator}^G = H(m_i, \hat{m}_i)$$

- Was + GP

  Let $L_{default}^G$ and $L_{default}^D$ represent the discriminator and generator GAN losses for Was + GP. The unique objective function of D aims to maximize discriminator output for real data and minimize it for generated data, encouraging the generator to produce more realistic data. The gradient penalty term stabilizes the training process.

  $$L = \underbrace{E_{\hat{x} P_g}\left[D\left(\tilde{x}\right)\right] - E_{x P_r}[D(x)]}_{original\,discriminator\,loss} + \underbrace{\lambda E_{\hat{x} P_{\hat{x}}}[(\|\nabla_{\hat{x}} D(\hat{x})\|_2 - 1)]^2}_{gradient\,penalty}$$

  For λ, the complete training objective is to minimize $L^G = L_{default}^G + L_{info}^G + L_{dstream}^G + L_{generator}^G$. The training loss for C is similar to the downstream loss for the generator, i.e., minimizing $L_{dstream}^C = E[|l(x) - C(fe(x))|]_{x \sim p_{data}(x)}$.

3.2.2. Conditional Vectors

In the context of EPA-GAN, the deployment of CGAN proves instrumental in managing imbalances within training datasets. Furthermore, a method of sampled training is employed to expand the dataset, incorporating patterns from both continuous and mixed columns. During the sampling of real data, the introduction of conditional vectors becomes pivotal for data filtering and the reestablishment of balance in the training dataset.

The conditional vector, denoted as V, takes the form of a binary vector. It is meticulously crafted by amalgamating the one-hot encoding β for all patterns across variables (pertaining to continuous and mixed variables) and the one-hot encoding γ for all categories (pertaining to discrete variables) as outlined in Formula (1). Each unique conditional vector corresponds to a specific pattern or category. To illustrate, consider Figure 4, which presents an example featuring three variables—a continuous variable (C_1), a mixed variable (C_2), and a discrete variable (C_3)—C_3 belonging to category 2.

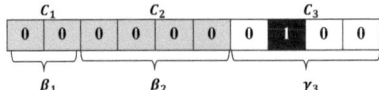

Figure 4. Example of Conditional Vector.

In instances where a conditional vector is requisite during the training phase, a variable is initially chosen at random with uniform probability. Subsequently, the probability distribution of each pattern (or category, in the case of discrete variables) within that chosen variable is computed based on its frequency. The sampling of patterns ensues, guided by the logarithm of their respective probabilities. The adoption of logarithmic probabilities, as opposed to raw frequencies, is deliberate, fostering a more equitable representation of rare patterns or categories and mitigating the challenges associated with their scarcity. The extension of conditional vectors to encompass both continuous and mixed variables serves to rectify the issue of imbalanced pattern frequencies inherent in the representation of these variables. Notably, the generator's imposition of conditions across all data types during the training process significantly enhances the inter-variable learning correlations.

3.2.3. Noise Vectors

In EPA-GAN, the utilization of Renyi differential privacy (RDP) imposes a more stringent control over privacy budgets, ensuring privacy protection during the data generation process.

RDP, an extension of differential privacy (DP), focuses on delivering heightened privacy protection in scenarios involving multiple queries or mechanisms. It introduces the concept of a budget, representing the cumulative upper limit of privacy leakage permitted across a series of queries or mechanisms. At the core of the RDP framework lies the composition theorem, allowing the synthesis of privacy losses from multiple queries into a comprehensive privacy constraint. This framework enhances the precision and robustness of privacy protection in the context of multiple queries or mechanisms, simplifying the management and control of privacy leakage in practical applications [21].

In this study, M corresponds to a GAN model equipped with a privacy budget (λ, ε). The (λ, ε)-RDP mechanism can be expressed as

$$\left(\varepsilon + \frac{\log\frac{1}{\delta}}{\lambda - 1}, \delta\right) - DP \tag{3}$$

if, for any adjacent datasets S and S', the Renyi divergence $D_\lambda(M(S) \parallel M(S')) = \frac{1}{\lambda-1} \log E_{x \sim M(S)}[(\frac{P[M(S)=x]}{P[M(S')=x]})]^{\lambda-1} \leq \varepsilon$ holds, where $D_\lambda(P \parallel Q) = \frac{1}{\lambda-1} \log E_{x \sim Q}\left[\left(\frac{P(x)}{Q(x)}\right)^\lambda\right]$ signifies the Renyi divergence.

Additionally, the Gaussian mechanism with parameter M_σ [7] is represented as $M_\sigma(x) = f(x) + N(0, \sigma^2 I)$, where f is an arbitrary function, and its sensitivity $\Delta_{2f} = \max_{S,S'} \parallel f(S) - f(S') \parallel_2$ encompasses all adjacent datasets S and S', adhering to a Gaussian distribution with mean zero and covariance matrix $\sigma^2 I$ (where I is the identity matrix) [21].

Furthermore, if M satisfies $(\varepsilon, \delta) - DP$, then $F \circ M$ holds, where F can be any arbitrary random function, $\circ$ denotes the composite operator. Consequently, in EPA-GAN, training only the generator in the generative adversarial network architecture ensures the entire GAN adheres to differential privacy, effectively curtailing privacy budgets.

4. Experimental Evaluation and Analysis

4.1. Parameterization and Evaluation Methods

This article's anonymized experimental dataset is derived from the actual measurement data of Irish smart meters released by ISSDA. The experiments were conducted in the Ubuntu 20.04 environment on a personal laptop with a 3.20 GHz CPU and 16.00 GB of memory, utilizing standard tools and software.

To substantiate the efficacy of the proposed EPA-GAN in this paper, a comparative analysis is performed against data generation models based on CWGAN [22], CT-GAN [19] and TableGAN [20]. The evaluation of EPA-GAN encompasses two key aspects: the machine learning performance on anonymized data and the assessment of statistical similarity. This scrutiny aims to determine the viability of EPA-GAN for safeguarding privacy in power data. Notably, decision trees and random forests have a maximum depth of 28, and MLP employs 128 neurons, with all algorithms implemented using scikit-learn 1.3.0 and default parameters.

For assessing the machine learning performance in classification tasks, the study employs five widely used algorithms (decision tree classifier, linear support vector machine, random forest classifier, multinomial logistic regression, and MLP) to measure performance metrics (accuracy, F1 score, and AUC) on both real and synthetic data.

Statistical similarity is evaluated using three metrics: Jensen–Shannon divergence (JSD), Wasserstein distance (WD), and correlation (corr.). JSD is used to quantify the difference between the probability mass distribution of individual categorical variables for real and synthetic data sets. However, the JSD measure is numerically unstable for evaluating the quality of continuous variables, so WD is used to quantify the correspondence of the distribution of variables between the synthesized data set and the original data set. And coor. is used to calculate the feature correlation between the original data and the synthesized data.

4.2. Experimental Results and Analysis

Table 1 and Figure 5 present quantitative data on the differences in machine learning performance during classification tasks on real and synthetically generated data using three models. The table reveals that the proposed EPA-GAN significantly reduces differences across all metrics. This suggests that the featurized encoding and loss feedback in EPA-GAN contribute to enhancing feature representation and GAN training.

Table 1. Differences of ML utility between raw and synthetic data.

Method	Differences of ML Utility		
	Acc	F1-Score	AUC
EPA-GAN	5.19%	0.089	0.037
CT-GAN	21.53%	0.284	0.256
TableGAN	11.44%	0.130	0.168
CWGAN	19.94%	0.346	0.287

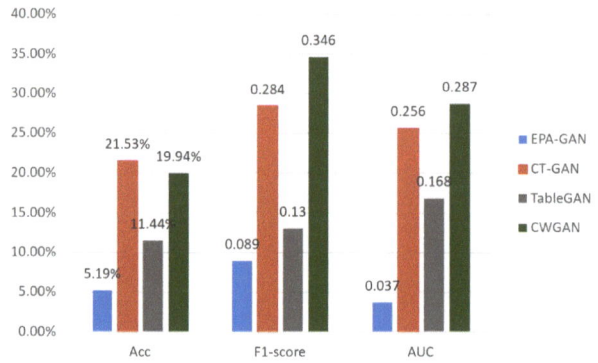

Figure 5. Differences of ML utility between raw and synthetic data.

Table 2 illustrates the contrast in statistical similarity between synthetic and real data. Notably, in JSD, the performance of EPD-GAN is 47.9% higher than CT-GAN, 53.2% higher than TableGAN, and 72.6% higher than CWGAN. Superiority is also evident in terms of WD and data correlation. This highlights EPA-GAN's effective preprocessing of imbalanced distribution data.

Table 2. Differences of statistical similarity.

Method	Differences of Statistical Similarity		
	JSD	WD (k)	Corr.
EPA-GAN	0.037	0.486	2.00
CT-GAN	0.071	1.771	2.68
TableGAN	0.079	2.122	2.29
CWGAN	0.135	237.089	5.72

The experimental results indicate high similarity in machine learning performance and statistical resemblance between the anonymized data synthesized by EPA-GAN and the original data. Therefore, these anonymized data can substitute for the original for mining analysis and interactive sharing, contributing to privacy protection in power system business data.

5. Conclusions

In response to potential privacy leaks during the interaction and mining analysis of power data in the marketing environment, this study proposes a generative adversarial network for anonymizing power data. This method synthesizes anonymized data with high practicality and similarity to the original, enabling its use in mining analysis and data sharing, effectively ensuring the privacy of original data. Initially, variables are encoded based on type and feature using different feature encoders to handle mixed categorical and continuous variables effectively. Subsequently, a generative adversarial network, improved through information, downstream, generator loss, and Was + GP, synthesizes anonymized data, incorporating random noise for privacy protection during data generation. Experimental verification demonstrates a significant reduction in machine learning performance and statistical similarity differences between the anonymized data synthesized by this method and the original data, compared to the current state-of-the-art models. The outcomes of this research provide an innovative privacy protection solution for the power industry, with broad application prospects. Therefore, in the future, we will research EPA-GAN to address scalability issues in handling large datasets, enhance its robustness against various types of attacks, and mitigate potential biases introduced during the anonymization process. Additionally, we will validate its effectiveness across different datasets and explore extending this method to other industries or domains, paving the way for its practical integration into real-world applications.

Author Contributions: Conceptualization, Y.Y.; methodology, W.S.; software, Q.G.; writing—review and editing, Q.S.; visualization, Y.C.; project administration, Y.S. All authors have read and agreed to the published version of the manuscript.

Funding: This work was supported by the Science and Technology Project of State Grid Anhui Electric Power Co., Ltd. [5400-202318221A-1-1-ZN].

Data Availability Statement: We confirm that the data supporting the findings of this study are available within the article. Additional data that support the findings of this study are available from the corresponding author upon reasonable request.

Acknowledgments: We would like to express our gratitude to the Science and Technology Project of State Grid Anhui Electric Power Co., Ltd. [5400-202318221A-1-1-ZN] for their support. We would also like to acknowledge the collective efforts of all the authors in this study.

Conflicts of Interest: The funders had no role in the design of the study; in the collection, analyses, or interpretation of data; in the writing of the manuscript; or in the decision to publish the results.

References

1. Alimi, O.A.; Ouahada, K.; Abu-Mahfouz, A.M. A review of machine learning approaches to power system security and stability. *IEEE Access* **2020**, *8*, 113512–113531. [CrossRef]
2. Yan, S. Research on Data Mining and Privacy Protection Methods in Smart Grid. *J. Microcomput. Appl.* **2019**, *35*, 101–104.
3. Guan, A.; Guan, D.J. An efficient and privacy protection communication scheme for smart grid. *IEEE Access* **2020**, *8*, 179047–179054. [CrossRef]
4. Jawurek, M.; Johns, M.; Kerschbaum, F. Plug-in privacy for smart metering billing. In *International Symposium on Privacy Enhancing Technologies Symposium*; Springer: Berlin/Heidelberg, Germany, 2011; pp. 192–210.
5. Kong, W.; Shen, J.; Vijayakumar, P.; Cho, Y.; Chang, V. A practical group blind signature scheme for privacy protection in smart grid. *J. Parallel Distrib. Comput.* **2020**, *136*, 29–39. [CrossRef]
6. Li, Y.; Yu, H. Research on Privacy Protection Scheme for Smart Grid Based on Group Blind Signature. *Autom. Instrum.* **2022**, *43*, 85–89.
7. Jang, E.; Gu, S.; Poole, B. Categorical reparameterization with gumbel-softmax. *arXiv* **2016**, arXiv:1611.01144.
8. Maddison, C.J.; Mnih, A.; Teh, Y.W. The concrete distribution: A continuous relaxation of discrete random variables. *arXiv* **2016**, arXiv:1611.00712.
9. Kusner, M.J.; Hernández-Lobato, J.M. Gans for sequences of discrete elements with the gumbel-softmax distribution. *arXiv* **2016**, arXiv:1611.04051.
10. Yu, L.; Zhang, W.; Wang, J.; Yu, Y. Seqgan: Sequence generative adversarial nets with policy gradient. In Proceedings of the AAAI Conference on Artificial Intelligence, San Francisco, CA, USA, 4–9 February 2017; Volume 31.
11. Zhao, J.; Kim, Y.; Zhang, K.; Rush, A.M.; LeCun, Y. Adversarially regularized autoencoders. In Proceedings of the International Conference on Machine Learning, Macau, China, 26–28 February 2018; pp. 5902–5911.
12. Choi, E.; Biswal, S.; Malin, B.; Duke, J.; Stewart, W.F.; Sun, J. Generating multi-label discrete patient records using generative adversarial networks. In Proceedings of the Machine Learning for Healthcare Conference, Boston, MA, USA, 18–19 August 2017; pp. 286–305.
13. Vincent, P.; Larochelle, H.; Bengio, Y.; Manzagol, P.-A. Extracting and composing robust features with denoising autoencoders. In Proceedings of the 25th International Conference on Machine Learning, Helsinki, Finland, 5–9 July 2008; pp. 1096–1103.
14. Mottini, A.; Lheriter, A.; Acuna-Agost, R. Airline passenger name record generation using generative adversarial networks. *arXiv* **2018**, arXiv:1807.06657.
15. Wei, N.; Wang, L.; Dong, F. A Hybrid Data Generation Method Based on Generative Adversarial Networks. *Comput. Appl. Softw.* **2022**, *39*, 29–34.
16. Mirza, M.; Osindero, S. Conditional generative adversarial nets. *arXiv* **2014**, arXiv:1411.1784.
17. Gulrajani, I.; Ahmed, F.; Arjovsky, M.; Dumoulin, V.; Courville, A. Improved training of wasserstein gans. In Proceedings of the Advances in Neural Information Processing Systems, Long Beach, CA, USA, 4–9 December 2017; Volume 30.
18. Hawes, M.B. *Implementing Differential Privacy: Seven Lessons from the 2020 United States Census*; Harvard Data Science Review: Cambridge, MA, USA, 2020; Volume 2.
19. Xu, L.; Skoularidou, M.; Cuesta-Infante, A.; Veeramachaneni, K. Modeling tabular data using conditional GAN. In Proceedings of the Advances in Neural Information Processing Systems, Vancouver, BC, Canada, 8–14 December 2019; Volume 32.
20. Park, N.; Mohammadi, M.; Gorde, K.; Jajodia, S.; Park, H.; Kim, Y. Data synthesis based on generative adversarial networks. *arXiv* **2018**, arXiv:1806.03384. [CrossRef]
21. Mironov, I. Rényi differential privacy. In Proceedings of the 2017 IEEE 30th Computer Security Foundations Symposium (CSF), Santa Barbara, CA, USA, 21–25 August 2017; pp. 263–275.
22. Rawat, S.; Shen, M.H.H. A Novel Topology Optimization Approach using Conditional Deep Learning. *arXiv* **2019**. [CrossRef]

Disclaimer/Publisher's Note: The statements, opinions and data contained in all publications are solely those of the individual author(s) and contributor(s) and not of MDPI and/or the editor(s). MDPI and/or the editor(s) disclaim responsibility for any injury to people or property resulting from any ideas, methods, instructions or products referred to in the content.

MDPI AG
Grosspeteranlage 5
4052 Basel
Switzerland
Tel.: +41 61 683 77 34

Electronics Editorial Office
E-mail: electronics@mdpi.com
www.mdpi.com/journal/electronics

Disclaimer/Publisher's Note: The title and front matter of this reprint are at the discretion of the Guest Editors. The publisher is not responsible for their content or any associated concerns. The statements, opinions and data contained in all individual articles are solely those of the individual Editors and contributors and not of MDPI. MDPI disclaims responsibility for any injury to people or property resulting from any ideas, methods, instructions or products referred to in the content.

www.ingramcontent.com/pod-product-compliance
Lightning Source LLC
LaVergne TN
LVHW072319090526
838202LV00019B/2310